Nanotechnology in Green Energy Generation

Nanotechnology in Green Energy Generation provides a comprehensive review of modelling, processing, and applications of all major categories of green energy generation materials. It explores different areas of green energy generation including hydrogen, solar, and wind energies, covering aspects such as synthesis, morphology, materials, and characterization.

Presenting the fundamental principles in the design and utilization of green energy generation materials, the book discusses the construction and equivalent circuits of traditional and new green energy cells. In addition, it provides thermal analysis and comparative studies with traditional power generation, including operation and cost-efficiency of new generation cells and modules. The book also includes many case studies, laboratory experiments, and research results throughout the chapters.

This book will be a valuable reference for applied researchers, academic researchers and graduate students studying advances in energy engineering, nanotechnology, and materials and composites.

Emerging Materials and Technologies

Series Editor: Boris I. Kharissov

The *Emerging Materials and Technologies* series is devoted to highlighting publications centered on emerging advanced materials and novel technologies. Attention is paid to those newly discovered or applied materials with potential to solve pressing societal problems and improve quality of life, corresponding to environmental protection, medicine, communications, energy, transportation, advanced manufacturing, and related areas.

The series takes into account that, under present strong demands for energy, material, and cost savings, as well as heavy contamination problems and worldwide pandemic conditions, the area of emerging materials and related scalable technologies is a highly interdisciplinary field, with the need for researchers, professionals, and academics across the spectrum of engineering and technological disciplines. The main objective of this book series is to attract more attention to these materials and technologies and invite conversation among the international R&D community.

Advancements in Nanomaterials for Energy Conversion and Storage
Edited by Piyush Kumar Sonkar and Vellaichamy Ganesan

2D Materials-Based Sensors
Technology and Applications
Vinod Kumar Khanna

Metal Organic Framework Derived Materials
Design Strategies and Applications
Gomathi Nageswaran, Varsha M V, Arun Kumar Rajasekaran and M Shashank Rao

Hydrogen Production, Storage, and Utilization
Technologies and Applications
Abbas Tcharkhtchi, Hamidreza Vanaei, Albert Lucas and Sedigheh Farzaneh

MXenes for Energy Storage Applications
Emerging Characteristics, Compositions, and Synthesis Methods
Muhammad Rafique, M. Bilal Tahir, and Saira Anwar

Multi-scale and Multifunctional Coatings and Interfaces for Tribological Contacts
Ajit Behera, Kuldeep K Saxena, Dipen Kumar Rajak and Shankar Sehgal

Nanotechnology in Green Energy Generation
Ahmed Thabet Mohamed

For more information about this series, please visit: www.routledge.com/Emerging-Materials-and-Technologies/book-series/CRCEMT

Nanotechnology in Green Energy Generation

Ahmed Thabet Mohamed

CRC Press
Taylor & Francis Group
Boca Raton London New York

CRC Press is an imprint of the
Taylor & Francis Group, an **informa** business

Designed cover image: Shutterstock

First edition published 2025
by CRC Press
2385 NW Executive Center Drive, Suite 320, Boca Raton FL 33431

and by CRC Press
4 Park Square, Milton Park, Abingdon, Oxon, OX14 4RN

CRC Press is an imprint of Taylor & Francis Group, LLC

ISBN: 978-1-032-84386-5 (hbk)
ISBN: 978-1-032-84387-2 (pbk)
ISBN: 978-1-003-51248-6 (ebk)

DOI: 10.1201/9781003512486

Typeset in Times
by Apex CoVantage, LLC

Contents

SECTION II Design and Economics of Green Power Generation

About the Author

Prof. Dr. Ahmed Thabet Mohamed received both BSc and MSc electrical engineering degrees from Aswan University, Egypt. He earned his PhD in electrical engineering in 2006 from El-Minia University, Egypt. Since 2006, Prof. Mohamed has been part of the Faculty of Energy Engineering at Aswan University, where he is currently a professor. His research interests lie in the areas of analysis and developing electrical engineering models and applications, investigating novel nanotechnology materials via addition nano-scale particles and additives for usage in industry, electromagnetic materials, electroluminescence, and the electrical and thermal aging of industrial polymers. In 2009, he established the first Nano-Technology Research Centre in Upper Egypt for developing industrial nanotechnology materials for AC and DC applications. Prof. Mohamed has over 90 publications in cited international journals and conferences, and he has written several books. In September 2022, he was ranked within the top 2% of scientists worldwide in the "Updated science-wide author databases of standardized citation indicators." Since 2018, Prof. Mohamed has held full professor positions at various different international colleges of engineering at Saudi Arabian universities.

About the Author

[illegible]

Aim of the Book

This book has the vital role of promoting the effectiveness and reliability of nanotechnology in green energy generation and industrial applications via the following aims:

- Studying the effects of NPs on the generation of hydrogen energy, solar energy and wind energy.
- Specifying the main theories and parameters for physical and electrical properties of fabricated nanocomposites for green energy generation.
- Specifying optimal NPs for industrial green energy applications.
- Fabricating experimental individual and multiple nanocomposites, membranes, and coatings for applications in green energy sources.

Green energy has become increasingly popular for serving large numbers of consumers. Therefore, it has become imperative to locate advanced frameworks that can securely move enormous amounts of energy in financially sustainable ways that meet the requirements for safety and security.

In recent years, nanotechnology has opened a wide range of applications in a variety of fields, with the energy sector being one of the most important among them in the form of nanostructured energy storage materials, solar cells, and other nanomaterials. The goal of these revolutionary approaches is to conserve resources and preserve the environment without compromising resources. This book encourages the development of nanotechnology concepts for green energy generation and monitoring devices; nanotechnology can play a significant role in establishing a world of advanced energy.

Nanostructured materials have become key to advancements in green energy generation: Nanotubes, nanorods, doped nanosized metal oxides, and related nanostructured architectures are powering fuel cells, wind farms, and photovoltaic energy production. Modern life requires energy, but a large percentage of the environmental damage related to climate change that has occurred in recent years has been caused by excessive energy consumption. Consequently, we must change our lifestyles to phase out conventional energy sources, including gas and oil, and develop more renewable energy resources, commonly called green or clean energy. The most commonly used renewable energy sources are solar, wind, water, and hydrogen.

Preface

As nanotechnology allows nanoscale structures to be automated at a much lower cost and in an environmentally friendly manner, it can increase the rate of improving energy and industrial materials characterization, making it faster, cheaper and more versatile. Nanotechnology has gone beyond the realm of young imagination and has become one of the most prominent trends and priorities of scientific research all around the world, especially in the advanced countries of the world, where it continues to be a major driver of innovation and research.

Several laboratories in universities, research institutions and industries have seen their image strengthened and their investments increased as a result of recent developments and serious discoveries in this field. A functional mixture of science and technology, nanotechnology begins with the fundamental components of matter, namely atoms and molecules. Nanotechnology is being incorporated into green energy sources, industries, and products around the world in a frenetic race between research centers and industry, and many scientists agree that nanotechnology will play an important role in the future of our planet as part of a new green energy revolution. A key component of the green energy revolution is the Nanotechnology Plan; countries around the world recognize that nanotechnology will be the future of green energy generation, and the importance of this modern technology is that it is highly economical, in contrast to fossil-based traditional energy sources.

For instance, hydrogen energy is a fascinating topic with immense potential for our clean and secure energy future. Hydrogen is a naturally occurring gas that is abundant in the universe, and clean hydrogen is produced with very low or zero carbon emissions, so it can be a game changer for achieving sustainable energy transitions. Derivative products of hydrogen, including clean fuels, can also contribute to decarbonization.

In addition, solar energy is a powerful ally in our transition to a cleaner and more sustainable world. As nanotechnology advances and costs decrease, its impact will continue to grow. On the other hand, wind energy offers significant advantages, but addressing challenges is essential for its continued growth and contribution to a decarbonized future. Wind turbines can be standalone structures or clustered together in what is called a wind farm. Wind power is a sustainable and renewable energy source, but wind energy is variable, so it requires storage or other dispatchable generation sources for reliable electricity supply.

Moreover, nanotechnology will have a profound impact on the economy of the energy sector. Scientific developments as well as technological advancements have contributed to the development of advanced materials with high specifications that assist in the efficient and accurate generation of green energy. Furthermore, the dependence of green energy units on nanomaterials addresses the economic problem represented by its call for natural resources. Therefore, the industries that incorporate nanotechnology will replace the current traditional energy generation methods because the new materials and methods will be better, cheaper, and cleaner and will not cause environmental pollution.

THE CHALLENGES

Welcome to the amazing 'nano world'! God willing, you will find in this book scientific and laboratory applications of nanoscience in the field of industrial applications and in the scope of green energy engineering. This book will assist and encourage researchers and investors in following the recent technology related to green energy generation because it is now and will continue to be necessary to train new generations of researchers, engineers, and technologists in generating green energy by nanoscience. Making a major contribution to education in nanoscopes, this book combines the basic ideas necessary to understand modern developments based on nanoscience and nanotechnology in green energy engineering.

The nanotechnology revolution aims to develop systems for generating green energy at the atomic level. In such systems, physics intertwines and overlaps with the fields of chemistry, biology, electricity, electronics, and mechanics, and this interconnectedness drives many new innovations. However, there are challenges with all of the recent green energy generation systems.

For instance, the challenges associated with hydrogen energy are storage, infrastructure, cost, and safety. Specifically, because hydrogen is highly reactive, it is difficult to store it, and separately, establishing a robust hydrogen infrastructure will be essential for widespread adoption. Making hydrogen production more affordable will also be crucial for widespread adoption, and the safety of handling and storing hydrogen is a crucial part of the process.

Solar energy also faces the challenges of intermittency—as solar energy depends on daylight hours and weather conditions—and material scarcity given that some solar technologies require rare materials that may become unavailable as the technology develops. In addition to high initial costs, there are a number of other factors that can hinder the adoption of solar power. The cost of installation and setup is significant, although it has steadily decreased over time. Land use is another issue: Large-scale solar plants require a considerable amount of land, which impacts ecosystems.

The main challenges associated with wind energy are competition with other energy sources and transmitting the energy. First, wind energy must compete with other low-cost energy sources to remain competitive with them. It is difficult for wind projects to compete economically with natural gas and solar power because wind is intermittent; this needs to be systematically adjusted to balance the supply and ensure grid integration. Additionally, it is important to balance the needs of energy with conservation of land in wind farms, as they require space, impact landscapes, and have a visual impact on scenery. Balancing renewable energy goals and environmental conservation are essential considerations. Second, wind-generated electricity needs to be efficiently transmitted to demand centers to achieve these goals, and the developed transmission infrastructure must be robust.

In the future, nanotechnology will be the guiding force for the green energy and industrial revolution in the world; new products will be smaller, cheaper, lighter, and more able to perform the tasks assigned to them. In recent years, nanotechnology has become an extremely active and rapidly growing research field, and many scientists working in this field confirm that the use of nanotechnology will lead to a new generation of renewable energy, bringing with it fundamental changes in economics and technology as well. Several scientists, industrialists, and funders have taken an interest in nanotechnology, and the governments of developed countries have been investing more and more in nanotechnology research and development over the past few years.

SEARCHING FOR A SOLUTION

Today, with the growth of the global energy industry, the energy sector has attracted the attention of both consumers and producers because of its effect on every aspect of life. The amount of energy that a nation consumes plays a crucial role in determining its progress. There has been a significant increase in the demand for green energy. In addition to the advancement and increased production of industrial products, green energy has become an increasingly important kind of energy. It is, therefore, necessary to find advanced systems capable of transferring these enormous amounts of green energy with a high degree of safety and efficiency.

In the field of nanotechnology, researchers from governmental agencies, businesses, academic institutions and research laboratories study it at a rapid pace. Many scientists working in this field have confirmed that nanotechnology will bring about a new green energy revolution in the near future in various areas of life as well as radical technological and economic transformations. There has been a lot of interest in nanotechnology applications among scientists, industrialists, and funders.

Nanotechnology is aimed at generating green energy through the use of new types of atomic electronics that utilize quantum mechanics as well as the movement of individual particles, producing equipment that can be faster and smaller than anything we see around us today. Such systems combine engineering, electronics, mechanics, physics, and chemistry, intertwining, overlapping, and strengthening each other; new innovations and inventions can result from this interconnectedness and overlap for new green energy generation systems.

ORGANIZATION OF THE BOOK

The energy sector has created a great deal of interest among both the consumer and the manufacturer because of its huge impact on every area of life. Green energy has become increasingly popular due to its reliability and its low environmental impact, causing no pollution to environment; this means that nanotechnology should be supported as a means of safely moving large amounts of green energy into prevalent endeavors. It is also imperative that green energy systems meet specialized requirements and be low-cost with extended operational lives. This book clarifies this NEW vision of green energy generation in ten chapters; I describe backgrounds of different technologies, current technologies, challenges, and trends and recommendations for all topics:

Chapter One discusses the principles of hydrogen energy, hydrogen cell construction, hydrogen energy development, NPs and hydrogen cells, and nanocells in technology.

Chapter Two covers the principles of fuel energy, fuel cell construction and development, and nanotech fuel cell modules.

Chapter Three discusses principles of solar energy, solar cell construction and development, and nanotech solar cells.

Chapter Four covers wind energy construction and development, NPs and wind energy, and nanotech wind energy modules.

Chapter Five covers nanotechnology in industry, including needs assessment, efficient of nanomaterials, nanomaterials of green energy, modern industrial applications, energy storage devices, and applications in energy storage devices.

Chapter Six discusses nanostructures of green energy (hydrogen, solar, wind), design and theoretical models, effective NPs and matrices, nanomaterials and life models.

Chapter Seven explains synthesis of nanotech green energy (hydrogen, solar, wind), fabrication methodology, thin-film morphology, and nanomaterial characterization.

Chapter Eight examines nanotech modules for green energy power plants (hydrogen, solar, wind), design modules, synthesis of nanotech hydrogen and fuel cells, synthesis of nanotech solar cells, and synthesis of nanotech wind farm modules.

Chapter Nine discusses nanotech green energy generation systems (hydrogen, solar, wind), green energy power generation, sustainability of green energy generation, and hybrid applications in green energy systems.

Chapter Ten overviews economic studies for nanotech green energy (hydrogen, solar, wind), theoretical models for green energy systems, traditional economic studies, nanotech economic studies, economic efficiency and investments.

The End

At the end of this book, it is important to focus on the fact that there is no doubt that the ever-increasing demand for fossil fuels and various industrial processes has exposed our environment high levels of greenhouse gases, including carbon dioxide, methane, and nitrogen oxide. In recent years, the appetite for green energy has been increasing among industries and in vehicles and homes, which account for a substantial share of the overall natural energy demand. The demand for various forms of energy, such as liquid fuels and renewable energy, has led to the need for energy extraction. Considering the global scenario and the need for energy, a cleaner source of energy, with the aid of nanotechnology, is making significant progress for the modern society.

It is important to recognize that nanoscience and nanotechnology have fundamentally changed the way advanced technology can be developed in any field related to materials. The materials used in these areas are profoundly influencing green energy generation. Nanostructures can transport energy on a global scale and allow carriers to have different characteristics than their bulkier counterparts. As nanotechnology has grown in relevance, especially in the energy sector, a variety of energy generation methods have been developed for generating energy such as hydrogen, wind, and solar power. Globally, there is an increasing demand for cheap and reliable energy, which makes it imperative that innovative technologies are created that can be disruptive and revolutionary as well as ecofriendly. Today and tomorrow, nanotechnology will play a vital role in improving green energy generation.

Prof. Ahmed Thabet Mohamed
Founder & Head of Nano-Technology Research Center,
Electrical Engineering Dept., Aswan University,
Egypt

Section I

Construction and Development of Green Energy

1 Nanotechnology in Hydrogen Energy

Hydrogen is not only the smallest element on earth, but it also has the lightest weight: One gallon of gasoline weighs approximately 2.75kg, while one gallon of hydrogen weighs an extremely small fraction of that amount (at a pressure of 1 atm and a temperature of 0°Celsius). There is no doubt that hydrogen will become an important fuel in the near future, and that this fuel will be useful in resolving local air quality issues. There is considerable progress being made in the development of hydrogen-powered transport means within the automobile industry. Because hydrogen combustion does not produce carbon oxides but merely water as a byproduct, hydrogen is considered an important fuel in the future. The most common method of transporting large amounts of hydrogen is by liquefying it or pressurizing and shipping it as compressed gas. Generating hydrogen from solar energy can play a significant role in supplying humanity's energy needs while also reducing global climate change-related environmental concerns. A major challenge is to find a way to combine sun-harvesting photochemical modules with catalytic modules that generate H_2 at high rates and quantum yields when the sun shines.

1.1 INTRODUCTION

Among all the elements of the observable universe, hydrogen is the lightest gas in nature and is the most abundant element; its tendency to interact with other elements is attributed to its small atmospheric presence. It is composed of three main naturally occurring isotopes, hydrogen-1 (protium), hydrogen-2 (deuterium), hydrogen-3 (tritium); its other isotopes include hydrogen-4, hydrogen-5, hydrogen-6, and hydrogen-7. As the most common hydrogen isotope, hydrogen-1, is found in more than 99.98% of all hydrogen.

Hydrogen is gaseous at room temperature (which normally lies around 298 K) and normal atmospheric pressure (1 atm). In industry, R&D is being focused on hydrogen as a vector to produce energy through reactions with various elements. Hydrogen is an efficient energy source because it has a high specific energy; for example, it contains more energy than 25 kg of gasoline per unit mass. In fact, in terms of the ratio of mass to energy content, hydrogen is by far the most energy-dense fuel [1].

Hydrogen is being considered for automotive use based on factors including weight, encumbrance, cost, and safety. However, its availability in stationary systems is a key factor in its potential use for energy: There are practically no commercial hydrogen filling stations, in contrast with petroleum. Instead, traditionally, hydrogen is compressed, stored, and delivered to customers in steel cylinders. There is a pressure difference between 250 and 350 bar in the hydrogen after the compression stage of the process. In addition to cryo-compression, hydrogen can also

DOI: 10.1201/9781003512486-2

be stored in two different ways that, combined with the ability to produce electric energy, make it applicable to both stationary and mobile systems (for electric energy generation) [2–4].

Hydrogen is categorized according to its production and impacts. Green hydrogen is derived from water using electrolysis powered by renewable electricity; a carbon capture and storage system is used to reduce emissions. Blue hydrogen emits no greenhouse gases. Grey hydrogen is produced from fossil fuels without the use of carbon capture and storage. It is preferable to store compressed hydrogen in containers with a special spherical shape when the quantity of gas is higher than 15000 Nm3. However, this has a higher specific cost (per unit volume unit) due to the special shape of the container.

Currently, modern tanks are capable of withstanding pressures from 350 to 700 bar; for vehicles, the problem of gas accumulation on board is caused by their low energy value per volume unit, which makes them more vulnerable to leakage. In a case of fire, the flame-proof fuses will prevent the hydrogen from bursting. Furthermore, an explosion-proof cut-off valve will be utilized if a collision occurs and the tanks feature robustness. All these factors contribute to the safety of compressed hydrogen storage in the liquid state; hydrogen is available at a temperature of –253°C, which must be maintained through use of specialized cryogenic tanks. One of the most significant challenges to storing liquid hydrogen is that it is technically complex to manage, which is one of several reasons why it is not common yet. Moreover, it is more expensive to compress and store hydrogen than compressed gas, which makes the notion less attractive.

1.2 PRINCIPLES OF HYDROGEN ENERGY

The availability and the high cost of hydrogen gas make it a noneconomic fuel to use because it is unreliable and difficult to produce [5]. A great deal of different feedstocks can be utilized as sources of hydrogen, including water, coal, natural gas, biomass, hydrogen sulfide, boron hydrides, and others through thermal, electrolytic, or photolytic processes.

A large amount of hydrogen is currently produced from fossil fuels by means of well-established technologies, such as steam reformation, gasification, partial oxidation; while these are well established technologies, they are criticized for their high energy consumption, high carbon emissions, and high energy use. In addition to carbon mitigation and the reduction of CO_2 emissions, there are other routes like biomass production which can substitute fossil fuels with sustainable biomass fuels. Even so, there is still quite a bit of work to be done before those technologies can be widely implemented across all industries.

In a clean and sustainable way, hydrogen can be produced from water via electrolysis when the electric power is generated using renewable resources. The hydrogen is produced in a compressed form and can be stored in gas or liquid form, which are commonly used in industry depending on the application. Recent researchers have been studying how to increase the density of hydrogen storage in order to increase its suitability for vehicular applications. The hydrogen atoms or molecules can attach to the surface or be integrated into the lattice structure of materials so that they can be attached to or integrated with the surface. Storage of large amounts of hydrogen is highly dependent upon the development of novel materials that can be used to store and release the hydrogen at a reasonable temperature and pressure for long periods of time.

Hydrogen can be combined with oxygen to generate electricity and heat, with only water being generated as a byproduct of the process. Methanation can be used to synthesize gas and inject it into the natural gas grid without limit, or hydrogen mixtures can be added directly into the natural gas grid with a limited amount of restrictions. There are two ways in which hydrogen can be sold locally, either selling directly to the national grid (in watt-hours) or selling to the public as compressed hydrogen bottles for use in hydrogen cars or microgrids. As a result of this power-to-gas technology, electricity grids and gas grids can be connected, increasing the storage capacity of the electricity grid, which is normally less capable and more expensive to run than the natural gas grid. It is possible that hydrogen technology will change our power infrastructure and even our lifestyle as a result of its use.

Academics and industry must work intensively to fully realize and harness the potential of the hydrogen economy. There are four main routes through which hydrogen can be produced: renewable, nonrenewable, nuclear, and biomass. These routes affect hydrogen production differently. A wide range of processes are available for producing H_2, generally categorized as conventional or renewable, including hydrocarbon reforming and pyrolysis. There are three chemical processes that are deployed in the hydrocarbon reforming process: steam reforming, partial oxidation, and autothermal steam reforming [6].

A majority of the hydrogen supply in the world is still derived from fossil fuels, since much of the costs involved in hydrogen production are influenced by fuel prices. Approximately 48% of hydrogen is produced from natural gas, 30% from heavy oils and naphtha, and 18% from coal [7–9]. A wide range of hydrogen production methods have been developed that can meet almost all of the demand for hydrogen. For instance, membrane reactors produce hydrogen from conventional fuels and are in many fields of the chemical and biochemical industries. Membranes are structures that are capable of mass transfer under varying gradients of driving forces (concentrations, pressures, temperatures, electric potentials, etc.) and are usually thin and wide [10]. Hydrocarbons are currently the primary feedstock used to produce hydrogen, but the need for renewable technologies to become increasingly prevalent in the field of H_2 production must be addressed in the near future as fossil fuels are becoming less abundant and greenhouse gas impacts are becoming more prevalent. Green technologies are expected to gain a greater share of the market in the near future and at the same time to dominate over conventional technologies over the long term [11–14].

Various processes have been developed to produce hydrogen from renewable resources. Thermochemical processes appear to be the most competitive for large-scale hydrogen production, and alternative energy sources include gas exhaust from gas turbines and concentrated solar energy. A number of issues need to be overcome regarding hydrogen's more widespread industrial and commercial use; if water-splitting methods are not further developed, hydrogen production will not become competitive and will not be affordable for the public. In the case of water electrolysis, the water is decomposed by electrical current by the following reaction: $2H_2O + \text{electricity} \rightarrow 2H_2 + O_2$; high efficiency depends in part on the types of materials used for the electrodes of the catalysts that are used. However, water electrolysis is less efficient because it requires a large amount of electricity [15].

There are four different categories of electrolyzers. The traditional alkaline electrolyzer ensures a balance between the electrolyte conductivity and corrosion resistance; it is the most effective of all electrolyzers. In advanced alkaline electrolyzers with high current density, conductivity increases as electrolyte temperature increases, matching the efficiency and electricity savings of hydrogen energy production. Solid polymer or membrane electrolyzers do not contain corrosive liquid electrolytes, have low maintenance requirements, and can operate at high current densities; for these reasons, they are far safer than their water-based counterparts. However, one issue is that the costs of the platinum and ion-exchange membranes play a major role in the final cost of the plant. Finally, high-temperature electrolyzers have up to 100% and use a solid ceramic electrolyte (oxygen-ion conductor) that is noncorrosive and leak-proof to both liquids and gases. A major disadvantage of the technique is its expense; renewable energy sources should be used at the beginning of the process to ensure that it is fully clean.

Although the possibility of producing hydrogen through a clean process is increasing every year, an increasing quantity of hydrogen is still being produced in the world by turning methane (CH_4) into hydrogen by a chemical reaction called reforming. In steam reforming, hydrocarbons (often methane) are converted into hydrogen by chemically reacting methane with water vapor at a temperature ranging between 700°C and 1100°Cunder pressure of 20 bar; the catalysts used in this process are usually nickel and alumina, which are able to withstand a high temperature. According to the following reaction, $CH_4 + H_2O(g) \rightarrow CO + 3H_2$ - 191.7 kJ/mol, a combination of carbon monoxide and hydrogen is produced in the process, and this mixture is called a syngas; with methane combustion typically supplying the thermal energy required for an endothermic reaction. Although the reaction is more favorable at low pressures, higher pressures reduce the amount of compression necessary [16].

1.3 HYDROGEN CELL CONSTRUCTION

Fossil fuels have long dominated global hydrogen fuel cells for energy production, but a number of contemporary technologies are being used to produce hydrogen today, of which hydrocarbon reforming and fuel pyrolysis and co-pyrolysis have been the most significant.

Steam hydrocarbon reforming has been the most significant technological advancement so far, and it has proven to be the most cost-effective and reliable method of producing hydrogen on an industrial scale, although it is still in the early phases of development. In terms of operating cost, partial oxidation is more effective and less expensive than steam reforming, but the subsequent shift makes the partial oxidation process more expensive. As far as investment cost is concerned, pyrolysis can be performed at acceptable levels and can yield oil yields that are satisfactory [17].

With ecofriendly, CO_2-neutral electrochemical water splitting, it can be produced hydrogen that can then be used to generate electricity via renewable sources such as solar and wind power. Water is decomposed into H_2 and O_2 by a process called electrolysis that uses direct current in an electrochemical cell as follows:

$$2H_2O \rightarrow 2H_2 + O_2 \quad (1.1)$$

This endothermic reaction requires an external source of energy and high voltages.

In electrolysis, an acidic or alkaline electrolyte is used to conduct ions between two electrodes. The reactions that occur on the negative electrode (cathode) and the positive electrode (anode) in a half-cell are the hydrogen evolution reaction (HER) and the oxygen evolution reaction (OER). There must be a separation of the gases produced to prevent the generation of explosive oxyhydrogen (stoichiometric 2:1 mixture of H_2 and O_2). This can be achieved by introducing a gas-impermeable membrane between the two electrodes, which allows the gas at each electrode to be withdrawn separately [18, 19].

Electrolysis can produce hydrogen carbon free by producing the electricity with nuclear or renewable energy. During electrolysis, water is split into H_2 and O_2 molecules in an electrolyzer. Small appliance-sized electrolyzers are well suited for distributing hydrogen in small quantities on a small scale, while larger, more powerful electrolyzers are well suited to large-scale, central production facilities that are directly connected to non-greenhouse-gas-emitting sources of energy.

It is important to keep in mind that different electrolyzers function in different ways, primarily because of the varying types of electrolyte materials involved and the varying ionic species that they conduct. A polymer electrolyte membrane (PEM) electrolyzer reacts with water at the anode to form positive hydrogen (H^+) and oxygen (O^+) ions; hydrogen gas is formed at the cathode when H^+ ions combine with electrons that flow through the external circuit. Alkaline electrolyzers produce hydrogen by moving H^+ through the electrolyte, while alkali ions (OH^-) circulate on the cathode side:

$$\text{Anode reaction: } 2H_2O \rightarrow O_2 + 4H^+ + 4e^- \quad (1.2)$$

$$\text{Cathode reaction: } 4H^+ + 4e^- \rightarrow 2H_2 \quad (1.3)$$

Since the early days of the electrolyzer, liquid alkaline solutions of sodium hydroxide (NaOH) or potassium hydroxide (KOH) have been used as the electrolyte and are considered to be safe and effective. One new approach that can be used on a laboratory scale is using solid alkaline exchange membranes as an electrolyte.

A solid oxide electrolyzer uses solid ceramic materials as the electrolyte to selectively conduct O^+ ions elevated temperatures while generating hydrogen using a slightly different type of electrolysis: Highly charged O^+ ions combine with H_2 gas at the cathode of a fuel cell to release electrons from the external circuit. When the O^+ ions pass through solid ceramic membranes, they react at the anode in a process known as redox (oxidation/reduction) reactions that form O_2 gas and release electrons into the external circuit. To ensure that solid oxide membranes are able to perform their functions, solid oxide electrolyzers typically operate between 700°C and 800°C, whereas PEM electrolyzers operate between 70°C and 90°C.

Advanced solid oxide electrolyzers based on proton-conducting ceramic electrolytes can operate at lab scale with the operating temperature lowered from 750°F to 500°F using solid oxide electrolyzes. Solid oxide electrolyzers produce hydrogen at higher temperatures to reduce the amount of electrical energy required. To keep a constant concentration of KOH or NaOH electrolyte during the reaction, it is important to replenish the electrolyte with water since it is consumed during the reaction. Several factors contribute to the stability of the water electrolyzer cell when an external current is applied to it, such as kinetic hindrances, losses through ohmic resistances, and irreversible electrochemical reactions [19].

There may be a synergistic potential for hydrogen production via electrolysis if some renewable energy technologies are able to achieve intermittent and dynamic power generation, which is typical of those technologies. There is a great deal of variability associated with wind power, which has made its use difficult even though its cost has continued to fall. However, it is also possible to use the surplus energy generated at wind farms to produce hydrogen fuel cell to achieve cross-over flexibility, which is better because today's grid electricity must be generated using methods that are responsible for greenhouse gas emissions and are energy intensive. In fact, it is the least preferred source of electricity for electrolysis. Hydrogen storage systems are an effective option for reducing power generation and demand imbalances and generate economic benefits through economizing on excess hydrogen and oxygen production.

1.4 HYDROGEN ENERGY DEVELOPMENT

Water, biomass, hydrogen sulfide, and fossil fuels are just a few of the substances that contain hydrogen. The process of removing hydrogen from fossil fuels must sequester all CO_2 and other pollutants that are typically released into the atmosphere to produce hydrogen at zero or low environmental impact. It is possible to generate electrical and thermal energy using fossil fuels (which must be processed in order to be deemed "clean"), renewable energies (such as solar, wind, hydro, wave, ocean, and thermal), biomass, nuclear, and recovered energy sources.

1.4.1 Industrial Electrolysis

Water electrolysis is currently one of the most common industrial processes for producing almost pure hydrogen, and it is expected to increase in importance; the process takes place when electrons are moved in a steady flow by an external circuit. The main electrochemical hydrogen production technologies are alkaline, polymer membrane, and solid oxide electrolyzers, and the production reactions are accelerated by applying homogeneous or heterogeneous catalysts to the electrode surfaces to increase the current density and reaction rate. Platinum is one of the most commonly used heterogeneous catalysts, but homogeneous catalysts are less expensive than heterogeneous ones due to their high turnover rates, up to 2.4 mol of hydrogen to one mole of catalyst every second [20].

Electrolysis requires the desalination and demineralization of the water because electrolyzers (especially PEM electrolyzers) are highly sensitive to the purity of the water. An electrolyzer, for example, is more likely to produce chlorine than oxygen if brine (or sea water) is fed to it since chlorine would be more likely to produce oxygen than brine. Using ion-selective membranes effectively removed salt from water without generating chlorine [21]. Under high-temperature electrolysis, steam disassociates into H_2 and O_2 by passing through a metal electrode at temperatures between 700°C and 7000°C. This method is considerably more efficient than conventional room-temperature electrolysis.

Using thermal energy, high-temperature electrolysis converts water into steam using either direct steam heating or indirect heat transfer to generate the heat. As a result, it requires less electrical energy than conventional electrolysis to achieve the same results. Clean external heat sources such as solar, geothermal, and nuclear can achieve zero greenhouse gas emissions. Although even though this process efficiently generates hydrogen, the system components must be designed to meet specific requirements due to the high temperature at which they operate [9].

1.4.2 Plasma Arc Decomposition

Plasma is characterized by the presence of electrons in an excited state and of atomic species; as a result of electrons in their excited state and the presence of electrically charged particles, plasma is capable of releasing high voltage electric currents. Plasma arc decomposition, also called high-temperature pyrolysis, breaks down methane in the absence of oxygen into hydrogen and carbon black (soot) under thermal radiation: While hydrogen is collected in the gas phase, carbon black remains in the solid phase at the bottom. Methane decomposes to carbon and hydrogen as follows:

$$CH_4 \rightarrow C + 2H_2,\ \Delta H = 74.6\ \text{MJ/kmol} \tag{1.4}$$

Using plasma cracking to make hydrogen is at least 5% cheaper than using steam methane reforming with carbon dioxide sequestration as a solvent [22].

1.4.3 Water Thermolysis

The reaction for water thermolysis, the thermal decomposition of water in a single step, is written as

$$H_2O \rightarrow H_2 + (½)O_2 \tag{1.5}$$

Adequate decomposition requires temperatures above 2500 K; at 3000 K and 1 bar, the decomposition rate reaches 64%; however, separating H_2 and O_2 from each other is one of the challenges of this production method. It is also important to cool the mixture before it is sent to the separation process because semi-permeable membranes can't be used at temperatures higher than 2500 K. The experimental solar thermolysis of water reaches 90% of equilibrium after only one minute of residence time at 2500 K [23].

1.4.4 Thermochemical Water Splitting

Thermochemical cycles that split water produce individual chemical reactions more efficiently than traditional chemical reactions because they require no catalysis. Additionally, water is the only material source of hydrogen in the thermochemical cycle that cannot be recycled. The sulfur–iodine (S-I) cycle is a viable method for generating electricity; however, it must be proven that these systems are commercially viable [24]. Thermal energy drives the first reactions of S-I cycles:

$$H_2SO_4 \rightarrow H_2O + SO_3 \tag{1.6}$$

$$SO_3(\text{heated to } 800\text{–}900°C) \rightarrow (1/2)O_2 + SO_2 \tag{1.7}$$

In the presence of water and iodine, sulfur dioxide (SO_2) undergoes an exothermic reaction before being separated from O_2, which occurs spontaneously under low temperatures at low temperatures:

$$SO_2 + I_2 + 2H_2O \rightarrow 2HI + H_2SO_4 \tag{1.8}$$

$$2HI\ (\text{heated to } 425\text{–}450°C) \rightarrow H_2 + I_2 \tag{1.9}$$

S-I cycles require high temperatures, which is why there are few effective and sustainable thermal energy sources. Gasification uses fixed, moveable, or fluidized bed gasifiers to produce hydrogen under autothermal or thermal processes, where autothermal gasification utilizes partial oxidation to provide the necessary heat.

A thermochemical process is used to produce hydrogen from ethanol and methanol. However, several variables need to be considered for photovoltaic (PV)-based electrolysis, including PV panels, DC bus bars, AC grids, accumulator batteries, the electrolyzer, and hydrogen storage canisters. PV electrolysis is one of the most expensive methods of producing hydrogen for fuel cells, 25 times more expensive than using fossil fuels. However, researchers are continually lowering the cost of this method, and it is expected that the factor will go as low as 6 by the middle of the century [25].

Photocatalysis is a process in which photonic energy is converted into chemical energy (hydrogen) using light from the sun. The energy in photons is proportional to the frequency of the radiation they are carrying; that particular amount of energy can be obtained from hn where h is the Planck constant and n is the frequency. An electron–hole pair is generated when a photon strikes the photocatalyst and the electric charge gained from the generated pair is used to decompose water when it hits the photocatalyst.

The photocatalyst is responsible for splitting water and generating hydrogen in the reaction. It should therefore have a band gap that is appropriate for redox and properly located conduction and valance bands. Additionally, it is essential for choosing the right photocatalyst that electron–hole pairs are generated and separated rapidly to make the reaction efficient. Moreover, complex supramolecular devices that have been engineered, modified, and chemically altered are also used to perform photocatalytic reactions. There are considerable studies on the potential of simple and complex photocatalysts based on their capacity to produce oxygen including on their efficiency and their impacts on the environment and human health [26].

In contrast with thermal water splitting, hybrid thermochemical cycles operate at lower temperatures and therefore are more efficient; thermal and electrical energy combine to meet the external energy requirements of each electrochemical reaction. Because hybrid cycles operate at lower temperatures than solar, nuclear, and biomass combustion, additional sustainable energy sources can be used to drive their associated processes including recovered waste heat from nuclear and geothermal facilities. Hybrid thermochemical cycles can operate at temperatures as low as 550°C and operate using Cu–Cl as the energy source, mainly for driving the cycle directly but also in part for generating electricity, which is one of the benefits of these cycles. In addition to taking place at lower temperatures, hybrid thermochemical cycles can be made sustainable by using energy that is produced through nuclear power, industrial heat, waste heat from power plants, concentrated sunlight, incineration of municipal waste, or geothermal heat.

For instance dark fermentation uses the biochemical energy stored in organic matter to produce H_2 in the absence of light; because the process requires no light, it is simpler and less expensive than photo fermentation bioreactors. Dark fermentation can also produce hydrogen from organic wastes potentially contaminated by biotoxins, for instance in wastewater treatment systems. Compared with petroleum sources, organic waste including wastewater is cheaper and more readily available [27, 28].

1.4.5 Coal Gasification and Fossil Fuel Reforming

Coal is currently most commonly used for large-scale hydrogen production. A high-temperature and high-pressure reactor is used to accelerate the gasification of coal by partially oxidizing it with steam and oxygen, producing H_2, CO_2, and syngas. In a shift reaction, hydrogen yield increases if the syngas is processed and cleaned using elemental S or SO_2. Gas turbines are then needed to further process the syngas before it can be consumed as energy and converted into electricity.

The CO_2 emissions from coal gasification are higher than those from other hydrogen production methods because of the high carbon content; work is taking place to develop technologies that can capture and store carbon to address the emissions. Additionally, producing hydrogen from coal gasification costs more than doing so by steam reforming natural gas, which also incorporates techniques that are better defined that those for coal gasification [29].

Steam reforming requires an external heat source, but it does not require oxygen. Its operating temperature is lower, and the ratio of H/CO is greater than with either partial oxidation or autothermal reforming. Under partial oxidation, hydrocarbons are oxidized and generate heat in the process. As opposed to steam and autothermal reforming, partial oxidation does not require a catalyst, and it is more S-tolerant. However, both autothermal reforming and partial oxidation do not depend on external heat sources; both also use pure oxygen as a feedstock and require oxygen separation units that increase the complexity and cost of the process. Although there are multiple technologies for producing hydrogen, steam reforming (particularly with methane) is still the most common and most economical [30].

1.4.6 Photo-Electrolysis, Bio-Photolysis, and Photo-Fermentation

Photo-electrolysis is a process in which heterogeneous photocatalysts are applied to either or both electrodes at the same time; a photo-electrolysis cell should be supported by an electrical current in addition to being exposed to solar irradiation for the success of the process. Photo-electrolysis converts both photonic and electrical energy into chemical energy in the form of hydrogen. Photo-electrolysis entails the following steps: (i) using a photon to generate an electron–hole pair that has a pen junction with a higher band gap than that of the anode, (ii) generating electricity via the flow of electrons from the cathode to the anode, (iii) decomposing water into H^+ ions and O_2 gas, (iv) reducing the ions at the cathode to form H_2 gas, and (v) separating the product gases, processing, and storage.

The performance of a photo-electrolytic system depends on the photon-absorbing material used: its crystal structure, surface properties, corrosion resistance, and reactivity. The photoelectrodes that efficiently convert photon energy into hydrogen are usually less stable in electrolytes, but the chemically stable photoelectrodes are less efficient for splitting water. Bio-photolysis can produce hydrogen via direct, indirect, or photo-fermentation. In the process, a photobioreactor uses light-sensitive bacteria to convert glucose to hydrogen [31]. The primary advantage of bio-photolysis is its ability to generate hydrogen from water at standard temperatures and pressures, which takes place via photoactivated enzymes and the following reactions:

$$6H_2O + 6CO_2 \rightarrow C_6H_{12}O_6 + 6O_2 \quad (1.10)$$

$$C_6H_{12}O_6 \rightarrow 6H_2O + 6CO_2 + 12H_2 \quad (1.11)$$

Many microorganisms can be used to produce hydrogen, but microalgae are the most suitable because they can be grown in closed systems and give high yields.

1.5 NPs AND HYDROGEN CELLS

A major attraction of nanomaterials for biohydrogen production is that they improve the physicochemical properties of the microorganisms. Specifically, the electron-to-electron transport pathway disrupts the incentive movement of microorganisms and the associated metabolic side effects, such as interfering with maturation pathways and adjusting hydrogenase quality through acetic acid derivatives and butyrate [23–34].

In keeping with the recent interest in using nanotechnology to increase dark fermentative hydrogen production, carbon nanotubes (CNTs) have shown tremendous effectiveness when they are combined with metal oxides such as copper (Cu), Au, palladium (Pa), silver (Au), iron–iron (F-F), nickel–nickel (N-N), silica (Si), and titanium (Ti) [35]. The high production of acetic acid, butyrate, and ethanol derivatives produces significantly high hydrogen yields, whereas low production rates cause the H_2 to explode. Au NPs have also been found to be appropriate for generating hydrogen [36, 37]. CNTs are in high use for forming biosensors and microbial power devices because of their decreased potentials, which are suitable for both the redox and the electron transfer reaction kinetics [38]. They have been widely used in biomedical and environmental fields, but they are increasingly being used to produce liquid biofuels such as biodiesel, biobutanol, and bioethanol. As much as there are concerns regarding the toxicity of NPs at higher concentrations, they can significantly increase the production of hydrogen under the right conditions by breaking down substrates that bacteria can convert into hydrogen [39].

Many new developments are being made in the field of nanotechnology not only in the academic community but also among investors, governments, and industry. For instance, every process that uses solar radiation as a source of energy uses nanotechnology in the process. The sun is available everywhere and is free, making it an ideal renewable energy source. It is possible to utilize solar energy in many different ways, such as via PV cells or solar collectors; for instance, passive solar involves building buildings to maximize their solar lighting and heating.

Biotechnology uses solar energy to drive chemical transformations in plants, resulting in complex carbs that are then used to produce energy, steam, or biofuels. PV solar cells convert the photons in the light from solar radiation into electrical current. Current solar panels are based on silicon (Si) wafers about 150-300 nm thick, forming the first-generation PV cells and the basis for the PV market. In terms of global solar cell market share, this technology accounts for over 86% [40, 41]. The second generation of PV cells use thin (1–2 nm) epitaxial layers that have been lattice matched. These cells make up around 90% of the market area but only a small fraction of the global PV market. However, despite their lower manufacturing costs, they unfortunately also have low conversion efficiency.

PV cells can now be constructed with nanoscale components, which first improves the flexibility and interchangeability that are provided by the ability to control the band gap energy and second improve the optical path and significantly decrease charge recombination probability. Researchers used thin-film solar cells based on nanocrystal quantum dots (QDs) coated on conductive transparent oxide substrates; QDs are NPs made of semiconductors with direct band gap and are particularly effective for PV technology. QDs and quantum wires are extremely effective light emitters that adjust to the wavelength of solar radiation coming into the device. They can emit multiple electrons per solar incident photon and have distinct absorption and emission spectra depending on the particle size and shape. PV energy directly decomposes water molecules into hydrogen and oxygen by photocatalytic electrolysis [42–44].

Sandwiching NPs between two transparent and conductive polymeric layers and irradiating them with light that are photonically irradiated releases electrons that are then carried through the polymer before being recombined with electrons before it degrades; an aqueous electrolyte fills the space surrounding the NPs. TiO_2 has a 3.2 eV band gap that can only be used for hydrogen production using UV light. NPs also substantially increase microbial metabolism in H_2 production, due to the transfer of electrons under aerobic conditions [45]. Smaller NPs have higher specific surface areas than larger NPs, which leads to competitive electron adsorption on the NP surface. There is growing interest in using metal and metal oxide NPs to improve the production of biological H_2 in the production media. Various organisms are also capable of producing H_2 from pure sugars or biowaste in the presence of

NPs, although their efficiency depends on the NP type and concentration [46, 47]. There is a direct correlation between the quantum size of NPs and the rate at which electrons are transferred between NPs and enzyme molecules, such as hydrogenase, which is known to convert hydrogen into protons and vice versa. This catalyzes the conversion of hydrogen into protons in order for electron sinks to be provided to the system or to provide reducing power from H2 oxidation as $H_2 \rightarrow 2H + 2e^-$.

1.6 NANOCELLS IN TECHNOLOGY

Graphene NPs are structurally integrated with 2D assemblies of electrocatalytic NPs to fabricate flexible electrodes with high performance. The demand for modern electronics, portable medical devices, and compact energy storage devices has made flexible electrodes one of the heated discussions in green energy generation today. A variety of applications have been investigated for hybrid electrodes anchored on carbon substrates and containing nanocrystals that can be used for sensing, energy conversion and storage, and catalysis [48–52].

Graphene, which is made up of a single layer of carbon atoms hybridized with sp^2, has emerged as a carbon-based scaffold for nanocrystals because of its unique structure and electronic properties that include large surface areas, chemical inertness, and superior electrical conductivity. Graphene oxide (GO) paper can transfer 2D NP arrays at oil–water interfaces onto monolayers of densely packed Au NPs of uniform sizes [53–56].

In a major discovery, GO paper was constructed and electrochemically reduced to restore the ordered structure and electron graphene transport properties; the resulting electrodes were robust and biocompatible and demonstrated excellent electrocatalytic capabilities. These electrodes exhibited excellent electrocatalytic properties by detecting glucose and hydrogen peroxide released by living cells with high sensitivity and selectivity. This approach is a modular one, but recently advanced nanocrystal synthesis and surface engineering methods have also provided the opportunity to examine the systematic relationship between structural parameters and the chemical composition of nanocrystals on the catalytic performance, a significant advancement in nanocrystal synthesis and surface engineering [57].

Owing to recent advances, the synthesis and surface engineering of metal and semiconductor nanocrystals with well-defined shapes and sizes have become extremely valuable for creating closely packed 2D nanocrystal arrays that have controlled shapes and sizes. Hybrid electrodes are produced in three primary steps: optimizing the GO paper and preparing 2D assemblages of Au NPs, loading the NP assemblages onto GO paper and fixing them to a solid substrate through dip-coating, and electrochemically reducing the 2D Au-GO paper into Au-rGO paper with a tightly packed monolayer of Au NPs.

The Au-rGO paper was tested as a freestanding working electrode in a three-electrode system. GO nanosheets were exfoliated using intensive ultrasonication to prepare hydrophilic GO paper that was then placed in a polytetrafluoroethylene (PTFE) mold to form an aqueous dispersion, and the GO paper could be easily peeled from the low-surface-energy PTFE substrate after the water had slowly and completely evaporated at room temperature [58–60].

1.7 TRENDS AND RECOMMENDATIONS

The use of hydrogen as a fuel alternative has been attracting special attention. However, unlike coal, gas, or oil, hydrogen does not have the potential to serve as a primary source of energy. In fact, hydrogen power is the same as energy from electricity in that both require first producing energy from another source before it can be used in its final applications and have particular transport and storage requirements.

Hydrogen for power can be produced from renewable (solar, wind, biomass, geothermal) and/or nonrenewable (coal, natural gas, and nuclear) sources. Electricity and heat can be generated and transported by fuel cells, internal combustion engines, and turbines. Hydrogen production produces only water as a byproduct, and hydrogen has outstanding potential to become a major catalyst in catalyzing the transition of our current global energy economy from a carbon-based one to one that relies on sustainable, renewable, clean power. Hydrogen has the potential to play a vital role in addressing issues like pollution, climate change, future energy supply availability and security.

Leaders in many countries have recognized hydrogen as a significant alternative energy source and a key technology for the transition to a sustainable future, and it is increasingly considered a viable alternative energy source for stationary power, transportation, industrial and residential applications. However, it is important to identify the technological, economic, and social bottlenecks of hydrogen storage systems. Measures of service and social impact are needed along with performance, market, environment, and social indicators and business models that can convert hydrogen storage system ideas into economic value [61]. For instance, the service, technology, organization, and finance (STOF) business model has proven to be particularly suitable for analyzing hydrogen storage systems [62]. Technology does not spread alone, so it is necessary to understand how a technical innovation is built and fits in a society. In light of this, it is important to pay attention to the alliances, competition, cooperation, and discrepancies between intersecting and interfacing logics [63].

REFERENCES

1. Di Sia P. Hydrogen and the state of art of fuel cells. *J Nanosci Adv Technol* 2(3):6–13. https://doi.org/10.24218/jnat.2017.27.
2. Zhevago NK, Chabak AF, Denisov EI, Glebov VI, Korobtsev SV. Storage of cryo-compressed hydrogen in flexible glass capillaries. *Int J Hydrogen Energ* 2013;38(16):6694–6703.
3. Di Sia P. On the use of hydrogen in the automotive sector. *Nano Energy Syst* 2017;1(2): 50–58. https://doi.org/10.24274/nes.2016.
4. Kampitsch TBM, Kircher O. Cryo-compressed hydrogen storage. In: Stolten D, Samsun RC, Garland N, editors. *Fuel Cells: Data, Facts and Figures*. Weinheim: Wiley-VCH Verlag GmbH & Co. KGaA; 2016. https://doi.org/10.1002/9783527693924.ch17.
5. Kapdan IK, Kargi F. Bio-hydrogen production from waste materials. *Enzym Microb Technol* 2006;38(5):569–582.
6. Zhang F, Zhao P, Niu M, Maddy J. The survey of key technologies in hydrogen energy storage. *Int J Hydrogen Energy* 2016;1–18.
7. Kothari R, Buddhi D, Sawhney RL. Comparison of environmental and economic aspects of various hydrogen production methods. *Renew Sustain Energy Rev* 2008;12(2):553–563.

8. Balat H, Kirtay E. Hydrogen from biomass—present scenario and future prospects. *Int J Hydrog Energy* 2010;35(14):7416–7426.
9. Dincer I, Acar C. Review and evaluation of hydrogen production methods for better sustainability. *Int J Hydrog Energy* 2015;40:11094–11111.
10. Koros WJ, Ma YH, Shimidzu T. Terminology for membranes and membrane processes. *Pure Appl Chem* 1996;68(7):1479–1489.
11. Hites RA. Persistent organic pollutants in the great lakes: An overview. *Handb Environ Chem* 2006;5(PartN):1–12.
12. Lund H. Renewable energy strategies for sustainable development. *Energy* 2007;32(6):912–919.
13. Ćosić B, Krajačić G, Duić N. A 100% renewable energy system in the year 2050: The case of Macedonia. *Energy* 2012;48(1):80–87.
14. Mathiesen BV, Lund H, Karlsson K. 100% Renewable energy systems, climate mitigation and economic growth. *Appl Energy* 2011;88(2):488–501.
15. Bessarabov D, Wang H, Li H, Zhao N, editors. *PEM Eelectrolysis for Hydrogen Production: Principles and Applications*. Boca Raton: CRC Press; 2015.
16. Angeli SD, Turchetti L, Monteleone G, Lemonidou A. Catalyst development for steam reforming of methane and model biogas at low temperature. *Appl Catal B-Environ* 2016;181:34–46.
17. Bičáková O, Straka P. The resources and methods of hydrogen production. *Acta Geodyn Geomater* 2010;7(2/158):175–188.
18. Smolinka T, Ojong ET, Garche J. Hydrogen production from renewable energies—electrolyzer technologies. In: *Electrochemical Energy Storage for Renewable Sources and Grid Balancing*. Amsterdam: Elsevier; 2014, pp. 103–128. https://doi.org/10.1016/B978-0-444-62616-5.00008-5.
19. Voitic G, Pichler B, Basile A, Iulianelli A, Malli K, Bock S, Hacker V. Chapter 10: Hydrogen production. In: *Fuel Cells and Hydrogen*. 2018. https://doi.org/10.1016/B978-0-12-811459-9.00010-4.
20. Karunadasa HI, Chang CJ, Long JR. A molecular molybdenum-oxo catalyst for generating hydrogen from water. *Nature* 2010;464:1329–1333.
21. El-Bassuoni AMA, Sheffield JW, Veziroglu TN. Hydrogen and fresh water production from sea water. *Int J Hydrogen Energy* 1982;7:919–923.
22. Gaudernack B, Lynum S. Hydrogen from natural gas without release of CO2 to the atmosphere. *Int J Hydrogen Energy* 1998;12:1087–1093.
23. Baykara SZ. Experimental solar water thermolysis. *Int J Hydrogen Energy* 2004;29(14):1459–1469.
24. Balta MT, Dincer I, Hepbasli A. Thermodynamic assessment of geothermal energy use in hydrogen production. *Int J Hydrogen Energy* 2009;34:2925–2939.
25. Rand DAJ, Dell RM. Fuels e hydrogen production: Coal gasification. *Encycl Electrochem Power Sources* 2009:276–292.
26. Acar C, Dincer I, Zamfirescu C. A review on selected heterogeneous photocatalysts for hydrogen production. *Int J Energy Res* 2014;38:1903–1920.
27. Koutrouli EK, Kalfas H, Gavala HN, Skiadas IV, Stamatelatou K, Lyberatos G. Hydrogen and methane production through two-stage mesophilic anaerobic digestion of olive pulp. *Bioresour Technol* 2009;100:3718–3723.
28. Das D, Veziroglu TN. Advances in biological hydrogen production processes. *Int J Hydrogen Energy* 2008;33:6046–6057.
29. *Royal Belgian Academy Council of Applied Science Report: Hydrogen as an Energy Carrier*. 2006. http://www.kvab.be/downloads/lezingen/hydrogen_energycarrier.pdf.
30. Holladay JD, Hu J, King DL, Wang Y. An overview of hydrogen production technologies. *Catal Today* 2009;139:244e60.
31. Kotay SM, Das D. Biohydrogen as a renewable energy resource—prospects and potentials. *Int J Hydrogen Energy* 2008;33:258–263.

32. Zhao W, Zhang Y, Du B, Wei D, Wei Q, Zhao Y. Enhancement effect of silver nanoparticles on fermentative biohydrogen production using mixed bacteria. *Bioresour Technol* 2013;142:240–245.
33. Han H, Cui M, Wei L, Yang H, Shen J. Enhancement effect of hematite nanoparticles on fermentative hydrogen production. *Bioresour Technol* 2011;102:7903–7909.
34. Hsieh P-H, Lai Y-C, Chen K-Y, Hung C-H. Explore the possible effect of TiO2 and magnetic hematite nanoparticle addition on biohydrogen production by Clostridium pasteurianum based on gene expression measurements. *Int J Hydrogen Energy* 2016;41:21685–21691.
35. Zhao W, Zhao J, Chen GD, Feng R, Yang J, Zhao YF, et al. Anaerobic biohydrogen production by the mixed culture with mesoporous Fe3O4 nanoparticles activation. *Adv Mater Res: Trans Tech Publ* 2011:1528–1531.
36. Zhang Y, Shen J. Enhancement effect of gold nanoparticles on biohydrogen production from artificial wastewater. *Int J Hydrogen Energy* 2007;32:17–23.
37. Pugazhendhi A, Shobana S, Nguyen DD, Banu JR, Sivagurunathan P, Chang SW, Ponnusamy VK, Kumar G. Application of nanotechnology (nanoparticles) in dark fermentative hydrogen production. *IJ Hydrogen Energy* 2019;44(3):1431–1440.
38. Mohan SV, Mohanakrishna G, Reddy SS, Raju BD, Rao KR, Sarma P. Self-immobilization of acidogenic mixed consortia on mesoporous material (SBA-15) and activated carbon to enhance fermentative hydrogen production. *Int J Hydrogen Energy* 2008;33:6133–6142.
39. Kumar G, Mathimani T, Rene ER, Pugazhendhi A. Application of nanotechnology in dark fermentation for enhanced biohydrogen production using inorganic nanoparticles. *Int J Hydrogen Energy* 44 (2019) 13106–13113.
40. Takabayashi S, Nakamuraa R, Nakato Y. A nano-modified Si/TiO2 composite electrode for efficient solar water splitting. *J Photochem Photobiol A Chem* 2004;166:107–113.
41. Serrano E, Rus G, Garcıa-Martınez J. Nanotechnology for sustainable energy. *Renew Sustain Energy Rev* 2009;13:2373–2384.
42. ISO 14040. *Environmental Management—Life Cycle 596 Assessment—Principles and Framework*. Geneva, Switzerland: ISO; 2006. https://www.iso.org/standard/37456.html (accessed on 10 February 2023).
43. Peppas A, Kottaridis S, Politi C, Angelopoulos PM. Taxiarchou M. Multi-model assessment for secondary smelting decarbonization: The role of hydrogen in the clean energy transition. *Hydrogen* 2023;4:103–119.
44. Lobo RFM. A brief on nano-based hydrogen energy transition. *Hydrogen* 2023;4:679–693. https://doi.org/10.3390/hydrogen4030043.
45. Beckers L, Hiligsmann S, Lambert SD, Heinrichs B, Thonart P. Improving effect of metal and oxide nanoparticles encapsulated in porous silica on fermentative biohydrogen production by Clostridium butyricum. *Bioresour Technol* 2013;133:109–117. https://doi.org/10.1016/j.biortech.2012.12.168.
46. Zhang Y, Shen J. Enhancement effect of gold nanoparticles on biohydrogen production from artificial wastewater. *Int J Hydrog Energy* 2007;32:17–23. https://doi.org/10.1016/j.ijhydene.2006.06.004.
47. Patel SKS, Lee J-K, Kalia VC. Nanoparticles in biological hydrogen production: An overview. *Indian J Microbiol* 2017. https://doi.org/10.1007/s12088-017-0678-9.
48. Khalap VR, Sheps T, Kane AA, Collins PG. Hydrogen sensing and sensitivity of palladium-decorated single-walled carbon nanotubes with defects. *Nano Lett* 2010;10:896–901.
49. Sanles-Sobrido M, Correa-Duarte MA, Carregal-Romero S, Rodríguez-González B, Alvarez-Puebla RA, Hervés P, Liz-Marzán LM. Highly catalytic single-crystal dendritic Pt nanostructures supported on carbon nanotubes. *Chem Mater* 2009;21:1531–1535.
50. Liao S, Holmes KA, Tsaprailis H, Briss VI. High performance PtRuIr catalysts supported on carbon nanotubes for the anodic oxidation of methanol. *J Am Chem Soc* 2006;128:3504–3505.

51. Hou Y, Cheng YW, Hobson T, Liu J. Design and synthesis of hierarchical MnO2 nanospheres/carbon nanotubes/conducting polymer ternary composite for high performance electrochemical electrodes. *Nano Lett* 2010;10:2727–2733.
52. Zanolli Z, Leghrib R, Felten A, Pireaux JJ, Llobet E, Charlier JC. Gas sensing with Au-decorated carbon nanotubes. *ACS Nano* 2011;5:4592–4599.
53. Novoselov KS, Geim AK, Morozov SV, Jiang D, Zhang Y, Dubonos SV, Grigorieva IV, Firsov AA. Electric field effect in atomically thin carbon films. *Science* 2004;306:666–669.
54. Geim AK, Novoselov KS. The rise of graphene. *Nat Mater* 2007;6:183–191.
55. Stankovich S, Dikin DA, Dommett GHB, Kohlhaas KM, Zimney EJ, Stach EA, Piner RD, Nguyen ST, Ruoff RD. Graphene-based composite materials. *Nature* 2006;442:282–286.
56. Geim AK. Graphene: Status and prospects. *Science* 2009;324:1530–1534.
57. Xiao F, Song J, Gao H, Zan X, Xu R, Duan H. Coating graphene paper with 2D-assembly of electrocatalytic nanoparticles: A modular approach toward high-performance flexible electrodes. *ACS Nano* 2012;6(1):100–110.
58. Guo HL, Wang XF, Qian QY, Wang FB, Xia XH. A green approach to the synthesis of graphene nanosheets. *ACS Nano* 2009;3:2653–2659.
59. Gao H, Xiao F, Ching CB, Duan HW. One-step electrochemical synthesis of PtNi nanoparticle-graphene nanocomposites for nonenzymatic amperometric glucose detection. *ACS Appl Mater Interfaces* 2011;3:3049–3057.
60. Shao Y, Wang J, Engelhard M, Wang C, Lin Y. Facile and controllable electrochemical reduction of graphene oxide and its applications. *J Mater Chem* 2010;20:743–748.
61. Chesbrough H. Open innovation: A new paradigm for understanding industrial innovation. In: Chesbrough H, Vanhaverbeke W, West J, editors. *Open Innovation, Researching a New Paradigm*. Oxford University Press; 2006, pp. 1–12.
62. Faber E, Ballon P, Bouwman H, Haaker T, Rietkerk O. Designing business models for mobile ICT services. *Paper Presented to Workshop on Concepts, Metrics & Visualization, at the 16th Bled Electronic Commerce Conference Transformation*, Bled, Slovenia, June 9–11, 2003.
63. Becherif M, Ramadan HS, Cabaret K, Picard F, Simoncini N, Bethoux O. Hydrogen energy storage: New techno-economic emergence solution analysis. *Energy Procedia* 2015;74:371–380.

2 Nanotechnology in Fuel Energy

The growing world population has brought about a surge in the use of liquid fuels, especially in the transport sector. It is becoming increasingly important to find sustainable, renewable energy sources that are efficient and cost-effective and produce fewer greenhouse gas emissions because of decreasing fossil fuel availability, increasing emissions of greenhouse gases such as carbon dioxide, and the rising price of petroleum fuels. Scientists across the world from government institutions, commercial businesses, universities, and research facilities are studying the field of nanotechnology, which is one of the new and rapidly developing fields of science and technology. In the near future, the nanotechnology revolution is likely to bring about radical changes in the economy and technology, according to many leading scientists working in green energy generation.

2.1 INTRODUCTION

Fossil fuels are the major energy source for most countries but have had extreme adverse impacts on the environment, including global warming, natural disasters, and health problems associated with air pollution, resulting in a series of social and economic problems. The phenomenon of global warming has been heavily criticized in recent decades; however, attempts have been made in many parts of the world to prevent its adverse effects, such as signing international agreements that in turn have led to a variety of local policies that are tailored to the needs of all signatories. There is no doubt that the circular economic strategy and sustainable development are both aligned on the preservation of resources, but it does not seem that this is happening right now [1].

Around the world, climate change is resulting in the emergence of a wide range of environmental problems, including pollution, hazardous materials, and a loss of ecosystems; technology progress has caused dramatic changes to the world that threaten to harm humans and the environment. Recently, there has been an increase in the awareness of environmental degradation and its risks. The growing evidence of environmental problems is due to a combination of several factors. During the last few decades, environmental concerns have expanded beyond conventional effluent gas pollutants like oxides (SO_X, NO_X), particulates, and CO to pollutants such as microplastics, although advances in environmental science reveal that the main greenhouse gas responsible for global warming is methane.

Rapid high increases in populations worldwide have driven increased industrialization and related increases such as in the road-based transportation of industrial goods and in personal automobile travel. With the increased energy demand, a gap is growing between the availability of energy and its spiraling consumption, and even

DOI: 10.1201/9781003512486-3

developed nations are having difficulties sustainably meeting the demand; in fact, coal is used for the majority of energy needs along with oil and natural gas. However, despite the huge popularity of these conventional fuels, all countries are faced with two major problems, the depletion of fossil fuels and the deterioration of the environment as a result of their use.

Industrial and academic researchers have for decades been exploring alternative energy sources that are cheap, sustainable, and nonpolluting, based on renewable sources such as solar, hydrogen, thermal, and biomass. There is no doubt that global warming and the depletion of fossil fuels have created an opportunity for the development of alternative forms of energy that can fuel the world's economy, and nanotechnology has a role to play in the transition to increasing green energy production. Achieving sustainable human development depends on the sustainable use of fuel energy to achieve the interdependent goals of economic, social, and environmental sustainability.

It is imperative that the researchers take action as soon as possible in order to reduce environmental damage and increase equity while ensuring the growth of global economics, and sustainable green energy production is the path to that process. Sustainability can be defined as the production and use of green energy in a manner that supports the full development of human beings in terms of their social, economic, and environmental dimensions. Engineering and science are currently confronted with one of the most important challenges of our time, which is producing green energy that is completely clean, efficient, and environmentally friendly.

2.2 PRINCIPLES OF FUEL ENERGY

Among the options available for green energy, fuel cells (FCs) are one of the most significant, and the cost-effective use of energy per capita has increased dramatically in recent years. A few decades from now, fossil fuels will be depleted, and the researchers will no longer be able to consume the same amount of fuel that the world uses today. In addition, the environmental effects have been detrimental for individuals and for industry including for the existing infrastructure for energy production. Academic, government, and industry scientists have long been predicting catastrophic consequences for contemporary levels of energy consumption; it is understood that to meet the energy needs of the global economy, alternative methods of producing energy must be developed and used.

In FCs, hydrogen fuel is transformed into electricity by electrochemistry rather than burning to create virtually no pollution; because of this efficiency, they are extremely important to future prosperity as well as a healthy environment for our entire planet. FC technology has made tremendous strides in recent years in terms of development and commercialization for a variety of applications owing to FCs' capacity to convert energy electrochemically rather than thermally. Static energy converts hydrogen or hydrocarbon fuel directly into electrical energy through chemical reactions in a clean and efficient process that creates water as a byproduct. This power source is more efficient at converting energy into work than internal combustion engines or most combustion systems.

FC systems generally operates efficiently and easily under load, and due to the high degree of modularity, they require short construction times; they also have low maintenance costs because they have no rotating parts except for auxiliary components. These devices show tremendous promise as possible alternatives to decentralized electricity generation as well as for combined heat and power production in small/medium sized communities.

2.3 FC CONSTRUCTION

A fuel cell is an electrical and thermal generator based on chemical conversion of hydrogen gas to a solid state by electrochemically reacting with oxygen or an oxidant gas in the presence of electrodes and across an ion-conducting electrolyte. The exhaust gases from this process are subjected to a chemical reaction that produces water.

In contrast to batteries, FCs do not run out of power or require constant recharging; they will continuously produce energy as long as they are continuously replenished with reactants (fuel and oxidizer) and their reactions products are discharged into the atmosphere. FCs directly convert chemical energy into electrical energy, which is a significant advantage over thermodynamic systems. FCs consist of two electrolytes, usually liquid sulfuric acid, sandwiched around two catalytic and relatively stable electrodes. The amount of fuel used is another important parameter, the primary determinant of hydrogen yield in fact. This process uses many different combinations of fuel and oxidant such as hydrogen, carbon dioxide, hydrocarbons, and alcohols as fuels and oxygen, chlorine, and chlorine dioxide as oxidants.

FCs produce energy by combining hydrogen atoms with oxygen atoms in an efficient manner, and there are no moving parts to wear out over time. Creating a basic FC entails only three steps [2]:

(1) Fuel is allowed into contact with the electrolyte through a porous anode, preventing consumption or corrosion. The device disperses hydrogen gas over its entire surface equally and uses the electrons freed from the hydrogen molecules as sources for the external circuits.
(2) Cathodes allow the oxidizer to enter contact with the electrolyte without consuming or corroding themselves in the process. In the electrolyte—a substance that conducts ions (negative or positive)—oxygen ions pass across the electrodes to form hydrogen, which after recombining with oxygen ions passes across the electrolyte to form water, also called a bicarbonate anhydride. This electrolyte also disperses oxygen onto its surface.
(3) The electrolyte (such as phosphoric acid, solid oxide, or an appropriate polymer) is used to block electrons escaping from the electrodes to avoid contact between the two electrodes, and this determines the operating temperature of the FC to prevent electrolysis. Moreover, it maintains a balance of electrical charge by allowing charged ions to flow between electrodes. The conductor used can conduct oxygen or hydrogen ions; the major difference between the two types of conductors is where water is produced in

the FC: On the proton-conductor side, there is an oxidant side, and on the oxygen-ion-conductor side, there is a fuel side. The catalyst on the anode converts hydrogen molecules into hydrogen ions (H^+) and electrons via oxidation. Electrons and protons (H^+ ions) follow different paths to reach the cathode based on their energies and create different currents when they pass through electrolytes.

Among what follows are some of the chemical reactions involved in the cathode and anode reactions, as well as its overall reaction:

$$\text{Anode reaction: } H_2 \Rightarrow 2H^+ + 2e^- \tag{2.1}$$

$$\text{Cathode reaction: } \tfrac{1}{2}O_2 + 2H^+ + 2e^- \Rightarrow H_2O \tag{2.2}$$

$$\text{Overall reaction: } H_2 + \tfrac{1}{2}O_2 \Rightarrow H_2O \tag{2.3}$$

In addition to emitting oxygen ions, the fuel cells also produce heat by reacting with atoms of another chemical in the electrolyte to generate power by generating water and heat.

As previously mentioned, the oxygen ions are produced at one electrode, which pass through the electrolyte to reach the other electrode. The only byproducts of this process are pure water molecules, as follows:

$$\text{Anode side: } H_2 + O_{2^-} \Rightarrow H_2O + 2e^-, CO + O_{2^-} \Rightarrow CO_2 + 2e^- \tag{2.4}$$

$$\text{Fuel-containing hydrogen: } CH_4 + 4O_{2^-} \Rightarrow 2H_2O + CO_2 + 8e^- \tag{2.5}$$

$$\text{Cathode side: } O_2 + 4e^- \Rightarrow 2O_{2^-} \tag{2.6}$$

$$\text{Overall reaction: } H_2 + \tfrac{1}{2}O_2 \Rightarrow H_2O \tag{2.7}$$

2.4 FC DEVELOPMENT

FCs can play a significant role in enabling a green energy transition by reducing emissions. In recent years, some cities have launched domestic energy policies and increased their use of hydrogen in producing renewable energy. Hydrogen technology refers to the collection of tools and processes used to generate, store, transport, and use hydrogen, and FCs are being promoted as renewable energy sources for both small- and large-scale hydrogen production. The efficiency of production, catalysts, and hydrolysis rate are also factors to consider as part of the H_2 generation process. Hydrogen produces zero carbon dioxide, whereas natural gas, the fossil fuel that produces the least carbon dioxide per unit of energy, still releases approximately 180 grams per kilogram of energy. A zone-by-zone conversion of the existing natural gas network to a hydrogen network is a feasible option with minimal disruption to customers during the conversion process and little impact on their total gas costs [3].

Carbonate FCs are one of the fastest-growing, most efficient, nonpolluting, and cleanest generation technologies available today. They can utilize various gases, liquids, and solid carbonaceous fuels for a variety of commercial and industrial applications. These FCs use the alkali metal carbonate mixture as a source of electrolytes and operate at temperatures up to 650 °C. Early challenges in their development relate to corrosion of the cell hardware and the strength of the ceramic components. Additionally, it is necessary to convert the carbonaceous fuel to a usable form for the FCs because both hydrogen and CO_2 are utilized at the anode. Steam reforming of a light hydrocarbon fuel in an external reformer is one of the most reliable and popular industrial processes for the production of hydrogen in FCs [4].

Anode FC reactions are exothermic, and steam-reforming reactions are highly endothermic. A fuel cell anode uses hydrogen as a reactant, but oxygen is used as the reactant at the cathode in the overall reaction. Using a carbonate FC system has the advantage of operating at such a high temperature that the kinetics of the reforming reaction for natural gas as well as other light hydrocarbons can be applied. Carbonate FCs and reforming reactions have unique thermal and chemical characteristics and can efficiently integrate both reactions into its anode compartment to generate the needed energy.

The overall fuel cell reaction is simply a combination of hydrocarbon fuel and oxygen that can be converted into electricity, heat, water, and CO_2, but it is necessary to introduce a hydrocarbon fuel such as natural gas into the anode compartment, along with steam, in order to produce electricity. As soon as the FC is switched on, the unburned fuel from the cell is oxidized by fresh air and introduced to the cathode side to produce hydrogen fuel. As a result of electrochemical reactions between hydrogen atoms at an anode of the fuel cell and carbon dioxide atoms at the cathode, water, CO_2, and heat are released (part of this heat is consumed in the reforming reaction, and the rest is removed from the fuel cell to achieve steady state operation). Internal reforming reactions consume two-thirds of the heat created by the reaction in the fuel cell, which results in a uniform fuel temperature in the cells. This process takes only a single step, resulting in a simpler, more efficient, more cost-effective system than external reforming fuel cells.

Carbonate FCs operate optimally between 600°C and 650°C; they do not require noble metal electrodes as do lower-temperature fuel cells or advanced ceramics as with solid oxide fuel cells that operate at higher temperatures. As with other FCs, carbonate cells use electro-catalysts that are less expensive and made of readily available commercial metals. Direct FCs incorporate several construction features; the bipolar plate and the corrugated current collector are made of 300-series stainless steel, while the injection plug is made of steel, and nickel-based porous material is used for making the electrodes. It is important to note that material properties such as creep strength, sintering resistance, low compaction, hot corrosion resistance, and low carburization will all affect the life of the cell. Anode and cathode electrochemical activity must also be stable [5].

In addition to providing ionic transfer, reactant gas separation, and perimeter sealing, an electrolyte matrix also provided ionic transport [6]. A composite paste-like structure formed by a layer of closely packed ceramic powder bed was impregnated with a dilute alkali carbonate electrolyte at an operating temperature, and the researchers examined FC performance, the stability of the matrix, and its robustness to thermal and mechanical stress. Tables 2.1 and 2.2 show technical characteristics and parameters for the current common FCs, such as the fuel source, electrolyte, operating temperature, cost, efficiency, and suitability for combined heat and power applications.

Engines were run at low to medium operating temperatures (50–210 °C) and by low electrical output efficiency (40%–50%) when operated with readily available FCs made from methanol, hydrocarbons, and hydrogen and by a maximum output of 50% when powered by pure hydrogen fuel. The latter three FCs are characterized by the facts that they operate at high temperatures (600–1,000 °C) and that methane directly takes advantage of their increased efficiency (50% to 60% for common fuels like natural gas and 90% for those with heat recovery) [7].

TABLE 2.1
Technical Characteristics of Different Fuel Cell Types [6]

Cell Type	Electrolyte	T(C°)	Fuel	Oxidant	Efficiency	Heat
Alkaline (AFC)	Potassium hydroxide	50–100	Pure hydrogen or hydrazine	O_2/Air	50%–65%	Very low quality
Proton-Exchange Membrane (PEMFC)	Polymer	50–80	Less pure hydrogen from hydrocarbons or methanol	O_2/Air	40%–50%	None
Phosphoric Asid (PAFC)	Phosphoric Acid	160–210	Hydrogen from hydrocarbon and alcohol	O_2/Air	40%–50%	Low quality
Sulfuric Acid (SAFC)	Sulfuric Acid	80–90	Alcohol or impure hydrogen	O_2/Air	40%–50%	None
Direct Method (DMFC)	Polymer	60–200	Liquid methanol	O_2/Air	40%–55%	Very low quality
Molten Carbonate (MCFC)	Molten salt such as nitrate, sulphate, carbonates	630–650	Hydrogen, carbon monoxide, natural gas, propane, marine diesel	CO_2/O_2/Air	50%–60%	High quality
Solid Oxide (SOFC)	Ceramic as stabilized zirconia and doped perovskite	800–1000	Natural gas or propane	O_2/Air	50%–60%	High quality
Protonic Ceramic (PCFC)	Thin membrane of barium-cerium oxide	600–700	Hydrocarbons	O_2/Air	45%–60%	High quality

TABLE 2.2
Parameters of Different Fuel Cell Types [6]

Parameters	AFC	PEMFC	PAFC	SAFC	DMFC	MCFC	SOFC	PCFC
Power Density (W/Cm^2)	0.2–0.35	0.35–0.36	0.2–0.25	0.2–0.3	0.04–0.23	0.1–0.2	0.24–0.3	0.2–0.3
Lifetime (103 h)	10–15	40–65	40–60	20–35	10–15	40–65	40–80	40–70
Cost/Price (US$/kW)	200	200	1000	800	200	100	1500	1500
Cell Voltage (V)	1	1.1	1.1	1.1	0.2–0.4	0.7–1.0	0.8–1.0	0.8–1.0
Capacity (kW)	10–100	0.03, 1, 2.5, 7, 250	100, 200, 1300	100, 200, 1300	0.001–1, 100–1000 (Research)	155, 200, 250, 1000, 2000	1, 25, 50, 100, 250, 1700	1, 5, 25, 100, 150, 1000

A typical stack configuration involves flat-plate stacks with hydrogen and air flowing down channels in the bipolar plates, exposing each electrode face on one side at the same time to the reactant gases flowing through the channels. When oxygen enters the cathode compartment, it is adsorbed, diffused, and reduced by the incoming electronic charge, which reduces its electronic gain by reducing the ratio of oxygen to oxygen, at the electrode-electrolyte interface. Due to the relative ease of manufacturing of a planar flat-plate stack and the fact that there are fewer energy losses, the configuration is of prime importance.

The bipolar plates have two main functions: transmitting electrons between the elementary cells on the plates and releasing heat to the external environment through the electrodes. An electrolyte, a cathode, and an anode are stacked to create the power section of a fuel cell power plant. Anodes are connected to cathodes by interconnecting plates between them. A common reason for using these plates is because of their high electronic conductivity, their stability in the fuel cell environment, and their compatibility with other components of the cell. Other types of FCs are less common today but could one day find a place in specific applications; these include regenerative cells, Na amalgam cells, air-depolarized cells, biochemical fuel cells, and alkali metal-halogen cells are examples of such cells [8].

As a result of the advances in materials science in the past few decades, current technology has made FC use possible in a few specialized applications. Alkaline FCs are one of the earlier FC systems, employed for NASA's space missions. The electrolyte used in this fuel cell system is an aqueous solution of KOH, which is used to transport positive charged ions in the electrolyte from anode to cathode, with water as the byproduct. In a phosphoric acid cell (PAFC), liquid phosphoric acid is used as the electrode, and the chemical reactions are very similar to those involved in a proton exchange membrane cell (PEMFC); pure hydrogen is used as the fuel source.

It is also possible to provide hot water along with electric power, depending on the heat and electricity load profiles of the facility [9].

Protonic ceramic FCs are made with ceramic electrodes, as in solid oxide FCs (SOFCs). This allows gaseous molecules of hydrocarbon fuels to be electrochemically oxidized without additional reformers at temperatures above 700°C. Additionally, SOFCs use a solid electrolyte, which keeps the membrane from drying out as with PEMFCs and liquid from leaking out like PAFCs [10]. Researchers recently compared stacks of SOFCs, PAFCs, PEMFCs, and molten carbonate FCs; PEMFCs are a very advanced type of fuel cell that is quite promising for a wide range of applications, including mobile carriers, transport systems, and cars. Overall, there is a great deal of potential for SOFCs in the medium power (kilowatts) area where there is a very large market [11].

Fuel processing, water management, and temperature play significant roles in the thermodynamic efficiency of the system; factors that affect the system's electrical efficiency include ohmic loss, stimulation loss, and concentration loss over the lifetime of the fuel cell. A linear region exists in this ohmic region because the voltage drops as the current density increases; in this region, the components exert an internal resistance which is referred to as ohmic polarization. As the output voltage of the device increases with a decrease in the current, the ohmic loss is less significant because of the increased activity of the chemical reactions (the time it takes for the reaction to warm up). Therefore, this region of the electromagnetic spectrum can also be called active polarization. This phenomenon is mainly due to the over flooding of water into the catalyst during the period of high current density [12].

Because there is no combustion as with conventional power generation, FC power-generation stations do not generate greenhouse gases, acid rain, or any other pollutants during operation, in addition to being efficient. In the case of hydrogen fuel cells driven by natural gas as a fuel, there are no net CO_2 emissions at all since any carbon released into the atmosphere is taken up by photosynthetic plants to replenish the oxygen in the atmosphere. There is no doubt that FCs, in general, are among the most efficient ways to generate nonrenewable electricity [13, 14].

It has been shown that FC devices are the cleanest and most efficient devices available for energy generation when combined with devices that use waste heat as a source of energy. FCs themselves can generate large amounts of high-quality waste heat that can be used to heat and cool homes without negatively affecting the environment. FCs are typically located in a distributed system, and, therefore, the optimal size will need to be selected to meet the peak load demands of various utility applications. Different FCs of various sizes are available on the market today, ranging in capacity from 0.5 kW to 2 MW. Scientists developed a novel stacking geometry that extended FC life span by more than 60,000 hours [15], and others used the same novel geometry to create a simple, cost-effective power plant powered by FCs directly on methane for the first time [16].

SOFCs use a thin-film electrolyte made using a relatively simple ceramic process. This process reduces cell resistance and doubles the power output while significantly reducing cost. FC industry professionals developed a nickel bis-di-phosphine catalyst that showed a 48.6% improvement over the basic FC design, the highest power density for commercial-sized FCs anywhere in the world, a significant step forward in their research and development [17].

Changes in cell composition and design has resulted in improved power density; FC systems with high power densities are generally lighter, bigger, and more affordable. FCs can be both modular and distributed to eliminate transmission lines. Furthermore, FCs operate quietly and are vibration free. Distributed applications can compete with combined-cycle gas turbines at low enough costs. Moreover, the benefits of coils include securing energy reliability, lowering operating and maintenance cost, and ensuring constant production of electricity [9, 18–20].

2.5 FC OPERATION

Recent decades have witnessed nanoscience and nanotechnology flourishing; nanomaterials have numerous applications in electrochemistry, electrocatalysis, optics, electronics, analytical devices, energy devices, and so on. The role of nanostructured materials for advanced energy conversion and storage devices has been extensively explored as functional components in these devices [21].

Separately, as I have described, FC type, operation, and output depend on the electrolyte and operating temperature. For instance, SOFCs exhibit some unique advantages over PEMFCs, namely their ability to operate at high temperatures and accept multiple fuels, and they also show fast response times to electrode reactions. All-oxide FCs are more effective than cells with liquid components in eliminating corrosion and requiring rigorous sealants. Electrolytes in SOFCs must provide ion transport pathways while being dense and electronically insulating. There are two types of SOFCs: oxygen ionic conductive and proton conductive [22].

The hydrogen molecules present in SOFCs with oxygen ion conductors dissociate and release two electrons. After the electrons reach the surface of the anode, they are transported along the external circuit to the cathode by the external circuit. Thus, the air oxygen molecules are dissociated and ionized as soon as they combine with the anode electron. Afterwards, a hydrogen ion transfers to the cathode/electrolyte interface, incorporates into the electrolyte, and migrates to the anode/electrolyte interface to reach the reaction site. When the water meets up with the proton at the TPB, it is converted into steam after being heated to a certain temperature; water then diffuses into the bulk gas of the exhausted fuel as it moves toward the anode's surface. An external circuit generates and supplies electricity using electron transport; additionally, heat is generated due to the migration resistance of ions and electrons as well as the reaction polarization loss that occurs during the reaction. This means that the SOFC is a combined heat and power generation system. Combined oxide conductivity results in a water generating anode and cathode chamber in a fuel cell because of mixed oxide ions and proton conductivity.

Several strategies have recently been developed with the aim of improving ionic conductivity of electrolytes by modifying their composition, microstructure, processing, and physical properties [23]. However, there is still a need for a single-phase electrolyte material with high stability and high ionic conductivity. To improve the ionic conductivity, it is necessary to vary the grain boundary conditions within the composite electrolytes by introducing a liquid phase between the boundary interfaces.

Meanwhile, there are two phases in a composite electrolyte: The first is a solid solution of doped cerium oxide, and the second is a salt or oxide phase that is usually insulating. A solid phase is formed when Li_2CO_3 and Na_2CO_3 are in solution at a temperature above 500°C or when Na_2CO_3 and some oxides are in solid state at temperatures below 500°C. There is excellent ionic conductivity at the interface as a result of the highway channel. Two factors should be considered when designing interfaces, first, the density of the interface or the area of contact between the two constitutions; second, it needs to be considered that the larger the interface, the higher the electrical conductivity. High densities of defects can be generated in multilayer nanostructures due to the larger area of interfaces and/or grain boundaries in the space-charge region, especially in nanostructured systems [24].

Researchers examined the interaction of doped cerium with the second phase and found that the effective conductivity of the second phase decreased when the doped phase increased without this interaction. No chemical reaction—no new phase or ion diffusion—takes place during the interaction; rather, it is a physical reaction. In addition, NPs with large interface areas are suitable for interacting with two different compounds due to their high energy surface. Consequently, to improve the conductivity of the composite electrolyte, it is necessary to reduce the size of the particles within the electrolyte [25].

Synthesized nanocomposite electrolytes showed a desirable combination of high ionic conductivity and a lower superionic phase transition temperature than electrolytes prepared at micro scale. There is no doubt that nanotechnology will work towards improving the ionic conductivity of conductive electrolytes at a large scale in the future [26, 27]. Direct methanol FCs (DMFCs) have been considered one of the most promising forms of substitute electricity for portable devices, military applications, and small performance vehicles [28–30]. DMFCs are characterized by electrochemical reaction at the interface, charge transfer along the interface, etc. The cause of overpotential in DMFCs is primarily polarization loss caused by the sluggish methanol oxidation reaction (MOR). Nanoscale catalyst structures showed high performance in catalytic reactions [31].

DMFCs are highly susceptible to poisoning by oxygen-containing carbon species such as CO, and researchers examined alleviating this intermediate adsorption on the anode catalysts to boost reaction kinetics. There is a growing body of research that reports the effectiveness of one-dimensional (1D) crystalline nanostructures, such as nanorods, nanowires, and nanotubes, in electro-catalyzing MOR [32, 33]. A nanocrystal catalyst showed improved reaction kinetics by generating numerous preferential catalytic facets or crystal defects, which resulted in a surface with a high degree of stress and uncoordinated structure, thus delivering sufficient reaction sites with increased activity [34].

Crystallizing 1D metal nanostructures prevents prevent corrosion and agglomeration of catalyst during electrochemical processes [35]. Removing carbon supports from 1D catalysts reduced catalyst degradation caused by carbon corrosion; the catalysts' durability and efficiency improved due to their physical and morphological architecture. Methanol and carbon dioxide were transported more rapidly, and intermediates were transferred more efficiently through the open pores and interconnected channels of the 1D structure [36].

Two-dimensional (2D) metal-organic framework graphene and hexagonal boron nitride nanosheets served as effective molecular sieves with a crystalline structure that offered ultra-high selectivity along with gas permeability. Monolayer graphene is believed to be impermeable against most molecules, including hydrogen, but this appears to be a misconception. Graphene monolayers could be penetrated by thermal proton, and this permeability increased with metal decoration and increased temperature [37, 38].

2.6 NANOTECH FUEL CELL MODULES

Developing cost-effective ways to fabricate SOFCs is an essential step toward sustainable energy, and there are two conventional methods for doing so: infiltration and calcination, both of which require many cycles as a result of the low catalyst load per cycle required for infiltration. Researchers fabricated nanostructured SOFC cathodes in just one loading step using chemically assisted electrodeposition (CAED) for the simultaneous deposition of multiple cations from diluted aqueous solutions of readily available salts. SOFCs are among the most efficient and cost-effective technologies since their performance is not limited by the fuel combustion efficiency limits, i.e., the Carnot cycle [39, 40].

During the Carnot cycle, the precursor metals of the cathode materials are introduced into a fully sintered backbone of the O_2-ion conducting phase in the form of a mixed metal salt solution to initiate the infiltration process. In the decomposition process, the metal salts are decomposed, and a NP is formed that is composed of the desired cathode material. A nanostructured cathode with low thermal expansion mismatch is achieved by infiltration nanofabrication, which provides the benefit of low-temperature manufacturing. On the other hand, infiltration is a tedious process since it is limited by the amount of precursor that can be loaded within the parameters of a single infiltration step; to reach the required catalyst loading, it is necessary to perform several cycles of infiltration and calcination [41, 42]. Researchers also fabricated SOFC cathodes through electrodeposition in a process that was similar to infiltration but provided the significant advantage that a single electrocatalyst (cathode or anode) could be loaded automatically in a single step [41–43].

Investigators used an ultrasonic bath to mix NiF ion containing a CNT-TiO_2 nanohybrid and then coated that mixture on both sides of carbon cloth (1.5 cm × 3 cm) with a loading of 5 $mgcm^2$. Then, it has been used the modified electrodes as anodes in methane FCs after they had been dried at room temperature, and they used the same procedure to prepare separate CNT- and TiO_2-based electrodes; the nanohybrid was synthesized without using any surfactants or creating harsh environmental conditions. The MFC anode electrodes showed significantly better performance when combined with the CNT-TiO_2 nanohybrid than when combined with only CNTs or TiO_2 NPs. TiO_2 and CNTs have unique properties that give them superior electrochemical performance, such as high surface area, good biocompatibility, good electrical conductivity, excellent stability, and good surface functionalization that is conducive to interfacial electrochemical reactions [44].

Microbial fuel cells (MFCs), which generate electricity from organic compounds through electrochemical reactions by microorganisms, are a continuous and cost-effective method for bringing together energy production and wastewater treatment. Researchers fabricated an H-shaped MFC with an anode of FeS_2 NPs and rGO (reduced graphene oxide) and evaluated its performance in the cell. Adding the NPs promoted surface interaction with the microbial biofilm, facilitating the electrode surface in exchanging extracellular electrons with electroactive bacteria, thus improving the performance of the MFC in a multichamber microfluidic device [45].

2.7 TRENDS AND RECOMMENDATIONS

Fuel cells have undoubtedly become a pivotal part of energy research in recent decades. The need for sustainable and efficient energy production is becoming increasingly pressing with the increasing global energy demands and the depletion of organic fuels. The high efficiency of FCs leads to low to zero greenhouse gas emissions as a byproduct in exchange for abundant fuel is available in abundance.

Among the current industry methods for converting stored chemical energy into electrical energy (an electrochemical reaction), FCs are the most efficient. For instance, solid oxide FCs oxidize not only hydrogen and hydrocarbons but essentially any fuel because the electrolyte transports oxygen ions, which are required for proton exchange membrane fuel cells to function. SOFCs provide a solution to the combustion efficiency limitations (e.g., the Carnot cycle) in a clean, efficient manner. However, although SOFCs can be used to burn conventional fuels (as well as hydrogen fuel) to generate electricity, they are limited by their high operating temperature (about 800 °C), and researchers are investigating the use of nanoscale structures to improve their application efficacy.

REFERENCES

1. Martins F, Felgueiras C, Smitkova M, Caetano N, Analysis of fossil fuel energy consumption and environmental impacts in European countries, *Energies* 12 (2019) 964, https://doi.org/10.3390/en12060964.
2. Boudghene A Stambouli, Fuel cells: The expectations for an environmental-friendly and sustainable source of energy, *Renew Sustain Energy Rev* 15 (2011) 4507–4520.
3. Fan L, Tu Z, Chan SH, Recent development of hydrogen and fuel cell technologies: A review, *Energy Rep* 7 (2021) 8421–8446.
4. Farooque M, Maru HC, Carbonate fuel cells: Milliwatts to megawatts, *J Power Sources* 160 (2006) 827–834.
5. Hoffmann J, Yuh C, Jepek AG, Electrolyte and material challenges, in: *Handbook of Fuel Cell: Fundamentals, Technology and Applications*, vol. 4, John Wiley and Sons, 2003 (Chapter 67).
6. The International Fuel Cells. *Fuel Cells Review*. A United Technology Company, 2009.
7. Galdo J, United State department of energy head quarter, in: *Bologna Conference on Fuel Cells*, Italy, May 31–June 01, 2001.
8. Boudghene Stambouli A, Traversa E, Fuel cells, an alternative to standard sources of energy. *Renew Sustain Energy Rev* 6(3) (2002) 295–304.
9. Refocus. *The International RE Magazine* (Suppl.), Elsevier, 2009.
10. Grover Coors W, Protonic ceramic fuel cells for high efficiency operation with methane, *J Power Sources* (2003) 150–156.
11. Cheng KWE, et al. Exploring the power conditioning system for fuel cell, in: *Proceedings of the 32nd IEEE Annual Power Electronics Specialists Conference*, 2001, pp. 2197–2202.
12. Boudghene Stambouli A, Fuel cells for large scale applications, in: *First International Workshop on Hydrogen*, Algiers, Algeria, 2005.

13. Boudghene Stambouli A, Traversa E, Fuel cells: Highly efficient and clean production of power and heat, in: *Proceedings of the Premier Séminaire International sur l'Implication de l'Energie Solaire et Eolienne dans le Développement Durable*, 2001, pp. 631–638.
14. Boudghene Stambouli A, Traversa E, Solid oxide fuel cells (SOFCs): A review of an environmentally clean and efficient source of energy, *Renew Sustain Energy Reviews* 6(5) (2002) 433–455.
15. Canha LN, et al. Optimal characteristics of fuel cell generating systems for utility distribution networks, in: *Proceedings of the 37th Intersociety Energy Conversion Engineering Conference*, 2004, pp. 597–602.
16. Farooque M, Maru HC, Fuel cells—the clean and efficient power generators. *IEEE Proceedings* 89 (2001) 1819–1829.
17. Le Goff A, Artero V, Jousselme B, Tran PD, Guillet N, Métayé R, Fihri A, Palacin S, Fontecave M, From hydrogenases to noble metal–free catalytic nanomaterials for H2 production and uptake. *Science* 326(5958) (2009 December 4) 1384–1387.
18. Yilanci A, Dincer I, Ozturk HK, A review on solar-hydrogen/fuel cell hybrid energy systems for stationary applications. *Progress in Energy and Combustion Science* 35 (2009) 231–244.
19. Boudghene Stambouli A, Fuel cells: The expectations for an environmental-friendly and sustainable source of energy. *Renewable and Sustainable Energy Reviews*15 (2011) 4507–4520.
20. Dincer I, Environmental and sustainability aspects of hydrogen and fuel cell systems. *International Journal of Energy Research* 31 (2007) 29–55.
21. Aricò A, Bruce P, Scrosati B, et al., Nanostructured materials for advanced energy conversion and storage devices. *Nature Materials* 4 (2005) 366–377.
22. Brett D, Atkinson A, Brandon N, Skinner S, Intermediate temperature solid oxide fuel cells. *Chemical Society Reviews* 37 (2008) 1568–1578.
23. Hui SQ, Roller J, Yick S, Zhang X, Deces-Petit C, Xie YS, Maric R, Ghosh D, A brief review of the ionic conductivity enhancement for selected oxide electrolytes. *Journal of Power Sources* 172 (2007) 493–502.
24. Arico AS, Bruce P, Scrosati B, Tarascon JM, van Schalkwijk W, Nanostructured materials for advanced energy conversion and storage devices. *Nature Materirials* 4 (2005) 366–377.
25. Fan L, Wang C, Chen M, Zhu B, Recent development of ceria-based (nano)composite materials for low temperature ceramic fuel cells and electrolyte-free fuel cells, *J Power Sources* 234 (2013) 154–174.
26. Thabet A, Allam M, Shaaban SA, Investigation on enhancing breakdown voltages of transformer oil nanofluids using multi-nanoparticles technique, *IET Gener Transm Dis J* 12(5) (March 2018) 1171–1176.
27. Thabet A, Allam M, Shaaban SA, Assessment of individual and multiple nanoparticles on electric insulation of power transformers nanofluids, *Electr Power Compon Syst J* 47(4–5) (2019) 420–430.
28. Acres GJK, Recent advances in fuel cell technology and its applications, *J. Power Sources* 100 (2001) 60–66.
29. Kamarudin SK, Daud WRW, Ho SL, Hasran UA, Overview on the challenges and developments of micro-direct methanol fuel cells (DMFC), *J Power Sources* 163 (2007) 743–754.
30. Kamarudin SK, Achmad F, Daud WRW, Overview on the application of direct methanol fuel cell (DMFC) for portable electronic devices, *Int J Hydrogen Energy* 34 (2009) 6902–6916.
31. Guo YG, Hu JS, Wan LJ, Nanostructured materials for electrochemical energy conversion and storage devices, *Adv Mater* 20 (2008) 2878–2887.

32. Koenigsmann C, Wong SS, One-dimensional noble metal electrocatalysts: A promising structural paradigm for direct methanol fuel cells, *Energy Environ Sci* 4 (2011) 1161–1176.
33. Lu Y, Du S, Steinberger-Wilckens R, One-dimensional nanostructured electrocatalysts for polymer electrolyte membrane fuel cells-a review, *Appl Catal B Environ* 199 (2016) 292–314.
34. Li M, Zhao Z, Cheng T, Fortunelli A, Chen CY, Yu R, Zhang Q, Gu L, Merinov BV, Lin Z, Zhu E, Yu T, Jia Q, Guo J, Zhang L, Goddard WA III, Huang Y, Duan X, Ultrafine jagged platinum nanowires enable ultrahigh mass activity for the oxygen reduction reaction, *Science* 354 (2016) 1414–1419.
35. Liu JG, Zhou ZH, Zhao XX, Xin Q, Sun GQ, Yi BL, Studies on performance degradation of a direct methanol fuel cell (DMFC) in life test, *Phys Chem Chem Phys* 6 (2004) 134–137.
36. Hu S, Lozada-Hidalgo M, Wang FC, Mishchenko A, Schedin F, Nair RR, Hill EW, Boukhvalov DW, Katsnelson MI, Dryfe RA, Grigorieva IV, Wu HA, Geim AK, Proton transport through one-atom-thick crystals, *Nature* 516 (2014) 227–230.
37. Guo Y, Jiang Z, Ying W, Chen L, Liu Y, Wang X, Jiang ZJ, Chen B, Peng X, A DNA-threaded ZIF-8 membrane with high proton conductivity and low methanol permeability, *Adv Mater* 30 (2018).
38. Peng Y, Li Y, Ban Y, Jin H, Jiao W, Liu X, Yang W, Metal-organic framework nanosheets as building blocks for molecular sieving membranes, *Science* 346 (2014) 1356–1359.
39. Zhang X, Chan SH, Li G, Ho HK, Li J, Feng Z, A review of integration strategies for solid oxide fuel cells, *J Power Sources* 195 (2010) 685–702, https://doi.org/10.1016/j.jpowsour.2009.07.045.
40. Kendall K, Kendall M, *High-Temperature Solid Oxide Fuel Cells for the 21st Century: Fundamentals, Design and Applications*, Elsevier, 2015.
41. Bidrawn F, Vohs JM, Gorte RJ, Fabrication of LSM–YSZ composite electrodes by electrodeposition, *J Electrochem Soc* (2010), https://doi.org/10.1149/1.3484096.
42. Zhao Y, Oh TS, Li Y, Vohs JM, Gorte RJ, Fabrication of MnCo2O4-YSZ composite cathodes for solid oxide fuel cells by electrodeposition, *J Electrochem Soc* (2016), https://doi.org/10.1149/2.0791608jes.
43. Jung SW, Vohs JM, Gorte RJ, Preparation of SOFC anodes by electrodeposition, *J Electrochem Soc* (2007), https://doi.org/10.1149/1.2790280.
44. Wen Z, Ci S, Mao S, Cui S, Lu G, Yu K, Luo S, He Z, Chen J, TiO2 nanoparticles-decorated carbon nanotubes for significantly improved bioelectricity generation in microbial fuel cells, *J Power Sources* 234 (2013) 100–106.
45. Wang R, Yan M, Li H, Zhang L, Peng B, Sun J, Liu D, Liu S, FeS2 nanoparticles decorated graphene as microbial-fuel-cell anode achieving high power density, *Adv Mater Communication* (2018) 1800618.

3 Nanotechnology in Solar Energy

A great deal of research and industrial work has been dedicated to improving the efficiency of solar cells and making them competitive with traditional energy sources in different aspects and performance levels with a common goal of increasing efficiency. PV solar cells are both energy efficient and cost-effective, but their performance depends on different factors such as their design and the materials used. Today, nanotechnology has the potential to improve the quality, stability, and overall performance of solar cells, and solar cells based on nanomaterials are expected to reduce power generation costs and improve power conversion efficiency.

3.1 INTRODUCTION

Nanotechnology is one of the newest and most fast-growing fields of research, and a lot of scientists who are involved in research in this field confirm that nanotechnology will bring about a new industrial revolution in the near future in a number of different aspects of life. Significant economic and technological changes are taking place around the world right now. For instance, PV technology is currently contributing greatly to the global energy scenario.

The sun is a clean and nearly limitless source of energy, and solar energy, the radiant energy emitted by the sun, can be harnessed and converted into various forms of usable energy. There are three primary ways to produce solar energy: thermal, PV conversion, and thermoelectric. PV is the most widely used; it converts solar energy directly into electricity via panels. Solar cells have evolved through four generations of continuous improvement.

Solar power has proven to be one of the most promising solutions to the world's energy crisis, although PV generation systems have a few disadvantages, which include the high cost and the low efficiency with which they convert light into electrical energy, as well as the fact that weather conditions can change greatly over time. Storage is also an issue; the solar energy generated during the day needs to be stored night, and high-performance solar cells and modules are needed [1–3]. Solar cells have attracted significant attention and rapid development efforts because solar power has remarkable potential for meeting the world's energy demands in the future. However, as I've noted, solar fuel cells currently feature high cost and low conversion efficiency, which remain the major obstacles to wider application.

Thin-film solar cells reduce cost by reducing material consumption and play a major role in the future of PV markets by trying to overcome these limitations, although the efficiency of power conversion decreases increasing temperature, a separate drawback. Researchers have made a variety of attempts to improve thin-film

DOI: 10.1201/9781003512486-4

solar cells' light absorption via light trapping [3–6]. Graded refractive index and photonic crystals are novel structures to improve light harvesting in solar cells [7–9].

3.2 PRINCIPLES OF PV SOLAR CELLS

A PV system uses semiconducting materials to convert light into electricity, but many losses occur before that energy can be converted: A solar cell can convert sunlight into electricity with a maximum efficiency of 30% (and even higher if the cell design is highly complicated), but the typical efficiency ranges between 10% and 15%.

Cell technology is mostly concerned with enhancing efficiency while reducing costs. Cell efficiency can be reduced by certain physical processes: Some of those processes are inherent to the cell and cannot be changed, but many of them can be improved through proper design; major factors that affect the efficiency of cell systems are reflection from the cells' surface, light with insufficient or excess energy, light-generated electrons and holes (empty bonds), resistance to current flow, and self-shading [10].

PV power systems use PV cells that converts sunlight into electricity. A solar cell generates electric power when light shines on it, which translates into both currents and voltages generated by the cell. A solar cell needs a material in which light is absorbed to raise electron energy, and a circuit that allows the electron to move from the solar cell to an external circuit; this energy is then dissipated by the electron as it travels from the solar cell to the external circuit and then returns to the solar cell. A solar cell that is connected to an electrical load will have a closed power circuit, which means that when the load is connected, electrons will flow through the load and be directed back to the rear contact of the solar cell. The conversion of direct current (DC, flowing in one direction only) from solar cells into alternate current (AC) is accomplished by inverters. PV energy conversion requires a wide range of semiconductor materials and processes, but the p-n junctions are the most commonly used in these processes. P-type layers are also known as absorbing layers, whereas N-type layers are known as window layers. PV devices convert light energy into electricity via three main processes: charge generation, charge separation, and charge collection [11–13].

3.2.1 Charge Generation

A semiconductor generates a charge as a result of absorbing photons and causing electrons to be excited and thereby forming electron–hole pairs at high energies. There is a relationship between the size of the semiconductor band gap and the energy of the charge carriers: A larger band gap can lead to a higher possible open-circuit voltage (Voc) for the cell. Because photons below the band gap are not absorbed by the band gap, fewer photons are converted to excited electrons, resulting in less photocurrent flowing in the system. With the decreased band gap height that results from thermalization, the electron energy reaches a level that is equal to the energy of the band gap, thereby absorbing more photons. As solar cells absorb solar radiation from the sun and exchange thermal radiation with the environment, a balance is achieved that helps determine the cell's optimal band gap. The balance process provides an optimum semiconductor band gap near 1.4 eV when you combine the solar spectrum, the absorption of the cell, and the emission of the cell; this results in a maximum efficiency of 33% when you take these factors into account [12, 13].

3.2.2 Charge Separation

The fundamental principle of PVs is that de-excitation of electrons in the conduction band can be achieved by restoring thermal equilibrium during recombination. A successful photogeneration event begins with the efficient separation of the electron–hole pair formed by the excitation dipole; the central layer must be able to effectively transport electrons and holes from the point of generation to the point of contact.

A single substance can be used to make up the central layer of an inorganic solar cell that can absorb photons efficiently and transport both electrons and holes efficiently. In active layers made of composite materials, for example in a dye solar cell, an absorber is sandwiched locally between nanostructured electron and hole conductors. For a low permittivity absorber to function effectively, especially at low temperatures, it is crucial that the local charge separation take place immediately and geminately between the post-generated electron–hole pairs.

Photogeneration forms bound states of opposite polarity, such as exciton states, and those bound states pass through the charge separation process in order to form unbound charge carriers with opposite polarities. During initial recombination between holes and electrons in the valence band, nongeminate recombination can occur. Charge separation can be achieved very efficiently, even without applying an electrical field, in homogeneous semiconductors with high dielectric permittivity and low exciton binding energies. The photogenerated carriers become part of the conduction or valence band as quickly as they are generated since they are incorporated into the respective ensemble of free carriers [10–13].

3.2.3 Charge Collection

Following the initial separation of charges, it is necessary to establish a flux of both types of carriers towards separate contacts. The solar cell is designed with preexisting asymmetry because it injects and extracts only one carrier per external contact; this means that the contacts should be selected for specific carriers, and this can be achieved in various ways.

The amount of material that goes into the absorbing layer of a solar cell is essential for the effective absorption of the solar photons. A device's absorption coefficient and its light-scattering properties are used to determine the appropriate thickness range for maximizing the optical path length of weakly absorbed photons within the absorber. In an organic solar cell, this lifetime depends on the local microstructure of the donor–acceptor network, and in an inorganic solar cell, the lifetime is affected by impurities, dislocations, and grain boundaries. For instance, introducing contact layers to metal halide perovskites increases recombination and decreases Voc [1, 3, 11, 13].

Additionally, conventional homojunction technology suffers from limitations due to the doped emitter and the Ohmic contacts made from Si wafers. Silicon heterojunction (SHJ) is characterized by better contact passivation than classical technologies, which leads to higher Vocs. Ideally, a good SHJ has a low contact resistance that is capable of maintaining great open circuit voltage [14, 15].

3.3 SOLAR CELL CONSTRUCTION

To fabricate solar PV cells in one study, the fabric was cleaned and lightly calendared to produce a smooth surface without melting the fabric. An aluminum (Al) vacuum-evaporated metal surface was surrounded by two layers of PEDOT:PSS, a liquid-coated conducting polymer. However, the polymer coating on the Al was an insufficient conductor to act as an effective PV electrode, and it was also too flexible to allow microcracks.

As soon as the amorphous silicon (a-Si) was laid down on the silicon substrate, it was followed by photoactivated and triple a-Si layers (N-type, undoped, and P-type) laid down by RF PECVD, at low pressures, from silane and phosphine or diborane at 200 °C or slightly less. To achieve the first impact strength, an upper contact layer of sputtered transparent conductive oxide was used as the first encapsulating coat. Using metal tracks mounted on the fabric, upper and lower contacts of the cell were connected to metal studs pressed through the tracks [16].

Researchers designed a textile-based PV power source based on specific cell shapes and sizes and how they interconnected with each other. For sensors and portable electronic devices with low power requirements, thin batteries could be incorporated into PV textiles albeit with some loss of fabric handle and drape. The designed fabrics do have specific use conditions and washing and cleaning requirements, however [17].

Researchers in a different study attempted to fabricate flexible PV cells using the same techniques as for textiles, including electroless and electro-plated parts, liquid printing, dip-coating, vacuum evaporating, atomic layer deposition, sputtering, chemical vapor deposition, and polymerization. Using methods that are already in use allows for fabricating these materials and devices in a cost-effective manner [18].

3.4 SOLAR CELL DEVELOPMENT

3.4.1 PV Nanocomposites

Intensive PV systems use organic, amorphous, or monocrystal solar cells, but nanocrystal PV cells are now in use [19, 20]. For instance, CIGS NPs showed significant promise as a possible replacement for NP-based solar cells. In one study, CIGS NP solar cells showed PV efficiency of 20.3%, which was significantly higher than the efficiency of any other cell [21].

Typically, CIGS layers are 1.5 to 2 meters thick in thin-film solar cells, but using thinner CIGS layers reduces both material consumption and cost. CIGS thin-film solar cells are expected to be produced in large quantities (70 GWp/year) in the foreseeable future, and experts predict global depletions of indium (I) and gallium (G) if CIGS thin-film solar cells are produced in large volumes. Decreasing the deposition time of the CIGS layer would reduce the production cost in a direct and immediate manner. In one study, CIGS layer thickness lower than 0.5 m triggered both electrical shuddering in devices and roughness. Thinner CIGS layers generated electrons closer to the back contacts, which increased the recombination between them [22].

Given the world's energy crisis, harnessing the sun's radiation to generate electricity is proving to be one of the most promising energy solutions. However, for solar cells to be cost-effective and reliable, they must be able to convert sunlight into electricity across a broad range of wavelengths; this will allow them to compete with conventional sources of energy. In order to achieve high efficiency, reliability, cost-effectiveness, and reliability, several solar technology options have been researched with high success, including wafer, thin film, and organic. Wafers and thin films are highly efficient because they use minimal materials.

Researchers in one study produced thin-film solar cells in three types, each with advantages and disadvantages: hydrogenated a-Si, CdTe, and CIGS [23]. Most CdTe solar cell absorber layers are 5 to 10 m thick, but other research groups designed and manufactured thin-film solar cells consisting of CdTe with CdS, and efficiency improved by 13.4%, 7.9%, and 4.7% at respective thicknesses of 3.5 m, 1 m, and 0.5 m. At an adsorb layer of 0.5 mm, the efficiency of a different cell of CdS/PbS (lead sulfide) was 2.16%, and it increased to 4.13% when the absorbing layer was increased to 2 μm. Reducing the CdTe and PbS thicknesses to submicrometer level, however, could decrease the material consumption in solar cells, as well as the production time and cost, because in contrast with the 2 μm CdTe cell, the thinner 0.5 μm cell created a more straightforward shunting path without the need for lasers [24, 25].

In a different set of studies, thin absorbing layers increased the tunneling recombination near the interface. Additionally, as the absorbing layer thickness decreased, the open circuit voltage decreased due to the large leakage current and the carrier recombination. PbS absorbing layer thicknesses of 0.5 m and 2 μm caused, respectively, 82% and 67% optical and recombination losses [25, 26]. The current goals for thin-film solar cell development involve reducing material amounts and costs as well as cell processing time, and this development is closely related to efficient high light absorption [27, 28]. New-generation solar cells have been developed based on semiconductors with mid-range band gaps that absorb light in the visible region of the solar spectrum that allow for efficient light capture. Researchers recently studied heterostructure solar cells that included a layer of thin-film PbS; the optical energy gap at 300 K was 0.41 eV, and the silicon dioxide (SiO_2) light source layer had a direct and narrow optical energy gap [29].

Recent researchers are developing and investigating solar cells with PbS as a primary absorbing layer and finding a significant influence of layer thickness. For instance, CdS/PbS cells showed efficiency of 2.16% at an absorbing layer thickness of 0.5 μm, but at 2 μm, the efficiency increased by 4.13%; the cells also had the highest open circuit voltages, current density, and output power density with increasing thickness of the absorbing layer [30]. CdTe solar cells usually have an absorbing layer thickness between 5 μm and 10 μm, and CIGS thin-film solar cells typically have an absorbing layer thickness of 1.5–2 μm. Researchers found 15.9% efficiency with a CdS/CIGS thin-film solar cell thickness of 1.8 μm and found 8.1% at 0.36 mm and efficiencies of 13.4%, 7.9%, and 4.7%, respectively, at 3.5 mm, 1 mm, and 0.5 mm. In the back contact region of CdTe, there were high current leakage and carrier recombination losses with thinner films [31–34].

3.4.2 PV Quantum Dots

QDs represent a promising nanotechnology; changing NP size changes QD size, and reducing QD size can reduce costs and boost absorption and efficiency [35]. Changing NP size also changes the driving forces for electrons and holes to be injected into the surrounding phases, which in turn has a direct effect on the performance of the solar cell [36]. There is a wide variety of QD solar cells, including Schottky, HJ solar cells with depleted QDs, QD sensitized, and hybrid organic/inorganic [37–42]. Depleted heterojunction solar cells are further classified into quantum funnel (QF), quantum junction (QJ), or heterojunction (HJ) cells. These further distinctions are intended to improve the connection between QDs and wide-band-gap semiconductors.

Quantum funnel solar cells (QFSCs) had a power-conversion efficiency (PCE) of 2.7% in a study in which researchers used the layer-by-layer technique, varying the QD sizes and therefore the electron affinity values, to create two layers: The TiO_2 layer was coated with large QDs of colloidal PbS, and the upper layers were composed of smaller QDs. Photoelectrons were driven to the electron acceptor layer by quantum funnels as they were generated at their points of generation [43–49]. Photodetectors, optical switches, solar cells, and optical switches all have potential applications that may benefit from PbS QDs [50–53]. Indeed, researchers have synthesized PbS NPs in a variety of morphologies, under different conditions, and according to different methods to achieve a wide range of properties [54].

Tunable band gaps match the distribution of spectral wavelengths of the solar spectrum, which is crucial for minimizing the consumption of the absorbing layer material. QDs produce a tunable band gap due to the size variation and the creation of intermediate bands. Nanostructured solar cells can harvest more of the solar spectrum because QD band gap increases as QD size decreases [55].

In the presence of a single photon of energy exceeding the band gap, QDs produce multiple excitons (electron–hole pairs) after collisions with several photons. This means that a single photon absorbs multiple electron–hole pairs when it comes into contact with an electron–hole pair; this phenomenon is called multiple exciton generation (MEG). In one work, QDs absorbed UV photons more efficiently than near-IR photons and thus produced more electrons, three per photon, as opposed to only one in Si nanocrystal solar cells [56].

Researchers observed more confinement with smaller NPs because the band gap energy (Eg) increased. There are two major benefits to this process: Not only do high-energy photons of light from the solar spectrum absorb more light, they also deliver a greater driving force to inject electrons and holes into the surrounding phases, thereby improving the performance of the solar cell [57]. PbS-QD/CdS HJSCs were demonstrated on thin films [58]. QDs have been widely integrated into solar cells of different types, including Schottky [59], depleted HJ [60], hybrid organic/inorganic [61, 62], and QD-sensitized [63, 64].

A depleted heterojunction solar cell can be classified into different types including quantum funnels (QF), quantum junctions (QJ), and heterojunctions (HJ) QD solar cells, which were discovered most recently; in these cells, QDs and wide-band-gap semiconductors are connected in a way that improved their performance [65, 66]. The heterojunction quantum dot solar cell (HJ-QDSC) is the same as the p-n quantum dot solar cell (QJSC) except the QDs of the HJ-QDSC are composed of different semiconductor materials. Researchers synthesized PbS QDs using n-type CdS and Bi_2S_3 in a core/shell structure and studied two planar layers on top of each other [67, 68]. Other researchers studied metal-semiconductor core/shell NPs. Their improved optical properties make them a popular choice among nanostructures with core shells containing metals and semiconductors. The researchers used metal NPs as the excitation energy for surface plasmon resonances by matching the NPs with respective band gaps [69–71].

3.4.3 PV Core Shell

MEG is the process of generating electron–hole pairs from a single photon by its absorption. QDs generate multiple electron–hole pairs after colliding with a photon of energy that is greater than the band gap of the semiconductor. Solar QFSCs, QJSCs, and HJSCs improve the connections between the QDs and wide-band-gap semiconductors and thereby improve their performance. QFSCs and QJSCs are synthesized by the same group [72, 73]. In HJ-QDSC, p-n QDs are constituted using a different semiconductor material than that used in a QJSC. Researchers combined n-type materials CdS and Bi2S3 to form PbS QDs in the form of a core/shell structure, that is, two planar layers stacked upon each other [74–76], or they can be formed in bulk [77].

Metal-semiconductor core/shell NPs are among the most intensively researched core/shell nanostructures because of their improved optical properties over other nanostructures. Surface plasmon resonances transfer energy more quickly between semiconductor coatings and metal cores by matching the excitation energy from the NPs to the band gap energy of the semiconductor core [78–80]. The surface plasmon resonance properties of metal NPs that are formed through the synthesis of metal-semiconductor heterostructures are strongly correlated with the composition, dimensioning, and morphology of metal NPs [81, 82].

Nanostructures with multiple components are seeing a great deal of interest since multiple functions can be incorporated into a single system for wide application. Metals improved the separation of charges in noble metal-semiconductor nanostructures and the binding of light to semiconductors. Photocatalysis and light harvesting are two of the key functions of semiconductors, and both increased both components' performance including semiconductor luminescence. A further benefit of coupling semiconductor components and noble metal components is that it allows the plasmonic modes of the latter components to be excited and controlled for the purpose of capturing and transferring information [83]. HJ-QDSc metal-semiconductor core/multiple shell structures improved the performance of thin-film solar cells in terms of high efficiency and minimal material usage [77, 84, 85].

3.5 NANOTECH SOLAR CELLS

3.5.1 Nanocrystal Solar Cells

QD solar cells are also known as nanocrystal solar cells. This type of solar cell consists of a semiconductor composed of transition metal groups. QDs were developed as the promising third-generation solar cells to reduce costs and improve absorption and efficiency. Their performance is determined by controlling their size. During the production of conventional compound semiconductor solar cells, a photon excites an electron, forming an electron–hole pair. In contrast, many electron–hole pairs form when a photon strikes a QD made of the same semiconductor material [86–89].

3.5.2 Quantum Dot Solar Cells

The quantum confinement concept is used in nanotechnology to describe the change in the structure of an atom that occurs when a small scale affects the band structure of the energy of the atom; in semiconductors, the quantum confinement region is 1–25 nm. An electron's energy will adjust in response to geometrical constraints; this is known as the quantum size effect. Quantum confinement also shifts from continuous energy bands of bulk materials to discrete energy levels of these materials.

Quantum dot solar cells (QDSC) behave similarly to traditional solar cells but exhibit lower resistance. Semiconductors exhibit higher resistance than metals when they absorb light, including in conventional solar cells. Researchers studied Si, CdTe, and CIGS for moving electrons from the valance band to the conduction band. The electrons on the cell surfaces moved above the conduction band and relaxed, releasing phonons that heated the solar cell without actually producing any energy; this damaged the cell and decreased performance. A QDSC generates a large number of electrons with every photon it absorbs, in contrast with a conventional solar cell [89–93]. PbS is an essential binary semiconductor of the IV–VI group. Researchers applied PbS NPs with a band gap of 0.41 in a near-IR communication device, and it gave a large excitonic Bohr radius of 18 nm with a significant dielectric constant [90].

The absorption in an optical domain can be tuned by decreasing the size of the particle; PbS has a relatively narrow band gap, allowing for tuning the absorption in a wide domain. A research groups investigated the effects of controlling PbS particle size in quantum confinement and found that it significantly affected charge carriers, resulting in a large blue shift. PbS films have a p-type conducting nature, with an energy gap that ranges from 0.40 eV to 2.34 eV, making them capacitors that show excellent photoconductive properties [91–93].

3.5.3 Polymer-Based Solar Cells

In solar cells, an array of thin polymer layers is stacked on a polymer foil or ribbon encapsulated with thicker polymer layers that are connected in a serial manner. They are used in combination with a polymer donor and a fullerene acceptor. Sunlight can be absorbed by cells made of a variety of including organic materials like polymers that act as conjugates or conductors. The unique features of polymer solar cells are allowing for new applications such as stretchable solar devices for use in textiles, changing the face of solar technology [94–96].

3.5.4 Dye-Sensitized Solar Cells

Another cell type for solar energy generation is dye-sensitized solar cells (DSSCs). There are three components of the DSSC: a semiconductor electrode (n-type TiO_2

and p-type NiO) and a dye molecule between the electrodes. Using optically active dyes made of nanograined TiO_2 to coat DSSC surfaces made them more efficient than conventional solar cells, but the dye molecules degraded after being exposed to ultraviolet and IR radiation, decreasing their life spans. In addition, adding barrier layers also increases manufacturing costs [97].

3.5.5 Perovskite Solar Cells

In a recent investigation performed by a prominent independent lab, perovskite solar cells showed efficiency of 31%; they are predicted to play an important role in electric car batteries. However, two significant issues with perovskite solar cells are their stability and durability; this material degrades over time, degraded the efficiency of the cells. A unique property of perovskites is that they combine halide groups, a type of anion, with carbon atoms in a complex Class ABX_3 structure; the X addresses the halide group, whereas A and B relate to the carbon atoms as cations. Scientists fabricated a perovskite solar cell by loading Si p-i-n nanowire cluster with alkyl radical ammonium ion lead halides. The unique resource used to construct the device was based on the importance of perovskite solar cells by keeping pace with continued advancements in the perovskite solar cells and therefore lead-free salts perovskites [93].

3.5.6 Thin-Film Solar Cells

A thin-film cell has the following general advantages [95]:

1. They are less costly because there is less material used; however, for their size, thin-film modules require a lot more space.
2. Their performance is less affected by low-light radiation and shading than a-Si panels.
3. They are flexible due to their lack of rigid substrates, so they can be used for a wider range of applications.

Thin-film cells to power solar PV systems have pros and cons; although some technologies like GaAs and CIGS are more efficient, they cost more. As the market for thin-film solar PV technology continues to grow and production increases, many countries are taking notice of the potential of this renewable source of energy [95]. Si monocrystals represent the first generation of nanotechnology for PV social cells. Worldwide, the market was 56% at the end of 2010, while monocrystal Si accounted for 36%, with multicrystal leading the market with 60%. Thin-film PV technology is expected to be more competitive than crystalline PV technology if the efficiency and reliability are acceptable [97].

Recently, almost 140 GW of solar systems have been installed globally [94–96]; this can be attributed to the increasingly low cost of solar modules and batteries. Currently, thin-film cells account for 5% of global annual global PV production, 4.5 GWp. Maximum power point tracking can be used to improve the efficiency of solar cells by avoiding load mismatch, which has allowed for the development of ever more sophisticated cells. One-crystal Si formed the first generation of solar cells and were widely used with moderate efficiency. The second generation of thin-film solar cells featured significant reductions in operating costs. The many layers in the cells made them more flexible; they were also easy to handle and showed lower current loss. Third-generation cells are made from dye-sensitized polymer nanocrystals, and higher dye quality produces the best efficiency [101–103].

Researchers tested CdS/CdTe, CdS/CIGS, and CdS/PbS under standard test condition (STC) and nominal operating cell condition, and CdS/(CIGS + 20wt.%Al) + Ag), CdS/(CIGS + 20wt.%Al) + Cu), CdS/(CIGS + 20wt.%Al) + Li), CdS/(CdTe + 20wt.%Al) + Cu) produced highly efficient solar power substations that required fewer cells and had a total installed capacity of 64,516.8 kWP [101]. Others examined the I-V (current-voltage) and P-V (power-voltage) characteristics of thin-film solar cells based on CdS/CdTe, CdS/CIGS, and CdS/PbS with a QD window layer, QD absorber layer, and core/shell or core/multiple shell absorber layer under variant thermal conditions and constant solar irradiation based on the AM1.5 reference spectrum [103].

3.6 TRENDS AND RECOMMENDATIONS

Nanocomposite absorbent layers exhibit nearly optimal energy band gaps in thin-film solar cells obtained from Al, Ag, Li, and Cu, individually or as nanocomposites, to PbS, CIGS, or CdTe; the refractive index decreases as absorption increases. The absorption edge wavelength of these nanocomposites shifts to the invisible and near-IR regions of the spectrum band, indicating a deep absorption and a blue shift in the absorption edge wave length. Nanocomposites with better adsorption energy bands and dielectric constants and refractive indexes have improved electron–hole generation rates, resulting in lower reverse saturation current densities and larger absorption energy band gaps.

Facile, cost-effective methods have made it possible to manufacture thin films that can be optimized according to crystal size and optical, depression, and dielectric properties. Nanocomposites produce thin films with impressive absorption; transmittance; optical and electric conductivities; reflectance; volumes; surface energy losses; and refractive indices as well as nearly optimal absorbing layers and band gap energies [104–109].

REFERENCES

1. Lee T and Ebong A, "A review of thin film solar cell technologies and challenges", *Renewable and Sustainable Energy Reviews, Elsevier*, Vol. 12, No. 3, pp. 1286–1297, 2016.
2. Yan D and Xuan Y, "Effect of temperature on performance of nanostructured silicon thin-film solar cells", *Science Direct, Elsevier*, Vol. 11, No. 4, pp. 109–119, 2015.
3. Sharma S and Ali K, *Solar cells from materials to device technology*, Chapter 1, 2, Springer, pp. 1–50, 2020.
4. Braun A, Katz A and Cordon J, "Basic aspects of the temperature coefficient of concentrator solar cell performance parameters", *Progress in Photovoltaics*, Vol. 21, No. 3, pp. 1087–1094, 2013.

5. Da Y and Xuan Y, "Role of surface recombination in affecting the efficiency of nanostructured thin-film solar cells", *Optics Express*, Vol. 21, No. 6, pp. 1065–1077, 2013.
6. Bozzola A, liscidini M and Andreani L, "Photonic light-trapping versus Lambertian limits in thin film silicon solar cells with 1D and 2D periodic patterns", *Optics Express*, Vol. 20, No. 2, pp. 224–244, 2012.
7. Buencuerpo J, Munioz E, Dotor M and Postigo A, "Optical absorption enhancement in a hybrid system photonic crystal thin substrate for photovoltaic applications", *Optics Express*, Vol. 20, No. 4, pp. 452–464, 2012.
8. Jang S, Song Y, Yeo C, Park C, Yun J and Lee YT, "Antireflective property of thin film a-Si solar cell structures with graded refractive index structure", *Optics Express*, Vol. 19, No. 2, pp. 108–117, 2011.
9. Deceglie M, Ferry V, Alivisatos A and Atwater H, "Design of nanostructured solar cells using coupled optical and electrical modeling", *Nano Letter*, Vol. 12, No. 6, pp. 2894–2900, 2012.
10. Technical Information Office, "Basic photovoltaic principles and methods", SERI/SP-290–1448, Solar Information Module 6213, Energy Research Institute 1617 Cole Boulevard, Golden, Colorado 80401, U.S. Department of Energy, 1982.
11. Boxwell M, *Solar electricity handbook*, Green Stream Publishing, pp. 250–276, 2019.
12. Rau U and Kirchartz T, "Charge carrier collection and contact selectivity in solar cells", *Advanced Materials Interfaces*, Vol. 11, 2019.
13. Kirchartz T, Bisquert J, Mora-Seroc V and Garcia G, "Classification of solar cells according to mechanisms of charge separation and charge collection", *Journal of Chemistry Physics*, Vol. 17, pp. 4007–4014, 2015.
14. Daniel N, Ricardo T and Juni S, "The influence of solar spectrum and concentration factor on the material choice and the efficiency of multijunction solar cells", *Scientific Report, Nature Research*, Vol. 9, No. 1, pp. 1–10, 2019.
15. Belghachi A and Limam N, "Effect of the absorber layer band-gap on CIGS solar cell", *Chinese Journal of Physics*, Vol. 6, No. 4, pp. 1127–1134, 2017.
16. Diyaf AG, Mather RR and Wilson JIB, "Contacts on polyester textile as a flexible substrate for solar cells", *IET Renew, Power Gener*, Vol. 8, pp. 444–450, 2014.
17. Mather RR and Wilson JIB, "Fabrication of photovoltaic textiles", *Coatings, MDPI*, Vol. 7, No. 63, 2017, doi:10.3390/coatings7050063.
18. Shishoo R (Ed.), *Plasma technologies for textiles*, Woodhead Publishing Limited: Cambridge, UK, 2007.
19. Purwahyudi B, Kuspijani K and Ahmadi A, "SCNN based electrical characteristics of solar photovoltaic cell model", *International Journal of Electrical and Computer Engineering (IJECE)*, Vol. 7, 2017.
20. Satyanarayana M and Kumar PS, "Analysis and design of solar photo voltaic grid connected inverter", *Indonesian Journal of Electrical Engineering and Informatics (IJEEI)*, Vol. 3, No. 4, pp. 199–208, 2015.
21. Liao Y, Yikuo S, Tyng W, Lai F and Hsieh D, "Observation of unusual optical transitions in thin-film Cu(In,Ga)Se2 solar cells", *Optics Express*, Vol. 20, 2012.
22. lundberg O, Bodegard M and Stolt L, "Influence of the Cu(In,Ga)Se2 thickness and ga grading on solar cell performance", *Progress in Photopholatics: Research and Application*, Vol. 12, 2003.
23. Lee T and Ebong A, "A review of thin film solar cell technologies and challenges", *Renewable and Sustainable Energy Reviews, Elsevier,* Vol. 12, 2016, https://doi.org/10.1016/j.rser.2016.12.028.
24. Mannan M, Anjan MS and Kabir MZ, "Modeling of current–voltage characteristics of thin film solar cells", *Solid-State Electronics, Elsevier*, Vol. 6, 2011, https://doi.org/10.1016/j.sse.2011.05.010

25. Bai Z, Yang J, Depba S and Wang D, "Thin film CdTe solar cells with an absorber layer thickness in micro- and sub-micrometer scale", *Applied Physics Letters*, Vol. 4, 2011, https://doi.org/10.1063/1.3644160
26. Mohamed H, "Theoretical study of the efficiency of CdS/PbS thin film solar cells", *Solar Energy, Elsevier*, Vol. 10, 2014, https://doi.org/10.1016/j.solener.2014.07.017
27. Hui P, Elisabeth R and Vehse M, "Cost-effective nanostructured thin-film solar cell with enhanced absorption", *Applied Physics Letters*, Vol. 7, 2014, https://doi.org/10.1063/1.4901167
28. Krishnakumar V, Barati A, Schimper HJ, Klein A and Jaegermann W, "A possible way to reduce absorber layer thickness in thin film CdTe solar cells", *Thin Solid Films*, Vol. 535, p. 233, 2013, https://doi.org/10.1016/j.tsf.2012.11.085.
29. Rondiya S, Nitin K, Yogesh K, Jadkar S, Haram S and Kabir M, "Structural, electronic, and optical properties of Cu2NiSnS4: A combined experimental and theoretical study toward photovoltaic applications", *ACS Paragon Plus Environment*, Vol. 17, 2017.
30. Mohamed HA, "Theoretical study of the efficiency of CdS/PbS thin film solar cells", *Solar Energy, Elsevier*, Vol. 10, 2014.
31. Bai Z, Yang J, DePba S and Wang D, "Thin film CdTe solar cells with an absorber layer thickness in micro- and sub-micrometer scale", *Apblied Physics Letters*, Vol. 4, 2011.
32. lundberg O, Bodegard M and Stolt L, "Influence of the Cu(In,Ga)Se2 thickness and Ga grading on solar cell performance", *Progress in Photopholatics: Research and Application*, Vol. 12, 2003.
33. Mannan MA, Anjan MS and Kabir MZ, "Modeling of current–voltage characteristics of thin film solar cells", *Solid-State Electronics, Elsevier*, Vol. 6, 2011.
34. Lundberg O, *Band gap profiling and high speed deposition of Cu(In,Ga)Se2 for thin film solar cells*, Uppasala University, 2003.
35. Tisdale W, Williams K, Timp B, Norris D, Aydil E and Zhu X, "Hot electron transfer from semiconductor nanocrystals", *Science Direct*, pp. 1543–1547, 2010.
36. Emin S, Loukanov A and Wakasa M, "Photostability of water-dispersible CdTe quantum dots: Capping ligands and oxygen", *Chemistry Letters*, pp. 654–656, 2010.
37. Olson J, Rodriguez Y, Yang L, Alers GB and Carter SA, "CdTe schottky diodes from colloidal nanocrystals", *Applied Physics Letters*, Vol. 96, p. 242103, 2010.
38. Debnath R, Greiner M, Kramer I, Fischer A, Tang J, Barkhouse D, Wang X, Levina L, Lu Z and Sargent EH, "Depleted-heterojunction colloidal quantum dot photovoltaics employing low-cost electrical contacts", *Applied Physics Letters*, Vol. 97, p. 023109, 2010.
39. Liu CP, Wang HE, Ng TW, Chen ZH, Zhang WF, Yan C and Tang YB, "Hybrid photovoltaic cells based on ZnO/Sb2S3/P3HT heterojunctions", *Physica Status Solidi*, pp. 627–633, 2012.
40. Chang JA, Rhee JH, Lee YH, Kim H, Seok S, Nazeeruddin M and Gratzel M, "High-performance nanostructured inorganic–organic heterojunction solar cells", *Nano Letters*, Vol. 10, pp. 2609–2612, 2010.
41. Huang X, Huang S, Zhang Q, Guo X, Li D, Luo Y, Shen Q, Toyodo T and Meng Q, "A flexible photoelectrode for CdS/CdSe quantum dot sensitized solar cells (QDSSCs)", *Chemical Communications*, pp. 2664–2666, 2011.
42. Kamat PV, "Boosting the efficiency of quantum dot sensitized solar cells through modulation of interfacial charge transfer", *Accounts of Chemical Research*, pp. 1906–1915, 2012.
43. Lee H, Leventis H, Moon S, Chen P, Haque S, Zakeeruddin S and Nazeeruddin M, "PbS and CdS quantum dot-sensitized solid-state solar cells: Old concepts, new results", *Advanced Functional Materials*, pp. 2735–2742, 2009.
44. Kramer I, Levina L, Debnath R, Zhitomirsky D and Sargent EH, "Solar cells using quantum funnels", *Nano Letters*, pp. 3701–3706, 2011.

45. Tang J, Liu H, Zhitomirsky D, Hoogland S, Wang X, Furukawa M and Levina L, "Quantum junction solar cells", *Nano Letters*, pp. 4889–4894, 2012.
46. Gonfa B, Zhao H, Li J, Qiu J, Saidani M, Zhang S, Izquierdo R, Khakani A and Ma D, "Air-processed depleted bulk heterojunction solar cells based on PbS/CdS core–shell quantum dots and TiO2 nanorod arrays", *Solar Energy Materials and Solar Cells*, pp. 67–74, 2014.
47. Chang LY, Lunt R, Brown P, Bulović V and Bawendi M, "Low temperature solution-processed solar cells based on PbS colloidal quantum dot/CdS heterojunctions", *Nano Letters*, pp. 994–999, 2013.
48. Bhandari K, Roland P, Mahabaduge H, Haugen N and Grice C, "Thin film solar cells based on the heterojunction of colloidal PbS quantum dots with CdS", *Solar Energy Materials and Solar Cells*, Vol. 7, 2013.
49. Jiao Y, Gao X, Lu J, Chen Y, Zhou J and Li X, "A novel method for PbS quantum dot synthesis", *Materials Letters, Elsevier*, Vol. 72, pp.116–118, 2012.
50. Cademartiri L, Montanari E, Calestani G, Migliori A, Guagliardi A and Ozin GA, "Size dependent extinction coefficients of PbS quantum dots", *Journal of the American Chemical Society*, Vol. 128, pp. 10337–10346, 2006.
51. Konstantatos G, Howard I, Fischer A, Hoogland S, Clifford J, Klem E, et al. "Ultrasensitive solution-cast quantum dot photodetectors", *Nature*, Vol. 442, pp. 180–183, 2006.
52. Kane RS, Cohen RE and Silbey R, "Theoretical study of the electronic structure of PbS nanoclusters", *The Journal of Physical Chemistry*, Vol. 100, pp. 7928–7932, 1996.
53. Zhao N, Osedach TP, Chang LY, Geyer SM, Wanger D, Maddalena T, et al. "ColloidalPbS quantum dot solar cells with high fill factor", *ACS Nano*, Vol. 4, pp. 3743–3752, 2010.
54. Ebnalwaled AA, Essai MH, Hasaneen BM and Mansour HE, "Facile and surfactant-free hydrothermal synthesis of PbS nanoparticles: The role of hydrothermal reaction time", *Journal of Materials Science: Materials in Electronics, Springer*, Vol. 28, pp. 1958–1965, 2017.
55. Nozik A, "Nanoscience and nanostructures for photovoltaics and solar fuels. Nano", *Nano Letters*, Vol. 10, No. 8, pp. 2735–2741, 2010.
56. Ebrahim K, "Quantum dots solar cells", *InTech Solar Cells New Approaches and Reviews*, Vol. 31, pp. 303–331, 2015.
57. Emin S, Loukanov A and Wakasa M, "Photostability of water-dispersible CdTe quantum dots: Capping ligands and oxygen", *Chemistry Letters*, pp. 654–656, 2010.
58. Bhandari K, Roland P, Mahabaduge H, Haugen N and Grice C, "Thin film solar cells based on the heterojunction of colloidal PbS quantum dots with CdS", *Solar Energy Materials & Solar Cells, Elsevier*, Vol. 117, pp. 476–482, 2013.
59. Olson J, Rodriguez Y, Yang L, Alers GB and Carter SA, "CdTe schottky diodes from colloidal nanocrystals", *Applied Physics Letters*, Vol. 96, p. 242103, 2010.
60. Debnath R, Greiner M, Kramer I, Fischer A, Tang J, Barkhouse D, Wang X, Levina L, Lu Z and Sargent EH, "Depleted-heterojunction colloidal quantum dot photovoltaics employing low-cost electrical contacts", *Applied Physics Letters*, Vol. 97, p. 023109, 2010.
61. Liu CP, Wang HE, Ng TW, Chen ZH, Zhang WF, Yan C and Tang YB, "Hybrid photovoltaic cells based on ZnO/Sb2S3/P3HT heterojunctions", *Physica Status Solidi*, pp. 627–633, 2012.
62. Chang JA, Im SH, Lee YH, H-j Kim, Lim CS, Heo JH and Il Seok S, "Panchromatic Photon-Harvesting by Hole-Conducting Materials in Inorganic–Organic Heterojunction Sensitized-Solar Cell through the Formation of Nanostructured Electron Channels", *Nano Letters*, Vol. 12, pp. 1863–1867, 2012.

63. Huang X, Huang S, Zhang Q, Guo X, Li D, Luo Y, Shen Q, Toyodo T and Meng Q, "A flexible photoelectrode for CdS/CdSe quantum dot sensitized solar cells (QDSSCs)", *Chemical Communications*, pp. 2664–2666, 2011.
64. Kamat PV, "Boosting the efficiency of quantum dot sensitized solar cells through modulation of interfacial charge transfer", *Accounts of Chemical Research*, pp. 1906–1915, 2012.
65. Lee H, Leventis H, Moon S, Chen P, Haque S, Zakeeruddin S and Nazeeruddin M, "PbS and CdS quantum dot-sensitized solid-state solar cells: Old concepts, new results", *Advanced Functional Materials*, pp. 2735–2742, 2009.
66. Tang J, Liu H, Zhitomirsky D, Hoogland S, Wang X, Furukawa M and Levina L, "Quantum junction solar cells", *Nano Letters*, pp. 4889–4894, 2012.
67. Gonfa B, Zhao H, Li J, Qiu J, Saidani M, Zhang S, Izquierdo R, Khakani A and Ma D, "Air-processed depleted bulk heterojunction solar cells based on PbS/CdS core–shell quantum dots and TiO2 nanorod arrays", *Solar Energy Materials and Solar Cells*, pp. 67–74, 2014.
68. Chang LY, Lunt R, Brown P, Bulović V and Bawendi M, "Low temperature solution-processed solar cells based on PbS colloidal quantum dot/CdS heterojunctions". *Nano Letters*, pp. 994–999, 2013.
69. Chaokang G, Xub H, Parkb M and Shannona C, "Formation of metal-semiconductor core shell nanoparticles using electrochemical atomic layer deposition", *ECS Transactions*, Vol. 10, 2008.
70. Ma G, He J, Rajiv K, Tang SH, Yang Y and Nogami M, "Observation of resonant energy transfer in Au:CdS nanocomposite", *Applied Physics Letters*, Vol. 84, pp. 4684–4686, 2004.
71. Sharma A and Gupta B, "Metal–semiconductor nanocomposite layer based optical fibre surface plasmon resonance sensor", *Journal of Optics*, Vol. 9, 2007.
72. Ebrahim K, "Quantum dots solar cells", *InTech Solar Cells New Approaches and Reviews*, Vol. 31, 2015.
73. Lee H, Leventis H, Moon S, Chen P, Haque S, Zakeeruddin S and Nazeeruddin M, "PbS and CdS quantum dot-sensitized solid-state solar cells: old concepts, new results", *Advanced Functional Materials*, pp. 2735–2742, 2009.
74. Tang J, Liu H, Zhitomirsky D, Hoogland S, Wang X, Furukawa M and Levina L, "Quantum junction solar cells", *Nano Letters*, pp. 4889–4894, 2012.
75. Gonfa B, Zhao H, Li J, Qiu J, Saidani M, Zhang S, Izquierdo R, Khakani A and Ma D, "Air-processed depleted bulk heterojunction solar cells based on PbS/CdS core–shell quantum dots and TiO2 nanorod arrays", *Solar Energy Materials and Solar Cells*, pp. 67–74, 2014.
76. Chang LY, Lunt R, Brown P, Bulović V and Bawendi M, "Low temperature solution-processed solar cells based on PbS colloidal quantum dot/CdS heterojunctions", *Nano Letters*, pp. 994–999, 2013.
77. Bhandari K, Roland P, Mahabaduge H, Haugen N and Grice C, "Thin film solar cells based on the heterojunction of colloidal PbS quantum dots with CdS", *Solar Energy Materials & Solar Cells, Elsevier*, Vol. 7, 2013.
78. Chaokang G, Xub H, Parkb M and Shannona C, "Formation of metal-semiconductor core shell nanoparticles using electrochemical atomic layer deposition", *ECS Transactions*, Vol. 10, 2008.
79. Khon E, Mereshchenko A, Tarnovsky AN, Acharya K, Klinkova A, Hewa Kasakarage NN, Nemitz I and Zamkov M, "Suppression of the Plasmon Resonance in Au/CdS Colloidal Nanocomposites", *Applied Physics Letters*, Vol. 11, pp. 1792–1799, 2011.
80. Sharma A and Gupta B, "Metal–semiconductor nanocomposite layer based optical fibre surface plasmon resonance sensor", *Journal of Optics*, Vol. 9, 2007.

81. Liu X, Liang S, Nan F, Pan Y, Shi J, Zhou L, Feng S, Wang J, Feng X and Quan Q, "Stepwise synthesis of cubic Au-AgCdS core-shell nanostructures with tunable plasmon resonances and fluorescence", *Optics Express*, Vol. 6, 2013.
82. Zhou L, Fu XF, Yu L, Zhang X, Yu XF and Hao ZH, "Crystal structure and optical properties of silver nanorings", *Applied Physics Letters*, Vol. 94, p. 153102, 2009.
83. Sun Z, Yang Z, Zhou J, Yeung M, Hongkai W and Wang J, "A general approach to the synthesis of gold–metal sulfide core–shell and heterostructures", *Wiley Interscience*, Vol. 5, 2009.
84. Bai Z, Yang J, DePba S and Wang D, "Thin film CdTe solar cells with an absorber layer thickness in micro- and sub-micrometer scale", *Applied Physics Letters*, Vol. 4, 2011.
85. lundberg O, Bodegard M and Stolt L, "Influence of the Cu(In,Ga)Se2 thickness and Ga grading on solar cell performance", *Progress in Photopholatics: Research and Application*, Vol. 12, 2003.
86. Blachowicz T and Ehrmann A, "Recent developments of solar cells from PbS colloidal quantum dots", *Applied Science*, Vol. 10, No. 5, pp. 1–16, 2020.
87. Shkir M, AlFaify S, Ganesh V and Yahia I, "Facile one pot synthesis of PbS nanosheets and their characterization", *Solid State Sciences*, Vol. 5, No. 4, pp. 81–85, 2017.
88. Mamiyeva Z and Balayeva N, "Preparation and optical studies of PbS nanoparticles", *Optical Materials*, Vol. 5, No. 3, pp. 522–525, 2015.
89. Mohamed HA, "Theoretical study of the efficiency of CdS/PbS thin film solar cells", *Solar Energy*, Vol. 10, No. 2, pp. 360–369, 2015.
90. Sharma S, Dhanunjaya A, Jayarambabu N, Kumar N, Saineetha A, Kailasa S and Venkateswara K, "Micro-structural, optical and vibrational spectra analysis of lead sulphide, cadmium doped PbS and strontium doped PbS nanostructured thin films synthesized through successive ionic layer adsorption and reaction technique for solar cell and infrared detector sensor applications", *Materials Today: Proceedings*, Vol. 26, No. 1, pp. 162–171, 2019.
91. Ebrahim K, "Quantum dots solar cells", *InTech Solar Cells New Approaches and Reviews*, Vol. 31, No. 3, pp. 303–331, 2015.
92. Bhandari K, Roland P, Mahabaduge H, Haugen N and Grice C, "Thin film solar cells based on the heterojunction of colloidal PbS quantum dots with CdS", *Solar Energy Materials & Solar Cells, Elsevier*, Vol. 7, No. 2, pp. 476–482, 2013.
93. Srimath A, Innocenzo D, Petrozza G, Enrico D, Filippo D, Henry S and Alison W, "Chapter 4: Photo physics of hybrid perovskites unconventional thin film photovoltaics", *Royal Society of Chemistry*, pp. 107–140, 2016.
94. Khare V, Khare C, Nema S and Baredar P, "Chapter 1: Tidal energy systems: Design, optimization and control", *Science Direct, Elsevier*, pp. 1–39, 2019.
95. Venkateswari R and Sreejith S, "Factors influencing the efficiency of photovoltaic system", *Renewable and Sustainable Energy Reviews, Elsevier*, Vol. 101, pp. 376–394, 2019.
96. Richhariy G, Kumar A and Samsher, "Chapter 2, 3: Go to photovoltaic solar energy conversion on science direct: Technologies, applications and environmental impacts", *Science Direct, Elsevier*, pp. 27–78, 2019.
97. Srinivas B, Balaji S, Nagendra M and Reddy Y, "Review on present and advance materials for solar cells", *International Journal of Engineering Research*, Vol. 5, No. 3, pp. 178–182, 2015.
98. Sagadevan S, Pal K, Hoque E and Zaman Z, "A chemical synthesized Al-doped PbS nanoparticles hybrid composite for optical and electrical response", *Journal of Material Science: Material Electron*, Vol. 15, No. 28, pp. 10902–10908, 2017.
99. Li X, Hylton N, Giannini V, Lee K, Ekinsdaukes N and Maier S, "Bridging electromagnetic and carrier transport calculations for three-dimensional modeling of plasmonic solar cells", *Optics Express*, Vol. 19, No. 4, pp. 888–896, 2011.

100. Reyes L, Rangel R and Jose S, *Nanocomposites for photonic and electronic applications book*, Elsevier, pp. 175–183, 2020.
101. Thabet A, Abdelhady S, Ebnalwaled AA and Ibrahim AA, "Modules allocation of innovative industrial solar power substation based on CdS/CIGS nanocomposites thin film solar cells", *Indonesian Journal of Electrical Engineering and Informatics (IJEEI)*, Vol. 7, No. 2, 2019.
102. Thabet A, Abdelhady S and Mobarak Y, "Design modern structure for heterojunction quantum dot solar cells", *International Journal of Electrical and Computer Engineering (IJECE)*, Vol. 10, No. 3, pp. 2918–2925, 2019.
103. Villalva MG, Gazoli JR and Filho ER, "Comprehensive approach to modeling and simulation of photovoltaic arrays", *IEEE Transactions on Power Electronics*, Vol. 24, No. 5, 2009.
104. Thabet A, Abdelhady S, Ebnalwaled AA and Ibrahim AA, "Innovative industrial Cu(In,Ga)Se2 thin film solar cell with high characterization using nanoparticles structure", *Indonesian Journal of Electrical Engineering and Informatics (IJEEI)*, Vol. 7, No. 2, pp. 382–392, June 2019.
105. Thabet A, Abdelhady S, Ebnalwaled AA and Ibrahim AA, "Improvement optical and electrical characteristics of thin film solar cells using nanotechnology techniques", *International Journal of Electronics and Telecommunications (IJET)*, Vol. 65, No. 4, pp. 625–634, October 2019.
106. Thabet A, Mobarak Y, Salem N and El-Noby A, "Performance comparison of selection nanoparticles for insulation of three core belted power cables", *International Journal of Electrical and Computer Engineering (IJECE)*, Vol. 10, No. 3, pp. 2779–2786, 2020.
107. Thabet A, Abdelhady S and Mobarak Y, "Design modern structure for heterojunction quantum dot solar cells", *International Journal of Electrical and Computer Engineering (IJECE)*, Vol. 10, No. 3, pp. 2915–2922, 2020.
108. Thabet A, Abdelhady S and Mobarak Y, "Ultra-optical characterization of thin film solar cells materials using core/shell absorber layer", *International Journal of Electrical and Computer Engineering (IJECE)*, Vol. 12, No. 2, pp. 1377–1384, 2022.
109. Thabet A and Abdelhady S, "Strategy trends of core/multiple shell for quantum dotbased heterojunction thin film solar cells", *Australian Journal of Electrical and Electronics Engineering*, Vol. 19, No. 3, pp. 203–218, 2022.

4 Nanotechnology in Wind Energy

Green energies are contributing country and contributes significantly to reducing our dependence on fossil fuels, and wind is a safe, abundant, sustainable, valuable renewable energy resource that can allow for clean, sustainable, and diversified electricity supplies. Expanding markets have led to rapid advancement in wind turbine technology over the years, which has led to more sophisticated, more reliable performance. There is no pollution associated with wind energy, and it produces no hazardous waste, in contrast with other forms of energy. It is anticipated that these improvements will enable wind energy to compete more effectively with other renewable energy technologies in the future. The energy market recently became flooded with technological developments that allow for the generation of renewable and clean energy. As a renewable alternative with a great potential role, wind energy is one of the fastest growing renewable energy options in the world, and wind farms are becoming more widespread around the globe.

4.1 INTRODUCTION

Wind is an indirect form of solar energy because the sun replenishes wind via differential heating of the earth's surface. Wind energy has only minor adverse effects on the environment and is therefore a green power technology: Manufacturing and transporting the materials used to construct a wind power plant consumes the same amount of energy as the new energy generated by the plant within a few months of operation. Wind energy systems also contribute to ensuring national energy security as fossil fuel reserves are rapidly depleting around the globe, which poses a threat to the long-term sustainability of the global economy.

A wind turbine is a device used to generate electricity by converting the kinetic energy of moving air into electricity using the kinetic energy of moving air. They consist of blades attached to a rotor; when the wind blows, it spins the rotor, which in turn drives a generator to produce electricity. Wind turbines are classified as horizontal or vertical axis. Horizontal-axis wind turbines (HAWT) are the most common type; they consist of long, thin blades similar to airplane propellers. Vertical-axis wind turbines (VAWTs) have shorter, wider curved blades; they look somewhat like the beaters used in an electric mixer [1].

Wind turbines also come in various sizes. Small individual turbines can produce around 100 kilowatts of power, enough to support a home or water pumping station, but it is possible to produce as much as 1.8 megawatts of energy by bigger turbines perched on tall towers. The largest turbines, standing over 240 meters (787 feet) tall, can produce anywhere from 4.8 to 9.5 megawatts of power [1].

DOI: 10.1201/9781003512486-5

Despite their benefits however, wind turbines have a number of negative impacts on wildlife, noise, and visual environment; for instance, collisions between birds and wind turbines are responsible for significant mortality and for habitat disruption and displacement related to avoiding these collisions. Moreover, due to the magnetic forces turbines generate, it can be difficult or impossible to receive radar or television signals, and lightning can strike them. However, these environmental impacts are minor compared with other forms of energy [1]. There is growing focus on finding ways to minimize or prevent wind energy's impact on wildlife. Thoughtful location of power plants can reduce bird mortality. In fact, climate change poses far more significant threats to wildlife [2].

The sound pollution caused by wind turbines is by far the greatest environmental impact, and construction engineers must be aware of the different types of noise that they will be making; they are *very* noisy machines. Mechanical and aerodynamic noise are two types of wind turbine noise. Mechanical noise derives from the movements of components such as gear boxes, generators, and bearings, and it is impacted by a variety of factors, including regular wear and tear, component design problems, or a lack of preventative maintenance. Aerodynamic noise occurs from the air flowing over and past the turbine blades, and it increases as rotor speed increases. There is an association between blade tip speed and blade noise; lower blade tip speeds result in lower plane wave levels, and the turbulence in the atmosphere causes characteristic "whooshing" sounds [3, 4]. There are several methods of minimizing mechanical noise, such as designing side toothed gear wheels at the design stage or insulating the inside of the turbine housing with acoustic material, and operation noise can be reduced by using acoustical insulation curtains and antivibration support footings [5]. Controlling the distances of turbines and wind farms from residential areas would also mitigate the impacts of noise pollution on property values.

Meanwhile, the visual impacts of wind power depend on a variety of factors, including whether the turbine is moving or stationary, the size and distance from residences, and whether there is shadow flickering [6, 7]. Geographic information systems (GIS) are among the most commonly used tools for assessing potential visual impacts and their magnitudes at chosen sites [8]. Mapping potential impacts for transmission lines separately addresses wind turbine safety considerations [9].

Wind energy is a clean, environmentally friendly, cost-effective source of renewable energy that has an important role in reducing atmospheric pollution and water consumption because wind energy consumes less water than fuel-based power plants that generate energy from petroleum or coal. There are some negative impacts caused by wind turbine power, but these are minor compared with those from petroleum-based energy. However, if turbines are planned and handled carefully, these negative impacts can be limited [10].

4.2 PRINCIPLES OF WIND ENERGY

The blades of a wind turbine rotate to convert wind kinetic energy into electrical energy, and the amount of energy is determined by the square root of the wind speed. Here I focus on the relationship between wind speed and wind power and how it translates into changes in electricity generation. The kinetic energy of the air parcel is represented as

$$KE = \frac{1}{2} m v^2 \tag{4.1}$$

Thus, the power available in the wind is as follows:

$$P = \frac{1}{2} \rho A v^3 \tag{4.2}$$

where A is a given cross-sectional area of the wind in m^2, v is wind speed in m/s, and ρ is density of the air in kg/m^3 density, which varies little horizontally and vertically near the rotor. Even small errors in calculating these equations can cause large errors in wind power forecasting. There is an intuitive reason why wind power increases in proportion to air density: When more air molecules hit turbine blades and transfer their momentum to the turbine [11].

The Weibull distribution is widely used, accepted, and recommended for estimating wind energy potential and expressing wind speed frequency distributions. Based on the Weibull distribution, the probability density function is calculated as follows:

$$f(v) = \frac{k}{c}\left(\frac{v}{c}\right)^{k-1} e^{-\left(\frac{v}{c}\right)^k} \tag{4.3}$$

where f(v) is the probability of observing wind speed v, k is the dimensionless shape parameter, and c is scale parameter in units of wind speed. The shape parameter of the Weibull distribution ranges from 1.2 to 2.75 for most wind conditions in the world. The cumulative distribution function of Weibull distribution is given as follows:

$$F(v) = 1 - e^{-\left(\frac{v}{c}\right)^k} \tag{4.4}$$

Other methods of estimating Weibull parameters include graphic, maximum likelihood, moment, and power density [12].

Different technologies have been emerging for converting wind energy into electricity in recent years, all of which aim at lowering costs, increasing efficiency, and improving reliability. Wind energy conversion systems (WECS) convert renewable energy from wind to electricity directly connected to the grid at constant speed via a squirrel cage induction generator and gearbox. With new developments in power electronics technology, variable-speed WECS with multiple-stage gearboxes, dual-fed induction generators, and partial-scale converters have extended both operation ranges and efficiency [13, 14].

Multiple technologies exist for energy for use in wind power applications; evaluating and selecting one depends on, among other things, the device's operating principles, their components, and their most relevant features as well. For instance, pumped storage hydropower is a large-scale energy storage system. Water is pumped from a lower reservoir to an upper reservoir during periods of low power demand, thereby taking advantage of its gravitational potential energy. The water flows from the upper to the lower reservoir when there is a great demand for power, and this action activates the turbines to produce electricity. There is a direct relationship between the volume of water and the height of the waterfall in the upper reservoir in terms of the stored energy. Generally, pumped hydro is used to store energy for between 1 and 24 hours, but longer-term (days) energy storage is also possible [15]. hydrostorage is available in many areas worldwide and have great potential [16].

Most low-power devices are powered by batteries; however, due to limitations in size and the need to charge them, batteries cannot achieve autonomy; they are increasingly being replaced with more efficient power options that come with no lifetime worries. There are many practical applications in which energy can be harvested from ambient resources, but portability is essential to many of these applications. Device portability is associated with its dimensions and weight. Among the major methods for producing portable devices, micromachining is one of the most convenient and efficient ones. This technique involves exposing the designed patterns on the NP surface using lithography or etching to produce micro-sized or even smaller mechanical parts [17].

4.3 WIND ENERGY CONSTRUCTION

Offshore wind energy construction involves the design, installation, and maintenance of wind turbines to harness wind power for electricity generation. Researchers built a work breakdown structure that described all the processes involved in wind turbine construction planning that featured phases for site selection and assessment, turbine design and components, construction, challenges and innovations, and environmental considerations [18]. Each subproject moves through a series of phases during the construction period, and all phases to the master schedule for the construction of the project [19, 20].

Other researchers built a chart that illustrated the construction phase (pre-assembly, installation, and commissioning) but excluded design, production, and operations. It has been described the relationships between suppliers and customers between the different phases of construction [21] and a model that shows how materials are transported from one manufacturing plant to another and from one shipping port to the other to arrive at the port of preassembly through a chain of production [22]. Logistics networks are responsible for the delivery of prefabricated modules that are, due to their weight and size, shipped from their origin or manufacture to the pre-assembly port by trucks. Researchers considered a number of factors before shipping components on specialized installation vessels such as port configurations, capacity, and logistics arrangements [23–25].

Many wind turbines will likely need to be decommissioned in the near future because their blades, which are largely made of glass fiber reinforced polymer composites (GFRPs), will reach the end of their service life of roughly 20–25 years [26]. In the next few years, the industry is expected to have to stockpile millions of tons of composite wind blades [27] made of the GFRPs (in conjunction with a small amount of carbon fiber) embedded in epoxy, polyester, or vinyl ester resins.

Based on the blade weight divided by the capture area, the specific weight determines the economic performance of the blades. Researchers built an innovative carbon/glass hybrid blade provided excellent weight; the specific weights of the blades increased significantly over the analysis range for all blade diameters. The blade's structural performance improved with thicker trailing edges in the inboard region; its specific weight decreased by 15% over the baseline weight. However, although the innovative design produced a lower blade weight, the blades present significant manufacturing, price, and cost challenges [28–31].

Researchers used two simplified methods to estimate blade design loads when a vehicle is either parked or operating in extreme wind conditions: following the design recommendations of International Electrotechnical Commission (IEC) Class I and calculation. They analyzed structural characteristics of a representative blade at baseline, thicker, and thickest and found no significant differences in load estimates between the two approaches. The blade was constructed four primary components: two called low-pressure shells on the downwind side and high-pressure shells on the upwind side and two shear webs that are bonded between the two shells. The researchers took multiple measurements of cross-sections of the blade at the baseline, thicker, and thickest spots and then calculated the stresses and deflections using 2D beam theory. They observed strong correlation between blade weight and length in the range between 30 and 70 m for all three cross-sections [32].

The modern wind turbine uses wind power to drive a rotor that rotates around what is called either a vertical or a horizontal axis depending on the wind's direction. Typically, the rotor in a VAWT is mounted on top of a tall tower made of steel or concrete that is fixed to the rotor. A gearbox on the rotor converts the low-speed, high-torque rotation of the rotor into high-speed, low-torque rotation that is then converted by an onboard generator into electricity. The wind turbine usually starts to generate electricity at a speed of 3–4 m/s, and it reaches its max output when the wind speed reaches around 13 m/s; it shuts off at about 25 m/s to prevent damage to the generator and other components as the wind speed increases [33].

There is a common use for reinforced concrete spread footings for towers because they are a simple and economical solution. However, different types of foundations are used when the soil conditions are weaker or the site has unusual conditions, such as limited spaces, sloped terrain, or offshore, including pilings, drilled shafts, and caissons [34].

Wind turbine loads can be classified into two types: static and cyclic. Static loads are caused by centrifugal and by aerodynamic load caused by uniform, steady wind speeds, and centrifugal force produce stationary loads. Cyclic loads occur on the turning rotor due to the flow fields that are stationary yet uneven throughout the swept areas. One issue is that rotating rotor blade weight mass force that constantly causes periodic, nonstationary loads. Apart from the stationary and cyclic loads driven rotor by the wind, the rotor is also exposed to random and nonperiodic loads that are generated by wind turbulence. An example of a typical concrete tower for a 1 MW wind turbine is given in [35].

4.4 WIND ENERGY DEVELOPMENT

Early wind turbine blades were made from stainless steel, but these stopped working within 100 hours of operation, the blades of the wind turbine stopped working. Since the early days, turbine blades have been built with materials including aluminum, steel, plastics, and wood, but composites have been the most common. The blades are the most expensive component of a wind turbine, and they are also some of the most complex: Blade materials must be stiff and light and have long fatigue life for optimal aerodynamic performance. Considerable research is being focused on developing lighter materials with a longer life cycle and lower cost such as carbon filament reinforced plastic, which are extremely strong as well as fatigue resistant [36–38].

The blades of wind turbines are becoming lighter with advances in research and technology. Since the blades of these wind turbines are lighter than conventional blades, they reduce inertial forces, which is what improves the efficiency of the turbine. In addition to the wheel shapes, researchers are experimenting with aerodynamically improving complex shapes for large blades, especially for the HAWT, using an algorithm that calculates the optimal twist angle; the ingenious internal blade structure makes it possible to fabricate long, large blades. Moreover, advanced construction techniques have been able to create tight strength-to-weight ratios [39].

Wind farms across the globe are employing large wind turbines with rotor diameters of as much as 100 meters to produce electricity; turbines are becoming larger even though the costs are high. These enormous blades primarily comprise a fiberglass reinforced epoxy resin, but carbon fiber and balsa are also part of their construction. Today, the blades consist of high proportions of carbon fibers to increase their durability against varying conditions; in addition to bending forces within the turbine blades, large wind turbines face aerodynamic loads, turbulent winds, dynamic thrusts and torques, and severe cyclic fatigue effects [40]; for instance, researchers explored CNT/polymer nanocomposites for blade materials. Multiwalled CNTs (MWCNTs) are created from a concentric arrangement of single-walled CNTs (SWCNTs), which comprise a single layer of carbon atoms. CNTs have a high aspect ratio (length-to-diameter ratio greater than 1000) that gives them large surface areas that allow the matrix and the fibers interact more effectively with each other at the interfacial level [41].

Turbines face many other challenges besides turbulent winds, such as fouling of rotor blades by insects, ice formation, and material erosion caused by dust particles or diffusion of moisture. There is a growing trend of installing wind turbines in cold-climate markets, which will increase ice formation on the blades; ice has multiple negative effects including higher cyclic loads and vibrations in rotation, reduced energy output, more noise, increased safety and health risks, and shorter life spans for turbine components. However, rotor blade fouling can be solved using nanocomposite paint [42].

The coating must be more durable for a wind turbine with a design life of up to 20 years if the researchers want to reduce the maintenance cost and the downtime of the turbine. The fatigue life of polymeric coatings has been carried out through experiments to overcome this problem. A nanocomposite filler in the coating can retard the initiation of fatigue at higher stress levels [43].

Wind energy harvesters are categorized by how they generate power: Piezoelectric harvesters utilize piezoelectric materials directly, electromagnetic harvesters use magnetic flux density changes, and electrostatic harvesters utilize capacitance variations. Rotational and aeroelastic mechanisms can be used to measure wind flow vibration; vibrations are produced by both vortices and movement in the aeroelastic mechanism. Portable wind energy harvesters can also be categorized by their physical dimensions in addition to their capabilities [44].

A number of desirable attributes need to be in place for wind power consumption to rise, including the creation of future-oriented, low carbon industries and the employment opportunities that these industries provide. Wind energy consumption patterns have changed over time, specifically increasing in the post-financial crisis period around the world; fiscal stimulus appears to have driven these increases. According to the World Wind Energy Association, wind energy consumption will continue to rise in the future if renewable energy participation continues to increase, which in turn is affected by energy policies designed to encourage the switch from fossil fuels to renewables [44].

4.5 NPs AND WIND ENERGY

A system's efficiency depends on what material it is made of; the quality of the material has a direct effect on the performance of the system. The development of nanotechnology has exponentially reduced the sizes and other specifications of renewable energy devices for converting and storing energy, monitoring the environment, and developing green sustainable materials. As a part of building affordable and clean energy, nanomaterials play an important role in the human life; in other words, they are now a globally marketed industry. In the field of energy production, wind energy storage, nanotechnology can significantly reduce the cost of components such as solar cells, fuel cells, hydrogen production, and wind energy storage. Researchers have recently developed composite materials that combine polymer, ceramic, and metal matrix components; these composite materials have shown great potential for applications related to energy [45–47]. Today, NP-based composites are being widely used in military devices because they possess high strength-to-weight and good operational safety and are affordable. NP-based composites have these features and other properties that make them useful for a wide variety of applications [48] including in green energy generation.

Successful applications of nanomaterials in wind turbine blades require choosing a polymer matrix and NPs that are optimized for the application in mind; NPs are often cleverly incorporated into polymeric materials in order to achieve specific properties. Clay is one of the most commonly used fillers due to its low cost and high surface-to-area properties; clay nanofillers have three basic classifications based on silica and Al content. Several factors influence the performance of composite materials, including the method of preparation. It has been used a wide variety of methods to disperse nanomaterials into polymers in combinations that give optimal outputs, including grafting, ultrasonication, mechanical, physical and chemical methods [49–51], injection molding [52, 53], compression molding [2, 54], extrusion molding [55], and resin transfer molding [56, 57].

The vital parameters for making polymer matrix composites with good properties are temperature, speed, time, and pressure. Polymer nanocomposites have been developed and used extensively in many different fields, including automotive and aerospace, medical applications, electrical and electronic, etc. There have also been significant advances in research and applications in the energy sector. Component preparation methods contribute to their performance; in fact, certain materials can only be prepared using a specific method [58].

4.6 NANOTECH WIND ENERGY MODULES

Nanotechnology improves the efficiency of wind turbines that generate electricity in a wide range of ways. Improving turbine performance and reliability results in longer turbine lifetimes, decreased fatigue failure, and lower generation costs. Among these nanotechnologies are NP-containing lubricants, nano-coatings for deicing and self-cleaning, and nanocomposites that make wind blades lighter and stronger than ever before [59]. Nanotechnologies also increase the energy storage capacity of advanced secondary batteries; a nanostructure electrode with ultrahigh surface area is based on CNTs and NP nanowires. Secondary batteries, including Li-ion, NaS, flow, and dry cell, benefit from nanoengineered ceramic electrolytes with well-controlled grains, grain boundaries, and crystal orientations [60].

Recently, there has been the aim for improving the efficiency of current energy sources through the knowledge of material properties at the nanometer level. Some materials exhibit different physical properties at the nanoscale from the properties of these same materials at the bulk scale. Many nanoscale materials can self-assemble and spontaneously change into ordered structures if they are given enough time. The surface to volume ratio of nanostructured materials is greater than that of non-nanostructured materials, and this makes them more suitable for interaction with other materials. Since a wide variety of chemical and electrical reactions occur at surfaces, this is an appropriate property since many of these reactions depend heavily on the shape and chemical compositions of the surfaces and the reactants involved.

It is possible to convert wind energy into useful forms of energy with turbines, windmills, and wind pumps; wind turbines convert wind energy into power, windmills use it to generate mechanical energy, and wind pumps rely on wind energy to pump water or drain away waste. There are many ways to use nanotechnology to increase the efficiency of wind energy conversions, such as in coatings, lubricants, and lighter materials; nano-sensors can monitor stability, possible damage, and any other potential problems that might occur, and protective. There is a great deal of nanotechnology with regard to the concern about wind turbines; one of those issues is corrosion of the blade, and one of the ways to prevent or decrease that corrosion is to use a protective coating, which will reduce maintenance cost and prolong the life of the wind turbines [61–67].

Recently, it has been created a nanocomposite of electroless nickel–phosphorus with glass fiber and reinforced plastic in a wind turbine blade that showed important corrosion resistance. Certain treatments were required before the deposition of the coating; between each step, it has been conducted electroless plating and rinsed the nanocomposite in deionized water: (a) 50 g/L NaOH for decreasing; (b) 200 g/L nitric acid for increasing; (c) a solution of 10 g/L stannous chloride sensitized for 2 minutes; 20 g/L sodium fluoride, and 40 g/L hydrochloric acid (HCl); and (d) 2.5 mL/L of a solution of palladium chloride and HCl activated for 1 minute [68].

Incorporating wind energy harvesters into the building skin maximizes the use of wind energy, which in one study was transformed into vibrations that could then be converted into electricity through microelectromechanical systems. Lubricating the turbine gearbox with a nano-coating also improves wind harvesting [69]. Other scientists created a nanocomposite composed of an epoxy and graphene oxide NP and coated turbine blades to significantly reduce light damage [70]. It is generally accepted that electromagnetic wind turbines are the most conventional application of wind energy harvesters [71], but today, harvesters can be installed on mobile and compact devices [72, 73].

There are four types of wind harvesters for converting wind energy: electromagnetic harvesters collect macroscale wind energy, whereas piezoelectric, electrostatic, and triboelectric harvesters collect microscale energy. A number of micro/nano wind sensors are available on the market; in case of communication and intelligent wind monitoring, passive and active wind sensors are exploited and improved to meet high precision and multifunction requirements. Micro/nano wind energy harvesters consist of fixed parts and moveable components. Moveable components are used as mechanical transducers for converting wind flow into mechanical rotation or vibration in the process of using wind energy. Wind energy harvesters can also be classified as windmill based or aeroelastic based on how they trap the wind. Windmill-based harvesters utilize blades to convert wind into rotating mechanical energy that is then converted to electrical energy by the coaxially connected components. Aeroelastic harvesters offer the advantages over windmill-based harvesters of low cost and simplicity. Aeroelastic devices are technically capable of harvesting wind energy from varying directions of wind flow, but since wind direction is random, neither direction is favored in terms of power generation; for arbitrary wind energy harvesting situations, the windmill-powered harvester can always be designed to achieve the optimal output performance. Vertical versus horizontal harvesters are arranged in different directions around an axis of rotation and have different wind speeds [74–76].

Micro/nano fabrication aims to achieve the optimum performance and robustness of triboelectric energy harvesters (TEH). The permittivity and surface morphology of triboelectric materials can be improved by optimizing their surface charge density, which is positively correlated with the output performance of TEH. Some of these methods include inductively coupled etching, transferring micro/nano structures, and doping nanocomposite materials. Optimizing the state of the friction pair formed by the two triboelectric films also increases robustness: Contact state optimization decreases the friction force between the two triboelectric films and the start-up wind velocity. Wind energy harvesters based on aeroelasticity tend to be smaller than windmill-based harvesters because they require less space for vibration [77, 78].

The thin-film membrane-based triboelectric nanogenerator (TENG) is a novel structure that offers the possibility of harvesting wind energy without blades. The technology uses a combination of contact electrification and electrostatic induction, and thin-film membranes allow for aggressively reducing the TENG's size. Nanotechnology for building materials that are more lightweight and durable, and it is being widely used to improve the efficiency of wind and ocean energy generators [79, 80].

Magnetic switch triboelectric nanogenerators (MS-TENGs) provide continuous electric power above the critical rate because when intermittent winds fall on the wind scoop, the energy modulation modules, which can store and release energy at any time regardless of wind speed, are responsible together with the magnetic force. This allows wind energy to be converted into electric energy on an almost continuous basis. The MS-TENG consists of a wind scoop, transmission gears, energy modulation modules, and a generation unit at its top. The scoop captures the wind energy, and the modulation modules and generation unit convert it into electric energy. Electrical energy can also be controlled by TENGs that operate via mechanical regulation, although these TENGs cannot provide continuous power because their outputs are intermittent. A wind scoop operates by converting the wind's kinetic energy into magnetic potential energy via transmission gears and the energy modulation modules. This process drives the generator to operate as long as the input speed exceeds the critical speed [81].

Wind turbines use various transmission components including gearboxes that can handle high-voltage fluctuations and rough conditions, which minimizes gearbox downtime and enables cost-efficient operation by monitoring the degradation conditions [82]. Assessing gearbox health of and computing prognoses can be divided into four steps: acquire the data, extract the features, construct health indicators, and predict remaining useful life [83]. Gearboxes are heavily impacted by noise; it is difficult to extract features for assessing their health because noise corrupts the vibration signals they acquire [84].

Wireless sensors can collect data instead of a traditional cabled sensor, but they can't completely replace the cabled sensors because they require batteries for long operation, and inefficient sensors rapidly deplete them; the wireless modem that links the wireless sensor and the outside world consumes the most energy, so it serves as the largest energy consumer. Hence, it is crucial to be selective when using the wireless communication channel in order to maintain the best possible quality of communication. In response to this fact, wireless sensors have actually been developed to process their own data in relation to the raw data, allowing for a relatively small amount of processed data to be broadcast instead of a massive amount of high-bandwidth raw data [85]. A novel approach was also developed that fused vibration signals from multiple sensors to examine the health status of gearboxes [86].

Design optimization is key in wind turbine system design, for both whole systems and individual parts, such as the tower, support structure, or even the mooring system in the case of floating offshore turbines. The best systems incorporate optimal features and outcomes for all areas such as cost and material consumption, loads, cycle times, and, ultimately, the overall response and performance of the system.

Wind turbine controllers measure wind speed at certain intervals. In a significant difference from onshore and bottom-fixed offshore controllers: Measurement speed is not constant in floating systems. Since measurements are made at such short intervals, the controller perceives decreasing wind speed (corresponding to a decreasing thrust) if the floating system moves with the wind; to prevent a reduction in the power output of the system, the controller pitches the blades directly into the wind. Using an onshore controller with a floating turbine causes damping, which destabilizes the system, making it necessary to adjust the parameters. The optimization goal is to determine the controller frequency at which the floating wind turbine system will maintain a stable state at a frequency lower than its lowest eigenfrequency, which puts it on the path toward stability [87, 88].

REFERENCES

1. Magoha P. Footprints in the wind? Environmental impacts of wind power development. *Fuel and Energy Abstracts* 2003;44(3):161.
2. Kuvlesky Jr William P, et al. Wind energy development and wildlife conservation: Challenges and opportunities. *Wind Energy and Wildlife* 2010;71:2487–2498.
3. Julian DBSM, Jane X, Davis RH. Noise pollution from wind turbine, living with amplitude modulation, lower frequency emissions and sleep deprivation. In: *Second International Meeting on Wind Turbine Noise*. 2007.
4. Oerlemans S, Sijtsmaa P, Mendez LB. Location and quantification of noise sources on a wind turbine. *Journal of Sound and Vibration* 2007;299:869–883.
5. Richard G. *Wind Developments: Technical and Social Impact Considerations*. Orkney Sustainable Energy Ltd., 2007.
6. Ian BD. Determination of thresholds of visual impact: The case of wind turbines. *Planning and Design* 2002;29:707–718.
7. Jacob L. Visual impact assessment of offshore wind farms and prior experience. *Applied Energy* 2009;86:380–387.
8. Juan PH, Joaquın F, Jorge PL, Eduardo B. Spanish method of visual impact evaluation In wind farms. *Renewable and Sustainable Energy Reviews* 2004;8:483–491.
9. Hadrian D, Bishop ID, Mitcheltree R. Automated mapping of visual impacts in utility corridors. *Landscape Urban Planning* 1988;16:261–282.
10. Saidur R, Rahim NA, Islam MR, Solangi KH. Environmental impact of wind energy. *Renewable and Sustainable Energy Reviews* 2011;15:2423–2430.
11. Archer CL. An introduction to meteorology for airborne wind energy. In: Ahrens U, Diehl M, Schmehl R. (eds) *Airborne Wind Energy. Green Energy and Technology*. Springer, Berlin, Heidelberg, 2013. https://doi.org/10.1007/978-3-642-39965-7_5.
12. Akdag SA, Dinler A. A new method to estimate Weibull parameters for wind energy applications. *Energy Conversion and Management* 2009;50:1761–1766.
13. Blaabjerg F, Teodorescu R, Liserre M, Timbus AV. Overview of control and grid synchronization for distributed power generation systems. *IEEE Transactions on Industrial Electronics* 2006;53(5):1398–409.
14. Blaabjerg F, Liserre M, Ma K. Power electronics converters for wind turbine systems. *IEEE Transactions on Industrial Electronics* 2012;48(2):708–719.
15. Dursun B, Alboyaci B. The contribution of wind-hydro pumped storage systems in meeting Turkey's electric energy demand. *Renewable and Sustainable Energy Reviews* 2010;14:1979–1988.
16. Díaz-González F, Sumper A, Gomis-Bellmunt O, Villafáfila-Robles R. A review of energy storage technologies for wind power applications. *Renewable and Sustainable Energy Reviews* 2012;16:2154–2171.
17. Nabavi S, Zhang L. Portable wind energy harvesters for low-power applications: A survey. *Sensors* 2016;16:1101. https://doi.org/10.3390/s16071101
18. Barlow E, Tezcaner Öztürk D, Revie M, Akartunalı K, Day AH, Boulougouris E. A mixed-method optimisation and simulation framework for supporting logistical decisions during offshore wind farm installations. *European Journal of Operational Research* 2018;264(3):894–906.
19. Ballard G, Tommelein I. Current process benchmark for the last planner system. *Lean Construction Journal* June 2016;89:57–89.

20. Daniel EI, Pasquire C, Dickens G. Development of approach to support construction stakeholders in implementation of the last planner system. *Journal of Management in Engineering* 2019;35(5):04019018. https://doi.org/10.1061/(ASCE)ME.1943-5479.0000699
21. Quandt M, Beinke T, Ait-Alla A, Freitag M. Simulation based investigation of the impact of information sharing on the offshore wind farm installation process. *Journal of Renewable Energy* 2017;2017:11.
22. Irawan CA, Akbari N, Jones DF, Menachof D. A combined supply chain optimisation model for the installation phase of offshore wind projects. *International Journal of Production Research* 2018;56(3):1189–1207.
23. Irawan CA, Song X, Jones D, Akbari N. Layout optimization for an installation port of an offshore wind farm. *European Journal of Operational Research* 2017;259(1): 67–83.
24. Barlow E, Ozturk DT, Day AH, Boulougouris E, Revie M, Akartunalı K. A support tool for assessing the risks of heavy lift vessel logistics in the installation of offshore wind farms. In: *Marine Heavy Transport & Lift IV*, 11–20. Royal Institution of Naval Architects, London, 2014a.
25. Paterson J, D'Amico F, Thies PR, Kurt RE, Harrison G. Offshore wind installation vessels—a comparative assessment for UK offshore rounds 1 and 2. *Ocean Engineering* January 2018;148:637–649.
26. André A, Kullberg J, Nygren D, Mattsson C, Nedev G, Haghani R. Re-use of wind turbine blade for construction and infrastructure applications. *International Symposium on Materials Science and Engineering* 2020;942:012015.
27. Liu P, Barlow CY. Wind turbine blade waste in 2050. *Waste Management* 2017;62:229–240.
28. Mishnaevsky L, Branner K, Petersen HN, Beauson J, McGugan M, Sørensen BF. Materials for wind turbine blades: An overview. *Materials* 2017;10:1285.
29. TPI Composites. *Parametric Study for Large Wind Turbine Blades. SAND2002–2519.* Sandia National Laboratories, Albuquerque, NM, 2002.
30. TPI Composites. *Cost Study for Large Wind Turbine Blades. SAND2003–1428*, Sandia National Laboratories, Albuquerque, NM, 2003.
31. TPI Composites. *Innovative Design Approaches for Large Wind Turbine Blades. SAND2003–0723.* Sandia National Laboratories, Albuquerque, NM, 2003.
32. Jackson KJ, Zuteck MD, van Dam CP, Berry D, TPI. Innovative design approaches for large wind turbine blades. *Wind Energy* 2005;8:141–171.
33. Fenwick S. Wind power in the UK. *International Cement Review*, May 2007, pp. 109–110.
34. Bromage A. Concrete for the next generation of wind farms. *Concrete*, January/February 2006, pp. 24–25.
35. Singh AN. Concrete construction for wind energy towers. *The Indian Concrete Journal* August 2007;43–49.
36. Mishnaevsky L, Branner K, Petersen HN et al. Materials for wind turbine blades: An overview. *Materials* 2017;10(11):1285.
37. Ancona D, McVeigh J. *Wind Turbine-Materials and Manufacturing Fact Sheet.* US Department of Energy: Princeton Energy Resources International for the Office of Industrial Technologies, 2001.
38. Griffith DT, Resor BR, Ashwill TD. Challenges and opportunities in large offshore rotor development: Sandia 100-meter blade research. In: *AWEA WINDPOWER Conference and Exhibition*, 2012.
39. Muzammil WK, Rahman MM, Fazlizan A, Ismail MA, Phang HK, Elias MA. Nanotechnology in renewable energy: Critical reviews for wind energy. In: Siddiquee S, Melvin G, Rahman M. (eds) *Nanotechnology: Applications in Energy, Drug and Food.* Springer, Cham, 2019. https://doi.org/10.1007/978-3-319-99602-8_3

40. Chou JS, Chiu CK, Huang IK et al. Failure analysis of wind turbine blade under critical wind loads. *Engineering Failure Analysis* 2013;27:99–118.
41. Ma PC, Zhang Y. Perspectives of carbon nanotubes/polymer nanocomposites for wind blade materials. *Renewable and Sustainable Energy Reviews* 2014;30:651–660.
42. Dalili N, Edrisy A, Carriveau R. A review of surface engineering issues critical to wind turbine performance. *Renewable and Sustainable Energy Reviews* 2009;13:428–438.
43. Zhou Y, Rangari V, Mahfuz H et al. Experimental study on thermal and mechanical behavior of polypropylene, talc/polypropylene and polypropylene/clay nanocomposites. *Materials Science and Engineering A* 2005;402(1–2):109–117.
44. Sadorsky P. Wind energy for sustainable development: Driving factors and future outlook. *Journal of Cleaner Production* 2021;289:125779.
45. Yu G, Xie X, Pan L et al. Hybrid nanostructured materials for high performance electrochemical capacitors. *Nano Energy* 2013;2:213–234.
46. Sun Y-K, Chen Z, Noh H-J et al. Nanostructured high-energy cathode materials for advanced lithium batteries. *Nature Materials* 2012;11:942–947.
47. Hussein AK. Applications of nanotechnology in renewable energies—a comprehensive over view and understanding. *Renewable and Sustainable Energy Reviews* 2015;42:460–476.
48. Yang Y, Han C, Jiang B et al. Graphene-based materials with tailored nanostructures for energy conversion and storage. *Materials Science and Engineering: R: Reports* 2016;102:1–72.
49. Zeng Q, Yu A, Lu G et al. Clay-based polymer nanocomposites: Research and commercial development. *Journal of Nanoscience and Nanotechnology* 2005;5:1574–1592.
50. Azeez AA, Rhee KY, Park SJ et al. Epoxy clay nanocomposites– processing, properties and applications: A review. *Composites Part B: Engineering* 2013;45:308–320.
51. Mittal G, Rhee KY, Mišković-Stanković V et al. Reinforcements in multi-scale polymer composites: Processing, properties, and applications. *Composites Part B: Engineering* 2018;138:122–139.
52. Hopmann C, Fischer T. New plasticising process for increased precision and reduced residence times in injection moulding of micro parts. *CIRP Journal of Manufacturing Science and Technology* 2015;9:51–56.
53. Ibrahim ID, Jamiru T, Sadiku ER et al. Mechanical properties of sisal fibre-reinforced polymer composites: A review. *Composite Interfaces* 2016;23:15–36.
54. Rubio-López A, Olmedo A, Díaz-Álvarez A et al. Manufacture of compression moulded PLA based biocomposites: A parametric study. *Composite Structures* 2015;131:995–1000.
55. Bouman J, Belton P, Venema P et al. The development of directnextrusion-injection moulded zein matrices as novel oral controlled drug delivery systems. *Pharmaceutical Research* 2015;32:2775–2786.
56. Van Velthem P, Ballout W, Dumont D et al. Phenoxy nanocomposite carriers for delivery of nanofillers in epoxy matrix for resin transfer molding (RTM)-manufactured composites. *Composites Part A: Applied Science and Manufacturing* 2015;76:82–91.
57. Zailinda A, Rivai A, Yuhazri M et al. Comparison of two different composite process using hand lay-up and resin transfer moulding for tensilenproperties. *Proceedings of Mechanical Engineering Research Day* 2017;2017:308–309.
58. Ibrahim ID, Jamiru T, Sadiku ER, Hamam Y, Alayli Y, Eze AA. Application of nanoparticles and composite materials for energy generation and storage. *IET Nanodielectrics* 2019;2(4):115–122.
59. Haldar P. The power of nanotechnology. *Nanotechnology Now*, 2007.
60. Eldada L. Nanotechnologies for efficient solar and wind energy harvesting and storage in smart-grid and transportation applications. *Journal of Nanophotonics* 2011;5:051704-(1-18).

61. Ashby MF, Ferreira PJ, Schodek DL. Nanomaterials and nanotechnologies: An overview. 2009;1–16, ISBN:978-0-7506-8149-0.
62. Kostoff RN, Koytcheff RG, Lau CGY. Global nanotechnology research literature overview. *Technological Forecasting and Social Change* November 2007;74(9):1733–1747.
63. Pitkethly MJ. Nanomaterials—the driving force. *Materials Today* December 2004;7(12):20–29.
64. Ashby MF, Ferreira PJ, Schodek DL. Nanomaterials: Synthesis and characterization. *Nanomaterials Nanotechnologies and Design* January 2009;257–290.
65. Ashby MF, Ferreira PJ, Schodek DL. Nanomaterials and nanotechnologies in health and the environment. *Nanomaterials Nanotechnologies and Design* January 2009;467–500.
66. Van Hove MA. Atomic-scale structure: From surfaces to nanomaterials. *Surface Science* June 2009;603(10–12):1301–1305.
67. Zäch M, Hägglund C, Chakarov D, Kasemo B. Nanoscience and nanotechnology for advanced energy systems. *Current Opinion in Solid State and Materials Science* June 2006;10(3–4):132–143.
68. Lee CK. Corrosion and wear-corrosion resistance properties of electroless Ni–P coating on GFRP composite in wind turbine blade. *Surface and Coatings Technology* 2008;202(19):4868–4874.
69. Du L, Fang Z, Yan J, Zhao Z. Enabling a wind energy harvester based on ZnO thin film as the building skin. *Sensors and Actuators A: Physical* June 2017;260:35–44.
70. Vryonis O, Andritsch T, Vaughan AS, Lewin PL. Improved lightning protection of carbon fiber reinforced polymer wind turbine blades: Epoxy/graphene oxide nanocomposites. In: *2016 IEEE Conference on Electrical Insulation and Dielectric Phenomena (CEIDP)*, 2016;635–638.
71. Joselin Herbert GM, Iniyan S, Sreevalsan E, Rajapandian S. A review of wind energy technologies. *Renewable and Sustainable Energy Reviews* August 2007;11(6):1117–1145.
72. Luong T, Seo Goo N. Use of a magnetic force exciter to vibrate a piezocomposite generating element in a small-scale windmill. *Materials and Structures* 2012;21.
73. Xie Y, Wang S, Lin L et al. Rotary triboelectric nanogenerator based on a hybridized mechanism for harvesting wind energy. *ACS Nano* August 2013;7(8):7119–7125.
74. Didane DH, Rosly N, Zulkafli MF, Shamsudin SS. Performance evaluation of a novel vertical axis wind turbine with coaxial contra-rotating concept. *Renewable Energy* 2018;115, 353–361.
75. Zhang Y, Zhao J, Grabrick B, Jacobson B, Nelson A, Otte J. *Journal of Fluids and Structures* 2018;80:316–331.
76. Bai CJ, Wang WC. Review of computational and experimental approaches to analysis of aerodynamic performance in horizontal-axis wind turbines (HAWTs). *Renewable and Sustainable Energy Reviews* 2016;63, 506–519.
77. Abdelkefi A. Aeroelastic energy harvesting: A review. *International Journal of Engineering Science* 2016;100:112–135.
78. Fu X, Bu T, Li C, Liu G, Zhang C. *Nanoscale* 2020. https://doi.org/10.1039/D0NR06373H
79. Meng XS, Zhu G, Wang ZL. Robust thin-film generator based on segmented contact-electrification for harvesting wind energy. *ACS Applied Materials & Interfaces* June 2014;6(11):8011–8016.
80. Ahmadi MH, Ghazvini M, Alhuyi Nazari M, Ali Ahmadi M, Pourfayaz F, Lorenzini G, Ming T. Renewable energy harvesting with the application of nanotechnology: A review. *International Journal of Energy Research, John Wiley & Sons, Ltd.* 2018;1–24.
81. Liu SH, Li X, Wang Y, Yang Y, Meng L, Cheng T, Wang ZL. Magnetic switch structured triboelectric nanogenerator for continuous and regular harvesting of wind energy. *Nano Energy* 2021;83:105851.

82. Leite GNP, Araújo MA, Rosas PAC. Prognostic techniques applied to maintenance of wind turbines: A concise and specific review. *Renewable and Sustainable Energy Reviews* 2018;81:1917–1925.
83. Lee J, Wu F, Zhao W, Ghaffari M, Liao L, Siegel D. Prognostics and health management design for rotary machinery systems-reviews, methodology and applications. *Mechanical Systems and Signal Processing* 2014;42(1–2):314–334.
84. Srikanth P, Sekhar AS. Wind turbine drive train dynamic characterization using vibration and torque signals. *Mechanism and Machine Theory* 2016;98:2–20.
85. Andrew Swartz R, Lynch JP, Zerbst S, Sweetman B, Rolfes R. Structural monitoring of wind turbines using wireless sensor networks. *Smart Structures and Systems* 2010;6(3).
86. Pan Y, Hong R, Chen J, Singh J, Jia X. Performance degradation assessment of a wind turbine gearbox based on multi-sensor data fusion. *Mechanism and Machine Theory* 2019;137:509–526.
87. Leimeister M, Kolios A, Collu M. Development of a framework for wind turbine design and optimization. *MDPI, Modelling* 2021;2:105–128.
88. Larsen TJ, Hanson TD. A method to avoid negative damped low frequent tower vibrations for a floating, pitch controlled wind turbine. *Journal of Physics: Conference Series* 2007;75.

5 Nanotech Green Energy in Industry

Nanotechnology has the potential to solve problems and make our lives more comfortable, but it also poses challenges and offers treatments. The nanotechnological industry is a multidisciplinary frontier technology that is able to provide numerous innovative solutions. Green technologies, including nanotechnology solutions, play a greater role in realizing sustainable development goals and eliminating the threats of technological advancement in development processes by combining ecofriendly nanotechnology solutions with green and sustainable solutions. Nanotechnology solutions are crucial to countering this threat in the field of green energy development. There is no doubt that nanotechnology has the potential to provide green energy sources, distribution lines, and systems that utilize green energy in a more efficient and effective manner because of its potential advantages in the generation and storage of green energy. In recent years, nanotechnology has been praised for its potential to be used to develop green energy as a ubiquitous energy system and to provide business opportunities and better environmental outcomes.

5.1 INTRODUCTION

Nanoscale materials are different from their bulk counterparts in terms of their physical properties (strength, conductivity, reflectivity, chemical reactivity, etc.); several nanoscale materials can spontaneously self-assemble into structures that are ordered by some intrinsic property. The surface area of nanostructured materials per unit weight or volume is enormous, which means that there is much greater areas are available for interacting with other materials. The shape, texture, and composition of the surface are all key factors in determining the occurrence and the intensity of many chemical and electrical reactions [1–7]. Green energy is being developed to maximize environmental efficiency, minimize secondary waste production, and meet social and economic societal needs using clean and sustainable sources of energy. Energy resources can be roughly categorized as fossil fuels, renewable resources, and nuclear [8]. Renewable energy sources, such as solar, hydrogen, wind, biomass, and geothermal, are often referred to as alternative energy sources [9].

Developing renewable energy systems allows us to solve some of the most crucial challenges of the present time as well as improve energy supply reliability, organic fuel economy, and water supply problems. Green energy helps in achieving sustainable development in remote regions, improving standards of living and employment opportunities for local populations [10]. In rural areas, renewable energy projects can create job opportunities as well as minimize migration to urban areas [11]. Rural and small-scale energy needs can be met reliably, affordably, and environmentally sustainably by harvesting renewable energy in a decentralized manner [12].

DOI: 10.1201/9781003512486-6

Hydrogen, one of the oldest known molecules, has shown an impressive range of applications and uses in many different industries, from chemical and refining processes to glass and electronics; its wide applicability is based on its reactivity rather than its physical properties. A combination of factors connected with changes in crude oil prices led to the rapidly growing use of hydrogen derivatives in petroleum refining. A variety of carbon nanomaterials, such as graphene, graphene oxide, graphene nanotubes, graphene monomers, fullerenes, g-C_3N_4, and nanotubes, have been synthesized that resolve band gaps, acknowledge and transfer electrons, and work as semiconductors and cocatalysts. A broad range of techniques can be used to increase the activity of carbon materials, including doping, functional groups, hybridization with other semiconductors, and interface engineering. The development and production of carbon nanomaterials have to do with their nonmetal nature, which makes them excellent as metal-free photocatalysts for water splitting [13].

The use of solar energy for industrial purposes is becoming a more accepted trend than ever before; industrial sectors increasing solar energy use is significantly reducing energy consumption in developed and underdeveloped countries alike [14–16]. It has become obvious that fossil fuels cannot support all of humanity's energy consumption needs throughout the foreseeable future, not only because they are not renewable and will not be available forever but also because they have had major detrimental impacts on the environment via global warming and climate change. Energy is one of the world's biggest concerns, especially since energy demands have increased rapidly with population growth and industry expansion [17, 18]. Energy storage has been a primary challenge in expanding the use of renewable energy; for instance, solar energy storage must synchronize the demand and supply of energy during new moons and equinoxes. Thermochemical energy, latent and sensible heat, or a combination of them, can be stored in well-insulated liquids and solids. A wide variety of technologies are now available for the storage of energy in different forms such as electrical, mechanical, and thermal [19].

5.2 INDUSTRY NEEDS ASSESSMENT

Many fuel cell products can be used to replace the fossil fuel-powered products that are currently used in many industries, although a series of development stages is needed to fully commercialize fuel cell motorcycles and backup power systems. Technology development, testing, verification, demonstration, and market introduction are some of the steps involved in bringing fuel cell products to the market. Developing such systems consumes considerable time and other resources. Fuel cell selection has become an important part of technology policies where governments have adopted clean energy [20].

Apart from technological advancements, applying fuel cells in products requires ensuring consumer safety and product performance; regulations and standards for testing and validation should ensure these. Furthermore, fuel cell products are currently much more expensive than traditional fossil fuel products because of their high cost of production. Therefore, governmental support for fuel cell product development plays a critical role in propelling their development [21].

A number of nations have implemented policies aimed at speeding up the rapid conversion of new energy technologies into products. These include incentives for research and pilot programs for new products. Moreover, various designs have been created for using environments for testing and verifying products; these environments are designed to reduce development costs through effective learning and to establish the potential for developing preliminary markets for product applications [22].

Hydrogen energy also recently received attention, even in terms of commercialization, for its application to stationary power generation. One significant improvement in hydrogen delivery has been using trucks and pipelines instead of producing hydrogen onsite at centralized locations [23]. It is possible to create hydrogen using existing liquid fuel infrastructure for the medium term until a pipeline can be built. Various hydrogen infrastructure scenarios are also available including still incorporating centralized hydrogen production with delivery options [24–26].

PV energy is another source that rose rapidly, largely because it does not produce air pollution or other harmful effects for the environment; therefore, it is the fastest-growing form of energy in many countries. In addition, no noise is produced, and solar energy is directly converted into electricity [27]. PV systems can be rapidly implemented and easily customized to meet local conditions and demands [28], features that have driven the emergence of micro-installations that are tailored specifically to meet the needs of prosumers [29].

Solar energy is most efficiently and economically used by the installation's user if its total consumption is accounted for [30]. This has made micro-PV installations a very attractive solution for firms that require small-scale, spatially distributed energy production [31, 32]. PV power plants, on the other hand, are large-scale installations and technical infrastructure devices that require large expanses of land, which is hugely expensive; the owners of these plants sell their entire output of electricity. In practice, this usually means that agricultural land is removed from use and not put back into production [33, 34].

A group investigating the profitability of renewal energy investments in northern Poland established that the following evaluation criteria should feature in any assessments of the profitability of a PV energy investment [35]:

- **Net present value (NPV)** allows investors to analyze investment projects in relation to their primary objective. NPV is a measure of the difference between future cash flow and project costs,
- **Internal rate of return (IRR)** shows the actual return rate on a PV farm over time. IRR tells us how much to discount all cash flows so that the researchers get a zero NPV;
- **Payback period** is the time needed for the initial commissioning costs and any negative flows of a project to be covered by positive cash flow. The investment is considered fully repaid as soon as cumulative cash flow is positive;
- **Profitability index** is similar to NPV and is used to analyze investment projects' profitability, among other things.
- **Accounting rate of return** is a financial indicator of the average turnover rate of an investment. Average annual net profit is calculated by dividing the net profit for a project by the number of years over which the project has run.

Among the parameters identified were electricity price increases and inflation levels. A combination of variables representing various economic scenarios was used in the sensitivity analysis. The investigators assessed the risks associated with solar plant operation via several scenarios, taking conservative approaches to cover uncertainties related to the future of the energy sector and possible economic downturns. In contrast, an optimistic scenario was characterized by averaged economic indicators that supported business growth [35].

5.3 NANOMATERIALS AND GREEN ENERGY

The nanoscale dimensions of nanostructured materials allow for large surface-to-volume ratios, favorable transport properties, altered physical properties, and confinement effects. Nanostructured materials now has massive applications related to energy, including solar cells, catalysts, thermoelectrics, Li batteries, supercapacitors, and hydrogen storage systems. Varying the nanoscale dimensions allows for controlling the band gap of the material; then, it can be used to tune the absorption and emission spectra in accordance with the specific requirements of the application. However, feasibility and cost remain significant prohibitors to large-scale nanomaterial synthesis and production, as does creating nanostructures that exhibit perfect morphology, structure, facets, surface chemistry, etc.

Earth-abundant and nontoxic materials have demonstrated high appeal for use in the development of renewable energy devices due to economic and environmental considerations. In case of using new mechanisms relying on nanostructures, solar cells have better optical absorption, less charge recombination, and fewer energy losses from electron transport, and Li-ion batteries tend to have better power and energy density. Indeed, developing new materials and structures for Li-ion batteries is always a priority; one of the key research themes in this field is finding materials and structures that can conduct Li-ion intercalation–deintercalation efficiently and with excellent cycling stability. Currently, composite and polymer-based materials are the most relevant materials for supercapacitors and hydrogen storage systems because both aim to reach high storage densities while maintaining low cost [36, 37].

There are two mature methods for storing hydrogen: compressed gas or liquefaction (storing in the form of a liquid), but there has recently been considerable interest in storage in a solid form. The hydrogen sorption capacity, kinetics, thermodynamic stability, and cycling stability of potential storage materials can be investigated using traditional characterization techniques like gravimetric, Sievert-type volumetric, and electrochemical methods. Even though gravimetry and volumetry are based on different mechanisms, they provide similar information. To overcome onboard difficulties, it is necessary to identify and develop lightweight, low-cost, solid-state hydrogen storage systems that are capable of storing large amounts of hydrogen with fast kinetics and high capacities.

Currently, there is considerable research pertaining to the development of nanoporous materials that could be used to store hydrogen in the future. These materials include carbon nanomaterials, metal-doped carbon nanomaterials, metal–organic frameworks (MOFs), covalent–organic frameworks, complex chemical hydrides, clathrates, amides, and zeolites, as well as metal or intermetal hydrides

[38, 39]. Under various conditions, hydrogen can be stored in solid state using these new nanomaterials via chemisorption or physisorption [40, 41].

Safety and cost-effectiveness are both advantages of using these processes over conventional hydrogen storage. There are, however, a number of disadvantages associated with chemisorption, such as high energy barriers, slow kinetics, and poor reversibility, whereas physisorption requires low temperatures (cryogenic) for reasonable uptake [42]. Physisorption can play a crucial role with regard to carbon-based nanomaterials and MOFs, whereas chemisorption occurs mainly with metal hydrides and complex hydrides.

Nanomaterials for hydrogen storage need to be accurately analyzed and characterized for have an efficient hydrogen economy. Conventional techniques evaluate hydrocarbon storage properties such as sorption capacity, kinetics, thermodynamics, and cycling stability. There are currently two techniques for characterizing hydrogen storage properties in ideal solid-state materials: gas sorption using Sievert's volumetric and gravimetric techniques and electrochemical approaches combine gas sorption and gravity [43].

Hydrogen storage continues to be one of the most major challenges in achieving a hydrogen economy, as conventional hydrogen storage systems (compressed and liquid) may not be suitable for applications on board vehicles because of inefficiency and safety. Chemical storage systems, such as metal hydrides and complex hydrides, offer both potential and limitations. Unfortunately, their poor reversibility and low cost-effectiveness are obstacles to their application in practice. Physical adsorption has also commanded in carbonaceous materials and other porous materials with large surface areas due to its simplicity in reversibility and rapid kinetics; a very low temperature is necessary to achieve high levels of hydrogen sorption [43].

In contrast, metal-PbS nanomaterials show good desorption capacity for both IR and visible light spectra, which makes them good material for fabricating thin films for PV applications. Researchers fabricated metal/PbS nanocrystal thin films by thermal evaporation under vacuum 10^{-6} torr using Al and Cu and used scanning electron microscopy (SEM) and energy dispersive analyzer x-ray (EDAX) to characterize their surface morphology and composition. Thin films of metal-PbS nanocomposites should analyzed by X-ray diffraction (XRD) to determine their properties, but it was possible to characterize these films with a UV spectrophotometer. Micro and optoelectronics, nonlinear optics, photocatalysis, and energy conversion industries are served by semiconductors as they are capable of tuning electrical, optical, and opto-electronic properties [44–46]; PbS has a direct band gap of ~0.41 ev [47, 48].

A variety of devices are made from PbS, including photodetectors, IR detectors, solar cells, etc. PbS emits radiation at wavelengths between 1 and 2.5 m, which are longer than those of the visible spectrum but shorter than those observed in everyday life. In addition to PbS, other materials that can be used to produce electroluminescent devices include solid-state lasers, biological sensing devices, and tunable near-IR detectors [49–52]. PbS layers have been deposited using a variety of deposition techniques, such as electrodeposition, laser pulse deposition, spray pyrolysis, and thermal evaporation [53–56], and, PbS NPs have been synthesized for a variety of applications [57, 58].

The development of nanotechnology has led to a great deal of interest in nanomaterials due to their unique properties, such as their high aspect ratios, which have attracted considerable attention recently. The surface characteristics of PbS-based nanomaterials have special optical characteristics that make them suitable for a wide range of applications such as LEDs, nonlinear optics, optical switches, nonlinear optical absorption, detectors, and single electron devices. Nanocomposite materials offer a greater range of desirable characteristics than either of their constituent materials alone [59–63].

Nanocomposites can improve the characteristics of nanomaterials. For instance, nanocomposites composed of multiple materials significantly affect the electronic and optical properties of semiconductor materials. Researchers have used a number of techniques to fabricate noble metal/semiconductor nanocomposites, such as hydrothermal, droplet, and electrochemical processes and thermal evaporation, which resulted in photodetectors and photocatalysts [64, 65].

5.4 MODERN INDUSTRIAL APPLICATIONS

There is a growing concern over the environmental impact of fossil fuels, and hydrogen's natural energy potential is likely to increase as fossil fuel supplies decline and concerns about the environment grow. Hydrogen may eventually become the principal chemical energy carrier. Hydrogen and electricity would be the two predominant sources of energy for end-use services if the majority of the world's energy resources became non-fossil-fuel-based.

The use of hydrogen as a fuel for vehicles could also reduce the amount of CO_2 emissions from this sector, although hydrogen storage is one of the major barriers to hydrogen auto applications, especially in applications requiring low weight and volume, one of the primary challenges for hydrogen application, but several options have been proposed in the automotive industry for replacing current fuels with hydrogen fuels. As hydrogen fuel cell electric vehicles produce no pollution and are comparable in range with conventional internal combustion engine vehicles, they appear to be a viable transportation option in the future [66].

It is widely recognized that hydrogen, as a carbon-free and sustainable energy carrier, offers a viable approach to energy collection and storage. A unique integrated solution to global warming and energy related issues is available by combining green energy sources with hydrogen energy systems; green energy-based hydrogen energy is one of the most promising solutions to environmental issues. Environmental pollution possibly can be eliminated or reduced, especially if greenhouse gas emissions are reduced [67–69]. Solar energy is another promising alternative to fossil fuels to achieve zero emissions [70]. Petroleum-based electricity is not sustainable, and hydrogen and fuel cells are highly efficient; they are carbon-free energy carriers with high density [71, 72].

To generate electricity, hydrogen that has been generated and stored either flows directly into a fuel cell or is compressed and then run directly through a fuel cell. On the market today, PV solar energy production technologies can be categorized according to the raw materials that are used in the PV cells, such as crystalline silicon, hydrogenated a-Si, CdTe, CIS, and CIGS, colored modules and high-power modules [73]. All PV systems can be characterized into one of five groups as follows:

(1) Connected to the network: Typically installed on house roofs, PV systems are connected to the network by a PV panel that converts the sun's energy into electricity (direct current). In addition to high productivity, this type of system avoids the isolation phenomenon by not using batteries [74].
(2) Isolated: PV systems are typically installed in rural areas with little access to electricity grids; this situation only requires PVs and some battery storage [75].
(3) Hybrid: PV generation can be combined with wind turbines or diesel engines; these systems can be connected to the network or stand alone [76].
(4) Solar power plants: Electricity is generated in a single point and is also connected to the network; plant size varies from hundreds of kilowatts to megawatts.
(5) Applied in consumer goods: Solar panels can be fitted on roofs, road and highway signs, poles, public phones, watches, calculators, toys, battery chargers, and poles.

The module is the main component in a PV system. It consists of multiple PV cells that generate electricity by converting the energy of solar radiation into electrical energy. At startup, it is very expensive to secure purified NPs and assemble large quantities of cells, but a factory needs to produce large volumes to be competitive. These factors pose a tremendous challenge for competitive and sustainable solar energy generation industry. In countries such as Brazil, for the solar energy sector to enter the electricity market productively and make a meaningful impact, specific lines of credit dedicated to solar electricity generation are vital. There is still a significant lack of attractive interest rates to finance distributed grid generation in the country. A lack of technology knowledge in the financial sector contributes to the limited number of alternatives that are available, thereby causing uncertainty and difficulties in understanding and measuring the risks associated with such assets [77].

During the last three decades, PV technology development and innovation has been focusing on reducing the overall system cost. The cost to finance a system is critical because most power generated by a system is prepaid via the system price. The operational life of a PV system plays a significant role in determining the system's economics, and PV panels can provide excellent performance (reliable and economic) after exposure to the sun for more than 20 years. PV plants can be operated for decades with no fundamental barriers; the systems are likely to last 40 to 80 years with periodic maintenance [78].

In the PV industry, the levelized electricity cost (LCOE) is the most commonly used to measure the economics of PV systems; in many cases, LCOE represents the levelized (average) cost of electricity generation over the life of the asset. In the context of comparing alternative technologies, LCOE can facilitate "apples-to-apples" comparisons of technologies when operating scales, investment levels, or operating time periods differ between companies, thereby making it easier to compare alternative technologies. In calculating LCOE, the major factors are system cost, capital cost, annual operating expenses, and the performance and the functional life of the system. In the last few years, the PV industry has witnessed remarkable improvements in all of its key LCOE drivers: New PV systems are much cheaper, have longer life cycles, are more efficient and perform better, and have lower operating expenses and capital costs, allowing LCOEs to approach competitive levels with conventional generation sources without incentives. The PV industry is expected to become increasingly competitive on LCOE over the next few years due to the continued scaling up of PV systems, and they may over the next few years become one of the most cost-effective sources of electricity generation [78].

Researchers designed and fabricated a novel nanotechnology called a triboelectric nanogenerator (MS-TENG) with a magnetic switch structure to harvest wind energy, another free and renewable energy source. Generators are driven by magnetic potential energy converted from kinetic energy by a wind scoop's transmission gears and energy modulation modules. An excellent example of the use of TENGs is their ability to turn random mechanical energy into a controllable source of electrical energy by using a mechanical regulation mode. It is important to note that TENGs do not have continuous outputs, so they cannot power electrical devices for long periods of time; the wind energy is harvested and generated continuously and regularly by the MS-TENG magnetic switch-structured triboelectric nanogenerator. As long as the input speed is above the critical speed, MS-TENG can maintain a continuous supply of electric energy. Whenever the input speed exceeds a critical speed, neodymium storage magnets are activated, and an electric output is created on the shaft made of stainless steel and machine lathed [79].

Researchers determined that for wind energy applications, nanostructured materials should have a structural unit length scale below 100 nm on any of its dimensions [80]. There are different types of nanostructured materials according to the dimensions of the materials; 0D nanosized powders, 1D nanocrystal multilayers, 2D nanostructured rods with nanoscale thicknesses, and 3D bulk materials with at least one nanosized phase [81]. Quite recently, the use of nanocrystalline and ultrafine-grained metals and alloys has seen a steady increase; these typically have grain sizes below 100 nm; their physical and mechanical properties are fascinating compared with coarse-grained materials. Several laboratory-scale processing techniques can be used to produce nanocrystalline metals and alloys, and there are processing-related challenges related to synthesizing nonagglomerated nanosized ceramic powders with the right densification route and selecting the right way to inhibit grain growth during sintering. However, many nanoceramic/nanocomposites have technologically significant relevance for wind energy applications in terms of improving strength and hardness [82].

5.5 GREEN ENERGY STORAGE

The energy storage system can perform a variety of functions to support the grid at all levels of electricity output. A centralized storage system can accommodate seasonal to hourly changes in supply and demand in the transmission system. It can also be used to support intermittent energy production from green sources.

Electrochemical technologies: As Li-ion batteries became increasingly common in different markets, including personal devices, electromobility, and industrial storage, they became the most common electrochemical storage systems [83]. Li-ion batteries lack the power density required for some applications [84].

Thermal energy storage (TES): TES is a technology that is used by a number of different companies, and it can be classified as low-temperature and high-temperature depending on the temperature range of the storage system. More precisely, it consists of sensible, latent, and concrete thermal storage, with the latter including liquid and cryogenic energy storage [85].

Electrical storage: The use of electrical storage devices as a means of storing energy can be an interesting alternative to using magnetic storage devices like superconducting magnetic energy storage (SMES), supercapacitors, capacitors, and hybrid supercapacitors. In this technology, electrical energy is stored using a coil and a cooling mechanism to minimize the application's losses during charging. SMES are designed to release the stored energy at the end of the loading phase using a power converters topology [86]. Future electricity networks will rely heavily on energy storage, which can balance demand and supply and support renewable energy source integration faster and more efficiently. Energy storage devices enable residential and commercial users to use green energy resources more efficiently. Hence, the storage capacity of energy is expected to increase, and to meet this need, a wide range of electric and thermal power technologies are being developed [87].

A number of improvements have been made to batteries with nanotechnology. Researchers experimented with carbon nanomaterials and found some favorable results, even though they were less electrochemically effective than some other materials such as the aforementioned metal alloys. Using the Li–SO_2 battery model, nanostructured carbon cathode materials were demonstrated as possible candidate materials with reversible capacities of more than 1000 mAh/g, at a maximum working potential of 3 V, and even providing a theoretical energy density that would be approximately 70% greater than the current Li-ion battery (651W h/kg) [88].

Lithium iron phosphate ($LiFePO_4$) is a promising and widely used cathode material that has much potential. $LiFePO_4$ is an attractive choice even at the pre anoscale condition because of its chemical and thermal stability, low toxicity, and tenably high capacity (170 mAh/g); A123 Systems Li-ion battery company even uses it [89, 90]. Nanotechnology offers an opportunity for batteries of all sizes to benefit from its capabilities. This is true regardless of whether the batteries are intended for devices as small as hearing aids or as large as grid energy storage units. In smaller and lighter battery applications, improving the energy density is often a primary objective. However, when the energy density has to be increased for larger batteries, objectives vary [91].

Researchers expanding on battery storage developed an energy storage device with improved performance using rolled-up nanotechnology, an innovative method for self-assembling nanomembranes into 3D structures using strain engineering. Using the technology, they fabricated nanotechnology-based single tubes for micro-batteries and micro-supercapacitors that allowed for direct diagnosis of the purity of active units. They constructed lab-on-chip electrochemical energy storage (EES) devices by combining standard lithography and thin-film deposition and performed studies on Li storage and charge-transfer kinetics. Using the rolled-up nanotechnology allowed for fabricating ultracompact capacitors with superior performance, which opens possibilities for integrating power storage systems with autonomous microelectronic entities on the chip [92].

In spite of the successful demonstration of rolled-up nanotechnology used to construct modular EES components, micro-sized single-tube EES devices are still in their early stages of development. Developing solid or gel polymer electrolytes to replace liquid electrolytes is one of the biggest challenges, but a miniaturized rolled-up EES device will have less power density if these electrolytes have lower ionic conductivities than liquid electrolytes. Energy storage devices are different from large-scale energy storage technologies, such as the pumped stage, because they have to meet the demands of a micro-milliwatt power system. For green energy to be more effective, it needs more efficient storage mechanisms [93, 94].

The hybrid energy harvester and self-charging power system are still prototypes; the individual components of these devices include different types of energy conversion and energy storage. A hybrid self-charging power system can increase power output by trapping certain types of energy and storing it directly, thereby scavenging certain types of energy. Future smart home and transportation devices may integrate artificial intelligence (AI) through hybrid self-charging power systems [95].

The development of nanomaterials and their processing into electrodes and devices is necessary to improve the performance and/or development of existing energy storage systems. For a more effective method of improving Li-ion batteries, Sila Nanotechnologies and others have been implementing large-scale Si NPs into the anodes of Li-ion batteries. This is a compelling demonstration of the possibility to scale nanomaterials for large-scale battery production. Nanoscale additives such as MWCNTs are being used to construct and reinforce battery electrodes. The capacity and size of small-volume or small-sized batteries and supercapacitors can be greatly improved by other nanomaterials such as SWCNTs and graphene. In the near future, nanomaterials will be able to be widely implemented in the energy storage technology because of their decreasing cost and increased confidence in their safety (on the healthy, ecological, operational levels). As nanomaterials gain full potential, they must be incorporated into sophisticated architectures allowing for multiple functions, including electron transport, ion transport, and interactions between devices [96].

5.6 APPLICATIONS IN ENERGY STORAGE DEVICES

Along with affecting electrical storage technologies, nanotechnology may also have an impact on electrochemical supercapacitors and batteries. At the laboratory scale, it has been shown possible to combine the high energy density of traditional batteries with the high electrical capacity and excellent performance of electrostatic capacitors by using redox-based supercapacitors with nanostructured electrode materials. It is important to note that the physical and chemical properties of the electrode material have a significant effect on both the energy density and the performance of rechargeable Li batteries. It is significant to note that nanomaterials are closely connected to electron transport due to the fact that they have a reduced dimension and are close to electrodes and electrolytes, respectively. Hence, electrons can move more freely and electrode-electrolyte contacts can be more comfortable, as well as providing an easy way to exert pressure and a higher degree of fracture resistance with nanostructures. There is a wide variety of high-capacity materials and conductive additives available to provide effective anode applications, which include CNTs, graphene-based nanostructures, and NP nanowires [97].

A supercapacitor provides high currents over a short period of time, making it a good energy storage and delivery technology. There are three main types of supercapacitors according to their energy storage capacity: electrochemical double layer capacitors (EDLC), pseudo-capacitors, and hybrid capacitors. Due to the excellent specific surface area, high electric conductivity, and the stability of carbon and carbon-based materials, they are receiving increasing attention as potential supercapacitor electrode materials.

The ultracapacitor is another type of electrochemical storage device that stores and delivers energy at a fast rate and provides a high current for a short period of time. Ultracapacitors are sometimes referred to as supercapacitors. They can store and deliver energy very rapidly and offer high currents in a burst, so they find applications in electric vehicles, uninterruptible power supply, and memory backups in IT systems. There are many other advantages of ultracapacitors, including their virtually unlimited cycle life, their high specific power, and their superior performance at low temperatures. EDLCs are commonly made with a liquid electrolyte such as KOH or H_2SO_4 as their electrolyte between each pair of electrodes [98].

A pseudo-capacitor is a hybrid between a battery and an EDLC; they also have an electrode and electrolyte, and they store charge chemically and electrostatically. Hybrid capacitors occupy a medium position between batteries and capacitors, and they are commonly used in the automotive industry. It is necessary to innovate in order to miniaturize energy storage devices (e.g., micro batteries) or integrate energy storage functionality into the device structures (e.g., structural batteries) so that they are flexible electrochemical energy storage devices (i.e., batteries and supercapacitors). In the current energy storage marketplace, Li-ion batteries are considered to be the most versatile energy storage devices [99].

A discharge involves the release of Li ions from the oxidized anode, their migration, and their integration into the cathode. A continuous flow of electrons through the external circuit causes the energy to be released at the same time; the reverse reaction occurs in the process of recharging. Li-ion batteries have electrodes that are nonflexible, such as oxides or phosphates containing L, graphite, Si, or Sn [100]. It is important for wearable applications to be aware of the flammability risk associated with Li-ion batteries, despite their high performance. The electrical energy stored in supercapacitors is stored in the electrical double layer as accumulated charges, which are higher than those in Li batteries [101, 102].

Pseudocapacitive material allows the supercapacitors to create a faradaic charge storage mechanism that increases their capacitance. The pseudocapacitive materials may be either metal oxides (such as RuO_2 [ruthenium oxide], MnO_2, V_2O_5, NiO, and Co_3O_4), conducting polymers (such as polyaniline, polypyrrole, and polythiophene), or some combination of these materials [100, 103, 104]. Supercapacitors are similar to Li batteries in terms of cell configurations; they require fewer materials to function. Planar design also allows for fabricating thin, mechanically robust batteries in addition to the layered design. Unlike traditional batteries, these planar batteries can be worn due to the flexible substrate encapsulating the electrodes. While planar batteries have an attractive structural design, they are also characterized by disadvantages such as high resistance and low capacity. A triangular pattern of carbon and metal electrodes can be used on planar batteries to increase their capacity and performance [105].

Developing energy-storage components small enough for wearable gadgets or wireless sensor networks requires developing components that can operate autonomously rather than needing to be recharged constantly. Some of these technologies become a reality with the development of miniaturized energy storage components. A miniaturized supercapacitor has a basic structure characterized by an electrode connected to a conductor electrolyte that functions as both a positive and a negative electrode. Originally, micro-supercapacitors were based upon thin-film micro batteries, which use a solid-state electrolyte sandwiched between electrodes (lithium phosphorus oxynitride) [106–108].

It is important to note that micro-supercapacitors (MSCs) need to be correctly sized and designed to be able to withstand the expected operating profile, as well as having a long-life expectancy and having low current leakage when they are used in a fully wireless sensor network (including an energy harvester, a sensor, and all related electronics) [109]. There is no doubt that microelectronic devices like MSCs fulfill the requirement of high-power delivery with a very short period of time due to the modern development of microelectronic devices. A particular limitation in the performance of the current MSCs is the use of aqueous gel electrolyte, which limits the working voltage of the device as a result of the lower energy density of the MSCs compared with micro-batteries [110].

5.7 TRENDS AND RECOMMENDATIONS

Recent developments in nanotechnology have provided a new avenue toward improving hydrogen-sensing performance; these technologies enable the synthesis and modification of different materials at a much faster rate. Materials being developing for use in sensing include conventional methods such as transit metals, metal oxide semiconductors, and their combinations as well as new technologies such as graphene and its derivatives [111].

In addition to metal oxide semiconductors, nanostructured transit metals have also been synthesized and dispersed over the sensor surface to reduce power consumption and increase the sensing response; this reduces power consumption and improves performance at the same time. On the other hand, it can lead to low conductivity, inhomogeneity of hydride reaction, and noise in the sensing signals; furthermore, metal-oxide semiconductors consume high amounts of energy and pose high thermal safety risks because they require external heaters to maintain high operating temperatures. During the sensing process, other materials are required to form a highly conductive surface; SnO_2, ZnO, and Cu_2O nanowires and nanorods with heterojunction structures show high sensitivity to a variety of gases, as do transition metal oxides such as RuO_2, NiO, MnO_2, Co_3O_4, and SnO_2 [111].

Several advances in nanomaterial preparation methods have made it possible to realize nano-objects with composite architectures that combine different functions at the nanoscale, such as catalysis, hydrogen storage, and oxidation protection. In addition to improving the hydrogen sorption properties of metal materials, this advancement also leads to a greater understanding of some of the fundamental mechanisms involved in metal hydrogen transformations at the nanoscale, which is contributing to the advancement of fundamental science. Especially in the context of materials science, the development of in situ microscopy and spectroscopy is essential as it opens exciting new avenues for high-resolution experiments that extend to other fields of the science of materials, including mechanical science [112].

Moreover, due to their unique morphology, nontoxicity, large surface area, and pore volume, these polymer materials make an excellent choice for producing hydrogen by photocatalytic means. In terms of water splitting, nanocarbon materials are quite advantageous since they are metal-free photocatalysts; it should be noted, however, that some metal-free catalysts produce very little hydrogen, thereby limiting their use. Since carbon nanomaterials have a nonmetal nature, their development for these materials should be the focus of the current trend [112].

Researchers took advantage of nanotechnology to CIGS composites and thin-film cells by adding cadmium (Cd) to individual nanofillers in a CIGS base matrix to determine the most suitable fillers for enhancing the optical properties of CIGS materials and the performance of CdS/CIGS thin-film cells based on submicron energy layer thickness. Al and Au NPs were more effective for improving the performance of PV cells than Au NPs. For multiple nanocomposites, CdS/(CIGS + 20wt.%Al) performed better than conventional nanocomposite cells when the absorbing layer was filled with 10wt.% Cd instead of the usual 20wt.%Al nanofiller. The metal nanofillers achieved high efficiency in the submicron CdS/CIGS absorbing layer of the model with less material, thereby saving both the production cost and time for CIGS thin-film solar cells [112].

In a different experiment, the dielectric constant and refractive index of CdTe and PbS nanocomposites decreased by increasing the concentrations of Au, Cu, Li, Al, or Cs nanofillers in the CdTe or PbS base matrix; in contrast, the energy band gaps increased with increasing the volume fraction of selected metal NPs. Adding Au, Cu, Li, Al, or Cs NPs to CdTe or PbS layers increased the absorption coefficient of the layers by many orders of magnitude; as the NP radii decreased below 10 nm, the absorption coefficient decreased as well. The increased absorption coefficient started to decrease the radii, and then the absorption coefficient of selected semiconductor layers remained constant with the increase in the radii of metal fillers. Adding the metal NPs also improved short circuit current density, open circuit voltage, output power density, and external quantum efficiency; increased the number of electron–hole pairs generated in the absorbing layer; and improved fill factor and efficiency. The CdTe and PbS absorbing layers contained 20wt.% Au, Cu, Li, Al, or Cs NPs [112].

Researchers who also examined nanocomposites for thin-film solar cells found that Cs NPs exhibited the best performance when incorporated as individual NPs into ITO/CdS/PbS/Al and SnO_2/CdS/CdTe/Cu nanocomposites. In semiconductor systems composed of CdTe, CIGS, CdS, and PbS, decreasing the grain size of the material to QD decreases the dielectric constant and refractive index of the material but increased the absorption coefficients and energy band gaps between layers. The high-temperature performance of the selected HJQD cells was improved by reducing the size of the QD window and absorb layer's materials [112].

In another study, the PbS shells made of QDs containing Au, Cu, Li, Al, or Cs cores showed lower quality and durability; increasing the concentrations of selected metal cores in the PbS shell increased the energy band gap. Increasing the concentrations of the metal cores also increased the absorption coefficient of the metal/PbS core/shell absorbing layers as well as the electron–hole generation rates. HJQD CdS/PbS thin films produced higher open circuit voltages, short circuit currents, maximum power points, and therefore better efficiency [112].

A significant challenge, however, is the development of next-generation Li-based rechargeable batteries that are highly efficient, cost-effective, and safe to use in portable electronics, electric vehicles, and grid-scale energy storage applications. These batteries have profound technological significance for the future. As part of the advancement of Li battery chemistry, there is a paradigm shift toward electrodes with a high Li-to-host ratio, which is achieved by utilizing a conversion or alloying mechanism; however, drastic changes in capacity are often accompanied by significant bond breakage, limited electronic/ionic conductivity, and unstable electrode/electrolyte interphase, which can be dangerous. Nano-structuring, surface protection, and nanocontainment stabilize advanced electrode materials that undergo drastic volumetric changes in their working environments. Adding nanofillers with fast-conducting surfaces to polymer solid electrolytes also significantly increases Li-ion conductivity [112].

Incorporating nanomaterials into both portable batteries and grid-scale energy storage systems can play an important role in improving battery safety and durability. The future is bright for nanotechnology as the future of the battery will depend on the design of environmentally friendly and low-cost methods of fabricating nanomaterials for use in batteries, and more efforts should be made in order to accelerate the large-scale application of nanomaterials in batteries. Nanotechnology also makes it possible to overcome the cracking threshold of the electrode material upon lithiation, which improve electron/ion transport within the electrode. Moreover, nanotechnology offers a variety of powerful methods for coating and functionalizing electrode materials to protect them from corrosion. Nanotechnology provides the opportunity for engineers to design each and every component inside a battery (separator, current collector, etc.) and perform novel battery functions unavailable by conventional methods [112].

REFERENCES

1. Thabet A, Theoretical analysis for effects of nanoparticles on dielectric characterization of electrical industrial materials. *Electr Eng (ELEN) J* June 2017;99(2):487–493. Springer.
2. Thabet A, Abdelhady S, Ebnalwaled AA and Ibrahim AA, Improvement optical and electrical characteristics of thin film solar cells using nanotechnology techniques. *Int J Electron Telecommun (IJET)* October 2019;65(4):625–634.
3. Thabet A, Abdelhady S and Mobarak Y, Design modern structure for heterojunction quantum dot solar cells. *Int J Electr Comput Eng (IJECE)* 2020;10(3):2915–2922.
4. Thabet A and Salem N, Experimental progress in electrical properties and dielectric strength of polyvinyl chloride thin films under thermal conditions. *Trans Electr Electron Mater J* 2020;21(2):165–174.
5. Thabet A, Abdelhady S and Mobarak Y, Ultra-optical characterization of thin film solar cells materials using core/shell absorber layer. *Int J Electr Comput Eng (IJECE)* 2022;12(2)1377–1384.
6. Thabet A, Al Mufadi FA and Ebnalwaled AA, Synthesis and measurement of optical light characterization for modern cost-fewer polyvinyl chloride nanocomposites thin films. *Trans Electr Electron Mater J* 2023;24(6). Nature Springer.
7. Guo KW, Green nanotechnology of trends in future energy. *Recent Pat Nanotech* 2011;5:76–88.

8. Demirbas A, Recent advances in biomass conversion technologies. *Energy Educ Sci Tech* 2000;6:19–40.
9. Rathore NS and Panwar NL, *Renewable energy sources for sustainable development.* New Delhi, India: New India Publishing Agency, 2007.
10. Zakhidov RA, Central Asian countries energy system and role of renewable energy sources. *Appl Sol Energy* 2008;44(3):218–223.
11. Bergmann A, Colombo S and Hanley N, Rural versus urban preferences for renewable energy developments. *Ecol Econ* 2008;65:616–625.
12. Panwara NL, Kaushikb SC and Kothari S, Role of renewable energy sources in environmental protection: A review. *Renew Sustain Energy Rev* 2011;15:1513–1524.
13. Singla S, Sharma S, Basu S, Shetti NJP and Aminabhavi TM, Photocatalytic water splitting hydrogen production via environmental benign carbon based nanomaterials. *Int J Hydrog Energy* 2021;46:33696–33717.
14. Ahamed JU, Saidur R, Masjuki HH, Mekhilef S, Ali MB and Furqon MH, An application of energy and exergy analysis in agricultural sector of Malaysia. *Energy Policy* 2011;39(12):7922–7929.
15. Saidur R, Mekhilef S, Ali MB and Safari A, Applications of variable speed drive (VSD) in electrical motors energy savings. *Renew Sustain Energy Rev* 2012;16(1):543–550.
16. Kelley LC, Gilbertson E, Sheikh A, Eppinger SD and Dubowsky S, On the feasibility of solar-powered irrigation. *Renew Sustain Energy Rev* 2010;14(9):2669–2682.
17. Mahlia TMI, Yong JH, Safari A and Mekhilef S, Techno-economic analysis of palm oil mill wastes to generate power for grid-connected utilization. *Energy Educ Sci Tech Part A: Energy Sci Res* 2012;28(2):1111–1130.
18. Mekhilef S, Saidur R and Safari A, Biomass energy in Malaysia: Current state and prospects. *Renew Sustain Energy Rev* 2011;15(7):3360–3370.
19. Mekhilef S, Faramarzi SZ, Saidur R and Salam Z, The application of solar technologies for sustainable development of agricultural sector. *Renew Sustain Energy Rev* 2013;18:583–594.
20. Eijndhoven JCN, Technology assessment: Product or process. *Technol Forecast Soc Change* 1997;54(2–3):269–286.
21. Chang PL, Hsu CW and Lin CY, Assessment of hydrogen fuel cell applications using fuzzy multiple-criteria decision making method. *Appl Energy* 2012;100:93–99.
22. Akai M, Stationary fuel cell programme in Japan. In *IEA EGRD workshop– transforming innovation into realistic market implementation programmes*, 27–28. Paris: International Energy Agency, 2010.
23. Tzimas E, Castello P and Peteves S, The evolution of size and cost of a hydrogen delivery infrastructure in Europe in the medium and long term. *Int J Hydrog Energy* July–August 2007;32(10–11).
24. Mulder G, Hetland J and Lenaers G, Towards a sustainable hydrogen economy: Hydrogen pathways and infrastructure. *Int J Hydrog Energy* July–August 2007;32(10–11).
25. Olateju K, Techno-economic assessment of hydrogen production from underground coal gasification (UCG) in Western Canada with carbon capture and sequestration (CCS) for upgrading bitumen from oil sands. *Appl Energy* November 2013;111.
26. Katikaneni SP, Al-Muhaish F, Harale A and Pham TV, On-site hydrogen production from transportation fuels: An overview and techno-economic assessment. *Int J Hydrog Energy* 2014;39:4331–4350.
27. Majewski J and Szymanek M, Technical, economic and legal conditions of the development of photovoltaic generation in Poland. *Acta Energetica* 2012;2:21–26.
28. Bigorajski J and Chwieduk D, Analysis of a micro photovoltaic/thermal—PV/T system operation in moderate climate. *Renew Energy* 2019;137:127–136.
29. Bartecka M, Terlikowski P, Kłos M and Michalski Ł, Sizing of prosumer hybrid renewable energy systems in Poland. *Bull Polish Acad Sci Tech Sci* 2020;68:721–731.

30. Kruzel R and Helbrych P, Analysis of the profitability of a photovoltaic installation in the context of sustainable development of construction. *E3S Web Conf* 2021;49:00061.
31. Abdmouleh Z, Gastli A, Ben-Brahim L, Haouari M and Al-Emadi NA, Review of optimization techniques applied for the integration of distributed generation from renewable energy sources. *Renew Energy* 2017;113:266–280.
32. Mitscher M and Ruther R, Economic performance and policies for grid-connected residential solar photovoltaic systems in Brazil. *Energy Policy* 2012;49:688–694.
33. Dinesh H and Pearce JM, The potential of agrivoltaic systems. *Renew Sustain Energy Rev* 2016;54:299–308.
34. Patel MT, Khan MR, Sun X and Alam MA, A worldwide cost-based design and optimization of tilted bifacial solar farms. *Appl Energy* 2019;247:467–479.
35. Brodziński Z, Brodzińska K and Szadziun M, Photovoltaic farms—economic efficiency of investments in North-East Poland. *Energies, MDPI* 2021;14:2087.
36. Zhang Q, Uchaker E, Candelaria SL and Cao G, Nanomaterials for energy conversion and storage. *Chem Soc Rev* 2013;42:3127.
37. Ghosh SK and Pal T, Interparticle coupling effect on the surface plasmon resonance of gold nanoparticles: From theory to applications. *Chem Rev* 2007;107:4797–4862.
38. Zacharia R and Ullah Rather S, Review of solid state hydrogen storage methods adopting different kinds of novel materials. *J Nanomater* 2015;1–18.
39. Sakintuna B, Lamari-Darkrim F and Hirscher M, Metal hydride materials for solid hydrogen storage: A review. *Int J Hydrog Energy* 2007;32:1121–1140.
40. Ren J, Musyoka NM, Langmi HW, Mathe M and Liao S, Current research trends and perspectives on materials-based hydrogen storage solutions: A critical review. *Int J Hydrog Energy* 2017;42:289–311.
41. Dillon AC, Parilla PA, Zhao Y, Kim YH, Gennett T, Jones KM, Zhang SB and Heben MJ, NREL activities in DOE carbon-based materials center of excellence overview: Timeline and budget timeline. *Time*, 2005. https://www.hydrogen.energy.gov/pdfs/review05/stp_63_heben.pdf (Accessed 18 June 2019).
42. Andersson J and Gronkvist S, Large-scale storage of hydrogen. *Int J Hydrog Energy* 2019;44:11901–11919.
43. Boateng E and Chen A, Recent advances in nanomaterial-based solid-state hydrogen storage. *Mater Today Adv* 2020;6:100022.
44. Rohom AB, Londhe PU, Jadhav PR, Bhand GR and Chaure B, Studies on chemically synthesized PbS thin films for IR detector application. *Mater Sci Mater Electron* 2017;28:17107–17113,
45. Londhe PU, Rohom AB and Chaure NB, CuInSe2 thin film solar cells prepared by low-cost electrodeposition techniques from a non-aqueous bath. *RSC Adv* 2015;5:89635–89643.
46. Rajashree C, Balu AR and Nagarethinam VS, Enhancement in the physical properties of spray deposited nanostructured ternary PbMgS thin films towards optoelectronic applications. *J Mater Sci* 2016;27:5070–5078.
47. Cabrera P and Arizmendi-Morquecho A, Thin film cds/pbs solar cell by low temperature chemical bath deposition and silver doping of the window layer. *J Non-Oxide Glasses* 2016;8:59–66.
48. Yucel E, Yucel Y and Beleli B, Optimization of synthesis conditions of PbS thin films grown by chemical bath deposition using response surface methodology. *J Alloys Compd* 2015;642:63–69.
49. Slonopas A, Alijabbari N, Saltonstall C, Globus T and Norris P, Chemically deposited nanocrystalline lead sulfide thin films with tunable properties for use in photovoltaics. *Electrochima Acta* 2015;151:140–149. Elsevier.
50. Seo J, Cho MJ, Lee D, Cartwright AN and Prasad PN, Efficient heterojunction photovoltaic cell utilizing nanocomposites of lead sulfi de nanocrystals and a low-bandgap polymer. *Adv Mater* 2011;23:3984–3988.

51. Safrani T, Kumar TA, Klebanov M, Arad-Vosk N, Beach R, Sa'ar A, Abdulhalim I, Sarusi G and Golan Y, Chemically deposited PbS thin film photo-conducting layers for optically addressed spatial light modulators. *J Mater Chem* 2014;2:9132–9140.
52. Kaci S, Keffous A and Hakoum S, Preparation of nanostructured PbS thin films as sensing element for NO2 gas. *Appl Surf Sci* 2014;305:740–746. Elsevier.
53. Mathewsa NR, Angeles-Chavez C, Cortes-Jacome MA and Toledo Antonio JA, Physical properties of pulse electrodeposited lead sulfide thin films. *Electrochimca Acta* 2013;99:76–84.
54. Atwa DMM, Azzouz IM and Badr Y, Studies on chemically synthesized PbS thin films for IR detector application. *J Appl Phys B* 2011;103:161–192.
55. Veena E, Bangera KV and Shivakumar GK, Studies on chemically synthesized PbS thin films for IR detector application. *Appl Phys A* 2017;123:366–378.
56. Castelo-Gonzalez OA, Sotelo-Lerma M and Garcia-Valenzuela JA, Effect of reaction time and temperature on chemical, structural, optical, and photoelectrical properties of PbS thin films chemically deposited from the Pb(OAc)2–NaOH–TU–TEA aqueous system. *J Electron Mater* 2017;46:393–400.
57. Wang Y, Minjie L, Jia H, Song W, Han X, Zhang J, Yang B and Zhao B, Optical properties of Ag/CdTe nanocomposite self-organized by electrostatic interaction. *Spectrochimica Acta Part A Sci Direct* 2006;64:101–105.
58. Lee Y, Kim E, Park Y, Kim J, Ryu W, Rho J and Kim K, Photodeposited metalsemiconductor nanocomposites and their applications. *J Materiomics* 2018;12.
59. Chen H and Wang L, Nanostructure sensitization of transition metal oxides for visiblelight photocatalysis. *Beilstein J Nanotechnol* 2014;5.
60. Xu M, Liang T, Shi M and Chen H, Graphene-like two-dimensional materials. *Chem Rev* 2013;113:376.
61. Neto AHC and Novoselov K, New directions in science and technology: Two dimension acrystals. *Rep Prog Phys* 2011;74.
62. Zhao S, Huang J, Huo Q, Zhou X and Tu W, A non-noble metal MoS2eCd0.5Zn0.5S photocatalyst with efficient activity for high H2 evolution under visible light irradiation. *J Mater Chem* 2016;4.
63. Perera S, Mariano R, et al. Hydrothermal synthesis of graphene-TiO2 nanotube composites with enhanced photocatalytic activity. *ACS Catal* 2012;2.
64. Viezbicke D, Patel S, Davis E and Birnie P, Evaluation of the Tauc method for optical absorption edge determination: ZnO thin films as a model system. *Basic Solid State Phys* 2015;252(8):1700–1710.
65. Hassanien A and Akl A, Effect of Se addition on optical and electrical properties of chalcogenide CdSSe thin films. *Superlattices Microstruct* 2016;89(89):153–169.
66. Rosen MA and Koohi-Fayegh S, The prospects for hydrogen as an energy carrier: An overview of hydrogen energy and hydrogen energy systems. *Energ Ecol Environ* 2016;1(1):10–29.
67. Sorgulu F and Dincer I, A renewable source based hydrogen energy system for residential applications. *Int J Hydrog Energy* 2017;xxx:1–10.
68. Yazici MS, Hydrogen and fuel cell activities at UNIDO-ICHET. *Int J Hydrog Energy* 2010;35:2754–2761.
69. Barbir F and Yazici S, Status and development of PEM fuel cell technology. *Int J Energy Res* 2007;32(5):369–378.
70. Tsatsaronis G, Kapanke K and Marigorta AMB, Exergoeconomic estimates for a novel zero-emission process generating hydrogen and electric power. *Energy* 2008;33: 321–330.
71. Midilli A and Dincer I, Hydrogen as a renewable and sustainable solution in reducing global fossil fuel consumption. *Int J Hydrog Energy* 2008;33:4209–4222.
72. Midilli A, Ay M, Dincer I and Rosen MA, On hydrogen and hydrogen energy strategies I: Current status and needs. *Renew Sustain Energy Rev* 2005;9:255–271.

73. Orduz R, Solorzano J, Aguilera MAE and Roman E, Analytical study and evaluation results of power optimizers for distributed power conditioning in photovoltaic arrays. *Prog Phot: Res Appl* 2013;21(3):359–373.
74. Urbanetz J, Zomer CD and Ruther R, Compromises between form and function in grid-connected, building-integrated photovoltaics (BIPV) at low-latitude sites. *Build Environ* 2011;46(10):2107–2113.
75. Villalva MG, Gazoli JR and Ruppert Filho E, Comprehensive approach to modeling and simulation of photovoltaic arrays. *IEEE Trans Power Electron* 2009;24(5):1198–1208.
76. Agencia Nacional de Energia Eletrica-aneel. *Brazilian atlas of electric power (in Portuguese)*. 2014. http://www.aneel.gov.br/aplicacoes/atlas/pdf/01-Introducao(3).
77. Ferreira A, Kunha SS, et al., Economic overview of the use and production of photovoltaic solar energy in brazil. *Renew Sustain Energy Rev* 2018;81:181–191.
78. Campbell M, The economics of PV systems. In *Photovoltaic solar energy: From fundamentals to applications,* First Edition. Edited by A. Reinders, P. Verlinden, W. van Sark and A. Freundlich. John Wiley & Sons, Ltd., 2017.
79. Liu S, Li X, et al., Magnetic switch structured triboelectric nanogenerator for continuous and regular harvesting of wind energy. *Nano Energy* 2021;83:105851.
80. Mukhopadhyay A and Basu B, Consolidation-microstructure property relations hips in bulk nano ceramics and ceramic nanocomposites: Areview. *Int Mater Rev* 2007;52(5):257–288.
81. Gleiter H, Nanocrystalline materials. *Prog Mater Sci* 1989;33:223–315.
82. Basu B, Alex S and Eswara Prasad N, Chapter 14: Lightweight nanostructured materials and their certification for wind energy applications. In *Nanotechnology for energy sustainability*. Wiley-VCH Verlag GmbH & Co. KGaA, 2017. ISBN:9783527340149.
83. Thielmann A, Sauer A, Isenmann R and Wietschel M, Technology roadmap energy storage for electric mobility 2030. *Fraunhofer Inst Syst Innov Res ISI* 2013;32.
84. Maddukuri S, Malka D, Chae MS, Elias Y, Luski S and Aurbach D, On the challenge of large energy storage by electrochemical devices. *Electrochim Acta* 2020;354:136771.
85. Luo X, Wang J, Dooner M and Clarke J, Overview of current development in electrical energy storage technologies and the application potential in power system operation. *Appl Energy* 2015;137:511–536.
86. Ali MH, Wu B and Dougal RA, An overview of SMES applications in power and energy systems. *IEEE Trans Sustain Energy* 2010;1:38–47.
87. Kebede AA, Kalogiannis T, Mierlo JV and Berecibar M, A comprehensive review of stationary energy storage devices for large scale renewable energy sources grid integration. *Renew Sustain Energy Rev* 2022;159:112213.
88. Jeong G, Kim H, Park JH, Jeon J, Jin X, Song J, Kim BR, Park MS, Kim JM and Kim YJ, Nanotechnology enabled rechargeable Li–SO2 batteries: Another approach towards post-lithium-ion battery systems. *Energy Environ Sci* 2015;8(11):3173–3180.
89. Bruce PG, Scrosati B and Tarascon JM, Nanomaterials for rechargeable lithium batteries. *Angew Chem Int Ed* 2008;47(16):2930–2946.
90. Vartanian C and Bentley N, "A123 systems' advanced battery energy storage for renewable integration. In *Power systems conference and exposition (PSCE)*, 20–24. IEEE/PES, March 2011.
91. Wong K and Dia S, Nanotechnology in batteries. *J Energy Resour Technol* January 2017;139:014001–1.
92. Deng J, Lu X, Liu L, Zhang L and Schmidt OG, Introducing rolled-up nanotechnology for advanced energy storage devices. *Adv Energy Mater* 2016;1600797.
93. Obrovac MN and Chevrier VL, Alloy negative electrodes for Li-ion batteries. *Chem Rev* 2014;114(23):11444–11502.
94. Bonaccorso F, Colombo L, Yu GH, Stoller M, Tozzini V, Ferrari AC, Ruoff RS and Pellegrini V, Graphene, related two-dimensional crystals, and hybrid systems for energy conversion and storage. *Science* 2015;347:1246501-9.

95. Zhang Q, Suresh L, Liang Q, Zhang Y, Yang L, Paul N and Tan SC, Emerging technologies for green energy conversion and storage. *Adv Sustainable Syst* 2021;5:2000152.
96. Pomerantseva E, Bonaccorso F, Feng X, Cui Y and Gogotsi Y, Energy storage: The future enabled by nanomaterials. *Science* 2019;366:969.
97. Ayalew ME, The role of nanotechnology for energy storage, conservation and post combustion CO2 capture in industry: A review. *Int J Mater Sci Appl* 2021;10(3):55–60.
98. Mensah-Darkwa K, Zequine C, Kahol PK and Gupta RK, Supercapacitor energy storage device using biowastes: A sustainable approach to green energy. *Sustainability* 2019;11:414. https://doi.org/10.3390/su11020414
99. Fu Y, Wei Q, Zhang G and Sun S, Advanced phosphorus-based materials for lithium/sodium-ion batteries: recent developments and future perspectives. *Adv Energy Mater* 2018;8:1702849.
100. Liu C, Li F, Ma LP and Cheng HM, Advanced Materials for Energy Storage. *Adv Mater* 2010;22:E28–E62.
101. Winter M and Brodd RJ, What are batteries, fuel cells, and supercapacitors? *Chem Rev* 2004;104:4245–4269.
102. Conway B, *Electrochemical supercapacitors: Scientific fundamentals and technological applications (POD)*. New York: Kluwer Academic/Plenum, 1999.
103. Simon P and Gogotsi Y, Materials for electrochemical capacitors. *Nat Mater* 2008;7:845–854.
104. Sumboja A, Foo CY, Yan J, Yan C, Gupta RK and Lee PS, Significant electrochemical stability of manganese dioxide/polyaniline coaxial nanowires by self-terminated double surfactant polymerization for pseudocapacitor electrode. *J Mater Chem* 2012; 22:23921–23928.
105. Sumboja A, Liu J, et al., Electrochemical energy storage devices for wearable technology: A rationale for materials selection and cell design. *Chem Soc Rev* 2018;47:5919–5945.
106. Yoon YS, Cho WI, Lim JH and Choi DJ, Solid-state thin-film supercapacitor with ruthenium oxide and solid electrolyte thin films. *J Power Sources* 2001;101:126–129.
107. Lim JH, Choi DJ, Kim HK, Cho WI and Yoon YS, Thin film supercapacitors using a sputtered RuO2 electrode. *J Electrochem Soc* 2001;148:A275–A278.
108. Kim HK, Cho SH, Ok YW, Seong TY and Yoon YS, All solid-state rechargeable thin-film microsupercapacitor fabricated with tungsten cosputtered ruthenium oxide electrodes. *J Vac Sci Technol B* 2003;21:949–952.
109. Kyeremateng NA, Brousse T and Pech D, Microsupercapacitors as miniaturized energystorage components for on-chip electronics. *Nat Nanotechnol* 2017;12.
110. Tyagi A, Tripathi KM and Gupta RK, Recent progress on micro-scale energy storage devices and future aspects. *J Mater Chem A* 2015;3:22507–22541.
111. Wang B, Sun L, Schneider-Ramelow M, Lang KD and Ngo HD, Recent advances and challenges of nanomaterials-based hydrog sensors. *Micromachines* 2021;12:1429.
112. Liu Y, Zhou G, Liu K and Cui Y, Design of complex nanomaterials for energy storage: Past success and future opportunity. *Acc Chem Res* 2017; 50:2895–2905.

Section II

Design and Economics of Green Power Generation

6 Nanostructures of Green Energy

Nanostructured materials are composed of atoms placed in the center of an original crystal of atoms so that they can form a new type of solid by integrating many features into the initial crystal of atoms. This type of solid can contain many different properties. Different nanostructured materials are produced based on the features that are incorporated into the material. There is little doubt that the strong cross-disciplinary nature of research conducted on nanomaterials has a significant role in the advances in our understanding of materials at the nanometer level. Traditional physical, chemical, and biological disciplines are converging at the nanometer level, and this convergence is an important principle for the advancement of fundamental as well as applied condensed matter science.

6.1 INTRODUCTION

Nanostructured materials can be produced using several techniques, including the condensation of inert gases or the condensation of chemical vapors [1, 2], pulse electron deposition [3], plasma synthesis [4], crystallization of amorphous solids [5], severe plastic deformation [6], and mechanical alloying or cryo-milling powders [7, 8]. New synthetic methods have been developed that have greatly assisted in the development of new nanostructured materials, including for fabricating materials that can be controlled in size, shape, and microstructure. Using high-intensity ultrasound to produce nanostructured materials is a straightforward, versatile, and convenient alternative to the conventional types. Ultrasound is a powerful tool in materials synthesis since two important physical phenomena are associated with it: cavitation and nebulization [9].

It is possible to synthesize nanostructured materials using ultrasound, including direct (so no chemical) reactions and ultrasonic spray pyrolysis, as well as incorporating ultrasonic vibrations into the synthesis process. Acoustic cavitation, created by ultrasonic irradiation, is a unique method of generating reaction conditions. Nanosize cells have increased energy, which increases cell voltage. Nanosized metastable phases can be stabilized by overlapping accumulated or depleted space charges; mesoscopic electrical conduction is referred to as the conduction of space charges. There is a wide range of applications for such interfacial controlled materials in areas such as fuel cells, solar cells, and molecular clotting mechanisms [10].

For instance, increasing the interfacial conduction of oxide ion conductors at moderate temperatures lowers the operating temperature of solid oxide fuel cells. The potential barrier at the contact of inorganic/organic conductors is optimized such that the charge recombination loss processes can be suppressed, thereby increasing

DOI: 10.1201/9781003512486-8

the efficiency of hybrid solar cells. Transition metal oxides and fluorides can display anomalous energy storage mechanisms that are attributed to nanocrystallinity. It is possible to reverse Li storage both through reversible insertion reactions and also by controlling the interfacial storage mechanism. The performance of supercapacitors improves significantly as a result of the mesoporous morphology and increased surface area of nanosized metal oxides. The conversion of waste heat into electricity is currently based on more exotic nanostructures of thermoelectric materials. When bulk materials are converted into nanostructures, exciting phenomena occur that provide new opportunities for the conversion and storage of energy [10].

6.2 DESIGN AND THEORETICAL MODELS

6.2.1 Hydrogen Energy

Hydrogen can be made with the help of photoelectrochemical water splitting (PEC); it is a promising approach for reducing greenhouse gas emissions through the production of hydrogen. A plasmonic photoelectrode fabricated of surface plasmon-sustaining material must be provided to sustain plasmonic photons on its surface. PEC depends also depends on the location of plasmonic nanostructures at multiple interfaces within the device, such as in the semiconductor or the electrocatalyst. Traditional materials used for plasmonic applications included Au, Ag, Al, and Cu. However, the plasmonic materials that are used for PEC water splitting must be low in cost, chemically stable, corrosion resistant, and mechanically strong and ultimately must have good electrocatalytic activity [11–14].

Using the Drude model [15], it is possible to conveniently obtain the optical properties of metals through the treatment of conduction electrons as a free electron gas. In this scenario, electrons with number density n and effective mass m* can oscillate in response to an applied electromagnetic field, while positive ions are treated as immobile. Above the characteristic frequency, referred to as plasma frequency (ωp), electromagnetic waves can be transmitted in the material, while at ω_p, a longitudinal collective oscillation of free electrons occurs. This collective oscillation is referred to as volume plasmon, and the corresponding frequency is given as follows [16]:

$$\omega_p = \sqrt{\frac{ne^2}{\epsilon_o m^*}} \tag{6.1}$$

where e is the elementary charge and ε_0 is the vacuum permittivity.

As was recently confirmed, this is especially the case with direct electron injection. In this regard, the role of Au and Ag can still be relevant, as they would be suitable for proof-of-concept to their well-known properties. PEC device development is highly dependent on direct observation of plasmonic effects under operational conditions. To make a decisive leap toward the integration of plasmonic components in real PEC devices for renewable energy storage, PEC microscopy, synchrotron techniques, and ultrafast optical spectroscopy methods will be crucial [17].

The recently developed nanocomposites are proving to be promising in terms of their high hydrogen uptake and release properties with a significant improvement in the kinetics of the reaction. A new generation of nano-coatings and nanocatalyst-dispersed alloys are being developed and showcased to create hydrogen storage materials for reversible energy storage, through nanoengineering and nanoconfinement of particle properties. In a nanocomposite material, at least one constituent possesses nanoscale features as a result of combining two or more elements, phases, or compounds. The unique properties of nanocomposite materials such as NPs indicate that they may prove useful for hydrogen storage [18, 19].

A high surface area is one of the primary factors that influences hydrogen sorption kinetics in nanoscale materials and nanostructured materials [20]. However, NP agglomeration during repeated cooling and heating together with contraction and expansion during cycling degrades performance over time. It is still important to note that the nanoscale systems provide the ability to control the properties of materials to a greater extent than can be done with bulk materials. Additionally, these nanoscale materials can be used to design and develop lightweight hydrogen storage systems that can be tailored to meet the requirements of an on-board application, which can be used as a fuel supply, or to provide power [21].

Nanostructured materials exhibit excellent properties for hydrogen storage due to desired and tunable specific surface area along with many others. Smart nanosystems have been designed to perform physical and chemical reactions more efficiently than ever before, with the greatest interaction on active surfaces, optimum bulk absorption, fast reaction kinetics, low-temperature hydrogenation and dehydrogenation, hydrogen atom dissociation from the surface of the nano-catalytic surface, and molecular division in the active polymeric chain resulting from the active polymer chain. Hydrogen dissociation can be facilitated electively by nanophase materials with their high surface to volume ratios and hydrogen adsorption properties. A granular composite consists of a small amount of NPs suspended in a medium volume of granular compound. This provides a short delusion path for hydrogen molecules to travel between the granules, thereby increasing their mobility. The use of nanosized catalytic doping agents results in greater dispersion of catalytically active species, which is more potent in the context of mass transfer reactions that require higher catalytic activity. Polyaniline nanofiber nanocomposites are unique nanostructures that can store hydrogen on-board vehicles by having a unique microstructure in addition to being functionalized with CNTs [21].

The development of several ternary nanocomposites, including $Fe_2O_3/EuVO_4/g\text{-}C_3N_4$ combining NPs of Fe_2O_3 and $EuVO_4$ with unique magnetic, optical, and structural properties, has also performed well for applications in hydrogen storage and water treatment. Several factors can be attributed to the electrochemical efficiency of these nanocomposites, such as a spillover mechanism, electrocatalytic effect, and optimized morphology. Controlling the weight of the magnetic NPs assists to modify the photocatalytic performance of magnetic nanocomposites prepared from magnetic NPs. Due to magnetic NPs properties, these nanocomposites can easily be decolored and recolored because they can be magnetically detached from the decolorization medium [22].

PMMA dissolved in water was converted to an oxygen radical in a highly stable solution. Organic Mg ions diffused into the PMMA solution and interacted with the acrylate ions of the MMA by electrostatic attraction. Second, Mg NPs reduced by Li-naphthalene were dispersed in a PMMA solution, and using oxygen atoms within PMMA to form an oxygen molecular layer on their surface prevented the aggregation of Mg NPs and the oxidation of Mg by O_2 or H_2O molecules [23].

An air-stable nano Mg-polymer composite was explored in a safe and low-cost manner, with organic Mg monomer being safer and less expensive. In addition, the prepared nano Mg-based composite did not oxidize in the air and was not affected by water molecules, and it remained stable longer than traditional composites. The Mg NPs in the PMMA substrates absorbed and released H_2 into the substrate more rapidly than they did in other substrates; the amount of Mg NPs directly relates to the amount of PMMA in the substrate. The presence of PMMA may interfere with the formation of Mg nanocrystals, or even compromise the reservoir's ability to store H_2 when the Li^+ ion is not able to completely mix with the Mg^+ ion [24].

Electric batteries are an essential source of energy for new vehicles; they are necessary for storing, converting, transmitting, and recovering energy, and they meet the growing demand for vehicle electric energy density. Battery systems that can charge quickly and have high energy density are urgently needed, as well as supercapacitor systems and nanoelectrode systems. Currently, there is a requirement for electrodes to have more channels so that charges will be transferred efficiently between them. In addition, electrodes must be designed in such a way that ions are efficiently transported in high-capacity electric batteries; Li-S batteries have improved [25, 26].

The key to optimizing the performance of batteries is to enhance the battery structures and nano-energy systems. In an electric vehicle for instance, Li batteries provide acceleration, and Li-ion batteries provide the power for the vehicle to cruise. Once an electric battery system has been designed, it needs to be tested to verify that it is meeting the specifications related to the system and to perform improvements. The electric battery industry has made great progress in its performance over the last few decades, but there is still a need for a greater joint effort among related fields, such as dynamics, chemistry, electricity, and nanoscience, to create a sustainable energy for the human society in the future [27].

6.2.2 Solar Energy

As thin-film solar cell nanocomposites continue to be developed, efforts are being made to further reduce the thicknesses of the absorbing layers to save material, processing costs, and time. Light management concepts need to be developed with a high absorption rate to achieve most absorption in the incoming light [28, 29]. To reduce the production costs of solar cells and become more competitive in the solar industry, it is suggested to decrease the thickness of CdTe, PbS, and CIGS, even to sub-micrometers [29, 30]. Solar cells with CdS/PbS absorbing layers have higher open circuit voltage, maximum current density, maximum voltage, output power density, and efficiency [31].

It is likely that manufacturing very large quantities of CIGS thin-film solar cells will result in substantial material savings, especially for indium and gallium, which might not be available in large quantities (70 GWp per year) if CIGS thin-film solar cells are manufactured in large quantities. In addition, thinner CIGS layers could also be deposited faster, lowering production costs directly by reducing the deposition time [32]. However, reducing absorber layer thickness is associated with technical problems [29–33].

When the absorbing layer is thinner, the carriers are generated closer to the back contact, which increases the likelihood of the back contact recombination. When thin absorbing layers occur on the surface, the recombination rate increases, enhancing tunneling recombination at the interface. Large current leakage and carrier recombination loss at the back contact also contribute to decreasing open circuit voltage as absorbing layer thickness decreases. During the formation of the 0.5 m thick absorbing layer of the CdTe solar cell, space-charge reaches the back contact of the solar cell. With an increase in CIGS layer thickness, electrons are generated closer to the back contact, resulting in a greater likelihood of back contact recombination as the layer thickness decreases to about 0.5 m. Tunnel recombination increases in the depletion region of CIGS layers thinner than 0.5 m [29–34].

A few theoretical models described the current–voltage (J-V) characteristics in thin-film solar cells. The following assumptions are made in the modeling to be analytically tractable: a) The thermal equilibrium concentration of charge carriers is negligible because of high band gap materials. b) The built-in electric field is nearly uniform across the absorber layer. c) Constant drift mobility (μ) and a single lifetime (t) are assigned to each type of carrier (holes and electrons). Toward the top of the absorbing layer, the voltage-dependent electric field is slightly higher than toward the bottom; the photogenerated carriers will drift with a slightly higher velocity near the top interface than that near the bottom interface. The calculations of charge collection are not significantly affected by assuming that the carriers drift uniformly throughout the absorbing layer; therefore, the assumption of average drift velocity does not make a significant difference [24]. The net external current density of solar cell [31, 35, 36–39] is as follows:

$$J(V) = J_{ph}(V) - J_{diode}(V) - \left(\frac{V + J(V)R_{ser}}{R_{sh}}\right) \tag{6.2}$$

where

$J_{diode}(V)$ is the forward diode current density

V is the applied voltage

R_{sh} is the shunt resistance

R_{ser} is effective series resistance including all contact resistances

$J_{ph}(V)$ is total photo generated current density which obtained by integrating over all incident photon wavelengths of the solar spectrum [35].

The open circuit voltage equation is expressed as follows [35–39]:

$$V_{oc} = \left(\frac{KT}{Q}\right)\ln\left[\frac{J_{ph}}{J_O} + 1\right] \tag{6.3}$$

The output power density of the cell is expressed as follows:

$$P_{cden} = J \times V \tag{6.4}$$

$P_{cmaxden}$ is the maximum output power density determined from $(P_{cden} - \mathrm{V})$ curve. P_o is the input power density. Fill factor and energy conversion efficiency are calculated as follows [40–45]:

$$F.F = \frac{P_{cmaxden}}{V_{oc} \times J_{ph}} \tag{6.5}$$

$$\eta = \frac{J_{ph} V_{oc} F.F}{P_o} \times 100 \;\; or \;\; \eta = \frac{P_{cmaxden}}{P_o} \times 100 \tag{6.6}$$

In the field of nanotechnology, a nanocomposite is a solid material that is multiphase and consists of one, two, or three phases. Nanocomposites possess a number of unique properties that cannot be matched by conventional components alone. An applicant for a solar energy application is evaluated based on three main aspects: the degree of optical absorption, the conductivity, and the environmental stability of the nanocomposite plastic [46–49]. It is preferable for the absorbing layer materials of solar cells to be semiconductors with direct band gaps. The band gap values should be selected to ensure that the solar spectrum can be effectively absorbed by the deposited films and that semiconductor/metal-based nanocomposites will exhibit properties derived from interfacial interactions between their component phases [50–52].

A further innovation in semiconductor particle research is the size-quantized optical and optoelectronic effects of semiconductor particles at (sub-)nanometer wavelength. These effects significantly modify the photonic characteristics of semiconductor particles. Because nanocomposites are composed of nanosized materials rather than bulk semiconductor materials, there have been tremendous developments in devices that serve as light-emitting diodes, solar cells, NP devices, and chemical/biological sensors [50–52].

In thin-film cells, the light scattering produced by metal NPs produces an additional optical absorption that can be intensified by coupling with the waveguide modes of the layer. As metal NPs have a resonant nature, they are ideal solutions for solar cells; they provide the necessary resonant feedback, thus making them attractive low-cost products. In semiconductor layers, the spectral profile can be modified by tuning the plasmon resonance frequencies (which depends on NP material, size, and distribution) [53].

The effective dielectric constant ε_{efj} of multiple nanocomposites absorber layer material using second type NPs with spherical metal inclusions calculated by the following equation:

$$\varepsilon_{efj} = \varepsilon_{efi} \frac{\left(\varepsilon_{mj} + 2\varepsilon_{efi}\right) + 2\mathcal{F}_j\left(\varepsilon_{mj} - \varepsilon_{efi}\right)}{\left(\varepsilon_{mj} + 2\varepsilon_{efi}\right) - \mathcal{F}_j\left(\varepsilon_{mj} - \varepsilon_{efi}\right)} \tag{6.7}$$

Whereas ε_{efi} is dielectric constant of base matrix. F_j is the volume fraction of the second NP, ε_{mj} is the dielectric constant of the second metal, as described using the Drude model.

The refractive index of a nanocomposite absorb layer using multiple NPs can be calculated as follows:

$$n_{fj} = \sqrt{\frac{|\varepsilon_{efj}| + \varepsilon_{refj}}{2}} \tag{6.8}$$

where ε_{refj}, $|\varepsilon_{efj}|$ are the real part value and the absolute value of the effective dielectric constant of nanocomposite absorb layer material using multiple NPs.

The absorption coefficient of metal/semiconductor multiple-nanocomposite material can be calculated using the following equation:

$$\alpha_{fj} = \frac{4\pi k_{fj}}{\lambda} \tag{6.9}$$

According to Tauc's argument, the direct energy band gap of metal/semiconductor multiple-nanocomposites material E_{gfj} can be obtained as follows [54–58]:

$$\alpha_{fj} \mathrm{h}\nu = \alpha_o (h\nu - E_{gfj})^{1/2} \tag{6.10}$$

The most commonly used polymerized vinyl chloride for cables and thermoplastics is polyvinyl chloride, which is made from ethylene and anhydrous hydrochloric acid. Interphase regions are characterized by their surface area, volume fraction, and thickness, and within each particle, an interphase zone surrounds the particle to form an interphase region [59–62]. Figure 6.1 graphically provides a better understanding of how to estimate the thickness of NPs from a closer perspective of the interphase area around the NPs [63–66]. There were recent efforts to explore the electrical characteristics and their effects on the electrical applications of a variety of industrial insulation materials from both theoretical and experimental perspectives

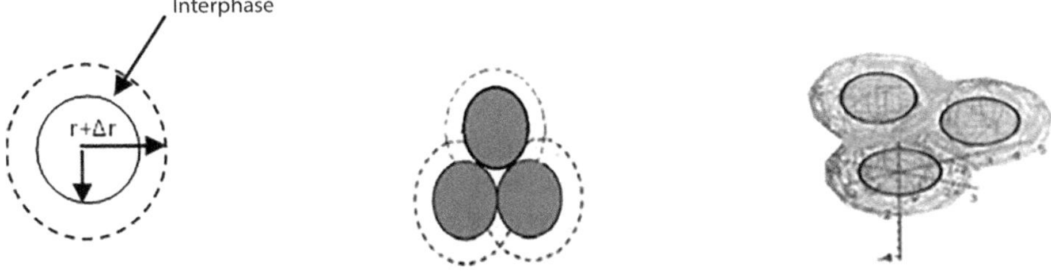

FIGURE 6.1 Interphase zone surrounds the particles in a nanocomposite system.

[67, 68]. Several nanocomposite models have been studied recently for their effective dielectric constants [69–71].

It is crucial for designers to be able to understand the magnetic properties of magnetic materials in order to design new composite thin-film solar cells based on nanocomposites which have strong J-V and I-V qualities based on their magnetic properties. Variant practical models used the parameters to achieve the efficiency and J-V characteristics of ITO/CdS/metal/PbS/Al, SnO_2/CdS/metal/CdTe/Cu, and ZnO/CdS/CIGS/Mo [72–76].

At nanoscale thicknesses, different metals doped with IV-VI nanocrystal thin films have shown impressive effects, and impure atoms barely penetrated the films. The semiconductor lattice of the host device is optimized for electron and hole absorption because the atomic levels on the semiconductor lattice are sharp. Nanostructured semiconducting materials gain desired properties and improved efficiencies based on dopant characteristics such as their nature and concentration. It is possible to modify the electrical and optical properties of a PbS structure by doping it with various metal ions such as zinc (Zn^+), silver (Ag^+), copper (Cu^+), platinum (Pt^+) and Al [72–76].

A strategy that has been adopted recently is doping PbS and CdTe films to tailor their physical properties to meet the needs of particular applications: Doping PbS and CdTe with the transition metals Cu, Zn, Cd, Al, and Hg had a profound effect on the morphological, optical, and electrical properties of PbS. Metal ion doping can also extend the band gap of a PbS semiconductor, increasing its visible photo response. Metal NPs are added to semiconductor absorbing levels to improve the performance of thin-film solar cells with SnO2/CdS/CdTe/Cu, ZnO/CdS/CIGS/Mo, and ITO/CdS/PbS/Al absorbers based on submicron absorbing layer thickness, improving efficiency and decreasing material usage as well as production cost and time [72–76].

Moreover, using new individual and multiple NP techniques in the absorbing layers of CIGS, CdTe, or PbS thin-film solar cells improved cell energy efficiency and fill factor by reducing the dielectric constant and improving the energy band gaps. The technique decreased the reverse saturation current, decreased recombination, decreased the space-charge region, increased the electron–hole generation rate in the absorbing layer, and increased the external quantum efficiency (EQE) and energy conversion efficiency of the cells, thereby increasing the absorption. Increasing concentrations of metal nanofillers in nanocomposite thin-film solar cells improved the optical characteristics of the absorb layers [72–76].

6.2.3 Wind Energy

Large-scale wind turbines are becoming increasingly efficient with today's ever-evolving technology, and they are skyrocketing in use. Scientists manufactured wind blades using an epoxy/SiC nanocomposite material selected because it has many technological advantages over its larger counterparts, including reduced noise and high aesthetic appeal. The blade was modeled using CATIA software, and the structural and model analyses were performed using ANSYS FEA software. The presence of NPs and fibers in the epoxy resin composite provided superior thermal stability and mechanical performance. The low specific weight of the epoxy/NP carbide nanocomposites reduced the mass of the wind turbine blades by 20% to 50% from the monolithic materials of conventional wind turbines [77].

There was a strong correlation between fiber orientation and the lay-up method, both of which increased the blade's mechanical strength. Furthermore, the composite material had higher impact resistance and damage tolerance, which increased its life span. Fiberglass is an insulator, which protects it against electric corrosion and galvanic corrosion. As little as 10% of nanofillers in a composite can have significant benefits in terms of strength and quality; the improvement is only attainable when the NPs are dispersed uniformly throughout the matrix [77].

The wind blade is the central component of an urban wind power plant, and numeric analysis is carried out to determine the optimum design of the blade under different wind loading conditions and different materials. In the case of steel, the blade is more rigid, and its deformation is minimal owing to its higher density; therefore, the specific strength of the blade will be lower. Due to the bidirectional fiber orientation of QI-GFR/epoxy/SiC blades, the stiffness toward unidirectional loads is low; they are suitable for massive wind turbines with heavy and bidirectional loads [78].

The wind speed in urban areas has decreased over the years since there are too many buildings and tall structures and not enough free space. As UD-GFR/epoxy/SiC nanocomposite blades are laminated parallel to the loading direction, they have higher load-bearing capacity than that with quasi-isotropic lamination. The natural frequency of the UD-GFR/epoxy/SiC blades was much lower than that of steel blades because of the lower mass of these blades; even though their deflection is higher than that of steel blades, their specific strength is still very high. In addition to its lightweight design, its high specific strength, stiffness to vibration, and low sound levels, the UD-GFR/epoxy/SiC nanocomposite blade is an ideal component for urban horizontal-axis wind turbines [78].

Among all of the approaches that have been investigated, one of the least has been capturing the waste heat from cooling wind turbine generators. A large wind farm, however, can dissipate a considerable amount of thermal energy, and this dissipated energy can be used to power several desalination units at a distance. As part of the wind turbine's cooling process, researchers applied five NPs, of Cu, copper oxide, titanium dioxide (TiO_2), aluminum oxide, and SiO_2, to the fluid that cools the generator to capture fresh water for drinking water. Based on the first and second laws of thermodynamics, they assessed the performance of the devised set-up for a number of selected nanofluids. It has been assumed that the NP diameters in the nanofluid were negligible before calculating the thermal properties of nanofluids. According to the thermal properties of the NPs (np) and base fluid (bf) at constant pressure, it was possible to calculate the mixture viscosity (μ), specific heat capacity (c), and the viscosity of nanofluids [78]:

$$\rho_{nf} = (1-\varnothing).\rho_{bf} + \varnothing.\rho_{nf} \tag{6.11}$$

$$c_{p,nf} = \frac{(1-\varnothing).\rho_{bf}}{\rho_{nf}}.c_{p,bf} + \frac{\varnothing.\rho_{np}}{\rho_{nf}}.c_{p,np} \tag{6.12}$$

$$\mu_{nf} = \mu_{bf}.(1 + 2.5\varnothing + 6.5\varnothing^2) \quad (6.13)$$

The Reynolds number through the cooling pipe of the WT's generator is expressed as:

$$R_e = \frac{\rho_{nf} V D_p}{\mu_{nf}} \quad (6.14)$$

where Dp is the pipe diameter. The Reynolds number should be set in a turbulence regime to increase the rate of heat transfer through the cooling process.

A hybrid system was developed based on WT/HDH (waste heat to humidification–dehumidification) and was organized into two subcycles: a cooling system for the wind turbine and a heat recovery system for the HDH. Due to its high cooling capacity and its popularity, water was used as the base fluid. In the wind turbine generator, variant NPs were used to improve the performance of heat transfer during the cooling process and ultimately improve the cooling efficiency [79].

The cooling of wind turbines is achieved largely by the circulation of water through the generator, or shaft, of the wind turbine. This is due to the large capacity of the wind turbines that have been installed in recent decades to guarantee reliable and durable operation of their generators. It is possible to utilize this thermal energy to produce water rather than releasing it into the environment. In some parts of the world, there is an urgent need for fresh water supply, especially for agricultural production. Using NPs in WT cooling pure water significantly increases the heat transfer rate, which significantly increases the amount of water that can be distilled using a high-tech desalination system such as HDH [79].

A hybrid WT/HDH system was modeled to address this issue using NPs of Cu, copper oxide, TiO_2, AlO, and SiO_2. The Cu/water mixture produced more fresh water than the SiO_2/water mixture, which had experimentally the lowest performance among all the NPs. Just a small increase in NP percentage increased the total energy efficiency of the devised WT/HDH setup. Increasing wind speed increased the distilled water rate, net electricity, and GOR, and decreasing the HDH unit decreased the overall energy efficiency. Using NPs to improve humidifiers rather than using dehumidifiers was significantly more effective [79].

6.3 EFFECTIVE NPs AND MATRICES

6.3.1 Hydrogen Energy

Doping can increase the efficiency of TiO_2 photocatalysis by either narrowing the band gap energy or increasing the charge separation. The most common dopants that are studied for TiO_2 are metals such as Cu and Au due to their superior work properties. The use of nitrogen as a dopant in structural materials has also been widely investigated in recent years, as it can reduce the band gap energy of TiO_2 and improve its photocatalytic activity.

Researchers synthesized doped TiO_2, explaining their synthesis methods and describing that they used sol-gel; they described that it has become one of the most widely used methods for producing doped TiO_2 due to its numerous advantages including high surface area, small NP size, and low calcination temperature. To maximize the possibility of photocatalytic hydrogen production in TiO_2, the amount of dopant incorporated is also important; otherwise, the highest concentration of dopant acts as a recombination center. The future requires more studies to focus on the development of a cheap dopant that can replace expensive noble metal dopants such as Pt and Au in the future. As a side note, photocatalytic chemists should also be focusing on finding a suitable sacrificial reagent that is used in the hydrogen production reaction to scavenge holes in the catalyst as part of the hole scavenging reaction [80].

A noble metal doped g-C_3N_4 is prepared; this material is an excellent photocatalyst; however, because of electron/hole recombination, it is not suitable for industrial applications. In order to increase the effectiveness of the catalyst, a number of valuable metal atoms, including Ag, Au, Pd and Pt, were used as co-catalysts to increase the separation of photoinduced charge carriers from the bulk to the surface and facilitate the evolution of hydrogen gas via the splitting of water molecules. By mixing $HAuCl_4$, H_2PdCl_4 and g-C_3N_4 nanosheets together after photochemically deposition precipitation of Au-Pt bimetal NPs with g-C_3N_4 samples, bimetal NPs were deposited and hydrogen was reduced by mixing $HAuCl_4$, H_2PdCl_4, and g-C_3N_4 nanosheets together [81].

Today, photocatalytic hydrogen production is one of the most promising solutions to the problem of global energy consumption and environmental problems. In this regard, it is extremely important to develop photocatalysts that are highly efficient, nonprecious, and stable. Graphitic carbon nitride (g-C_3N_4) has attracted considerable attention due to its advantages over metal-free n-type semiconductors such as a suitable band gap, 2D-layered structures, high thermal stability, low cost, ease of preparation, and good visible light response. As a potential source of photocatalytic hydrogen production in the presence of g-C_3N_4 composite surfaces, g-C_3N_4 represents a considerable prospect; however, pure g-C_3N_4 cannot perform as efficiently due to its inability to absorb enough visible light [82].

A g-C_3N_4 photo-generating electron–hole pair is prone to recombination because of the high content of photogenerated electrons and holes, resulting in low photocatalytic performance. Renowned candidates have been coupled with functional components to enhance their photocatalytic activity. Their unique electronic structure makes them excellent candidates for photocatalytic modification. Photocatalytic performance of g-C_3N_4 in visible light varies greatly, as it is highly dependent on the specific surface area, particle structure, active surfaces, particle size, and even extended light harvesting capability; nanocomposites were used to improve the photocatalytic performance of g-C_3N_4 to produce hydrogen under visible light [82].

It has been proven that mesoporous structures and ultrathin 2D g-C_3N_4 polymer matrixes can be synthesized with special structures and morphologies to achieve extended light-harvesting capacity. However, despite the extensive studies for the purpose of achieving the preferred optical properties of g-C_3N_4, additional efforts are still being made to achieve the desired outcome, although it is already known that energy band manufacturing can change the optical properties of g-C_3N_4. Its amazing photoactivity is not the only distinguishing feature of g-C_3N_4; it also easily reacts with transition metals, nonmetals, noble metals, semiconductor compounds, graphene, CNTs, carbon dots, and QDs. Combining these compounds significantly increases their adsorption capacities, surfaces, light absorption regions, and electron conductivity and in turn their performance [82].

Researchers performed photoelectrochemical water separation using NP-coated 3D core/shell nanostructures. An upper layer of TiO_2 was deposited over ZnO nanostructures comprising nanorods and nanosheets via liquid phase deposition. Then they performed thermal reduction on the ZnO NSs-NRs@TiO_2 surface to fabricate plasmonic Au NPs in situ in a three-step process: a) The ZnO nanostructures were prepared by electrochemical deposition and by bathing them in oil; b) a thin TiO_2 shell was formed on the ZnO nanostructure using cheap and easy liquid phase deposition; c) plasmonic Au NPs were deposited by thermal reduction on the core/shell nanostructure. This technique enabled creating the core/shell nanostructure and functionalizing the surface while maintaining the morphology and nanostructure of the ZnO surface. The characterization of the ZnO NSs-NRs@TiO_2 core/shell heterojunctions nanostructure showed that it facilitated charge separation even at low temperature. The surface plasmonic properties of the plasmonic Au NPs dramatically increased their photocatalytic activity. Photocatalytic water splitting is a method of producing green hydrogen that offers an effective way of dealing with environmental and energy challenges [83].

There is a newly emerging class of photocatalysts that utilize very highly p-conjugated polymer nanostructures. In addition to acting as electron sinks, metal NPs (such as Pt, Ni, and Pt-Ni) are also useful sites for significant proton reduction due to their low overpotential. There is a strong correlation between the activity of nickel-based NPs in catalyzing Hc recombination of H_2 to form H_2 [84, 85]. An efficient photocatalytic reaction exists between polypyrrole nanostructures modified with monometal NPs and hydrogen gas at high temperatures. As a generative process, radiolysis of polypyrrole NPs was utilized to create mono- and bimetal cocatalyst NPs (Pt, Ni, and Pt-Ni) that were used to mediate the reaction. A synergistic effect was observed with the co-modification of Pt and Ni to prepare Pt-PPy, Ni-PPy, and PtNi-PPy, demonstrating great photocatalytic activity for the formation of hydrogen gas. Pt plays a significant role in electron scavenging, ultimately forming more hydrogen cyanide, whereas Ni accelerated recombination, resulting in more hydrogen generation. In addition to being very stable in cycling, these Pt-Ni-PPy nanostructures may also be very durable and long-lasting [86].

However, the nanostructure of Mg/MG-H is a far more efficient hydrogen storage material, and Mg is becoming increasingly viable for hydrogen storage due to its cost-effectiveness, higher hydrogen capacity (both gravimetrically and volumetrically), cyclic stability, higher reversibility, and lower toxicity. A green and ecofriendly energy system should consider the production of hydrogen fuel as a source of fuel. In the future, the use of nanomaterials with novel sizes and structures is likely to be one of the most promising alternatives for the storage of clean, renewable, alternative, and sustainable energy. In Mg/MgH_2, high desorption temperatures and sluggish sorption kinetics can be overcome by using advanced synthetic techniques and different structural dimensions of Mg [87].

For instance, Mg nanowires with uniform diameters of 30–50 nm are significantly more efficient than nanowires and nanofilms with diameters larger than 30 nm.

Additionally, Mg/MgH_2 NPs are smaller than 5 nm in diameter offer optimal hydrogen storage properties and improved hydrogen composition. It is possible to synthesize tiny Mg NPs using nanopore-sized carbon aerogels and polymer encapsulation. There is a wide range of properties among different dimensional nanostructures that influence the hydrogen sorption kinetics and thermodynamic stability at different temperatures. There is potential benefit in using 1D Mg nanowires for hydrogen storage due to their properties; however, the nanowires collapsed after a few cycles of reaction and then converted to NPs [87].

6.3.2 Solar Energy

Thinner absorbing layers increase efficiency and reduce costs by saving material. Among the procedures that have been recorded for reducing a metal's radius to nanometal size with 1 nm radii, solar irradiation and nanometal fillers have been the most significant. Tables 6.1–6.3 give specifications and parameters of materials and nanocomposites that are commonly used in thin-film solar cells [3–5].

Figure 6.2 shows the comparisons of the EQEs of SnO_2/CdS/CdTe/Cu nanocomposite thin-film solar cells between adding individual NPs and multiple NPs. Adding Cu, Li, Al, and Au NPs (20wt.%) to the CdS/CdTe nanocomposite increased the EQE. Adding Cu as a second filler to the CdS/(CdTe + 20wt.%Al) film improved its EQE as well. The absorbing layer CdTe + 20wt.%Al is bonded with a thin-film absorbing layer CdTe + 10wt.% Li to be greater EQE than CdS/(CdTe + 10wt.%Al). In short, adding a second NP filler to the absorbing layer of CdS/CdTe thin-film solar cells significantly improved their EQE [3–5].

Figure 6.3 displays the EQEs calculated for thin-film solar cells using individual and multimetal NPs such as ZnO/CdS/CIGS and Mo nanocomposites. As the figure shows, the ZnO/CdS/CIGS/Mo cell had higher EQE than the ZnO/CdS/CIGS/

TABLE 6.1
Use Parameters for Thin-Film Solar Cell Models [3–5]

Parameter	SnO_2/CdS/CdTe/Cu	ITO/CdS/PbS/Al	ZnO/CdS/CIGS/Mo
Absorber layer thickness (nm)	500	500	500
Window layer thickness (nm)	100	100	100
Front layer thickness (nm)	100	100	150
Electron lifetime (s)	1×10^{-8}	10^{-9}	16×10^{-7}
Hole lifetime (s)	5×10^{-8}	10^{-9}	$1{\cdot}6\times10^{-5}$
$N_a - N_d$ the concentration of uncompensated acceptors (cm^{-3})	2.44×10^{12}	0.19×10^{10}	7.56×10^{13}
Diode quality factor	1.6	1.4	1.5
Series resistance (Ω. Cm^2)	1.08	2.1	2.5
Shunt resistance (Ω. Cm^2)	103	204	320

TABLE 6.2
Use Characteristics for Semiconductors in Thin-Film Solar Cells Layers [88–91]

Parameters	CIGS	CdTe	PbS	CdS
Band gap (eV) at 25° C	1.15	1.5	0.41	2.4
Electron mobility (cm^2/Vs)	100	320	1000	350
Hole mobility (cm^2/Vs)	25	40	80	50
Effective mass of electron	0.09	0.11	0.1	0.2
Effective mass of holes	0.75	0.35	0.1	0.7

TABLE 6.3
Dielectric Parameters for Using Materials as Window, Absorber, and Substrate Layers or Metal Fillers in Absorber Layers in Thin-Film Solar Cells [92–95]

Material	Plasma angular frequency $(\omega_p 10^{16} rad/s)$	Damping constant $(\gamma_m 10^{13} s^{-1})$	Fermi velocity $(\gamma_f 10^6 m/s)$
CdS	0.082	17.6	0.89
CdTe	0.052	8.88	0.59
PbS	0.014	1.76	8
CIGS	0.0386	19.5	0.45
Li	1.225	1.85	1.29
Cu	1.03	5.26	1.57
Ag	1.40	2.80	1.39
Al	1.09	12.4	2.03
Molybdenum	0.19	1.13	—

Mo cell. Additionally, the metal NP concentrations influenced the energy conversion efficiency and fill factor of ZnO/CdS/CIGS/Mo; increasing the concentrations of Al, Cu, Li, and Au in the absorbing layer of the cells improved the performance of CdS/CIGS, up to 20wt.% for Al. By increasing the energy conversion efficiency and fill factor, CdS/CIGS produced better performance. Adding Cu as second filler also improved the performance of the thin-film solar cells compared with the cells made from CdS/(CIGS + 20wt.%Al). The CdS/CIGS + 20wt.%Al nanocomposites performed better with Li added to the composition [3–5].

Figure 6.4 shows the EQEs of the CdS/(PbS + 20wt.%Al) nanocomposite thin-film solar cells within wavelength range 300–1100 nm at variant volume fractions. Adding NPs of 20wt.%Al + 10wt.% Cs to a PbS absorber layer base matrix produced a CdS/(PbS + 20wt.%Al) nanocomposites thin-film solar cell that showed markedly

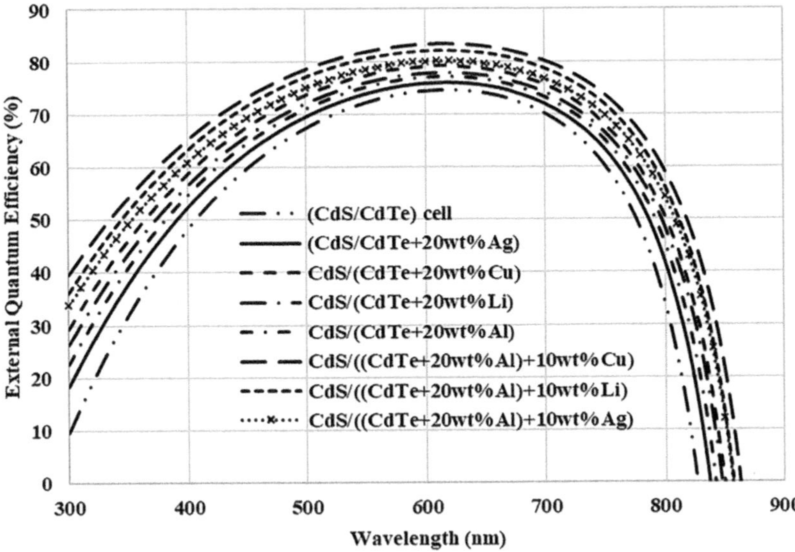

FIGURE 6.2 External quantum efficiency of SnO_2/CdS/CdTe/Cu thin-film solar cells using individual and multimetal NPs.

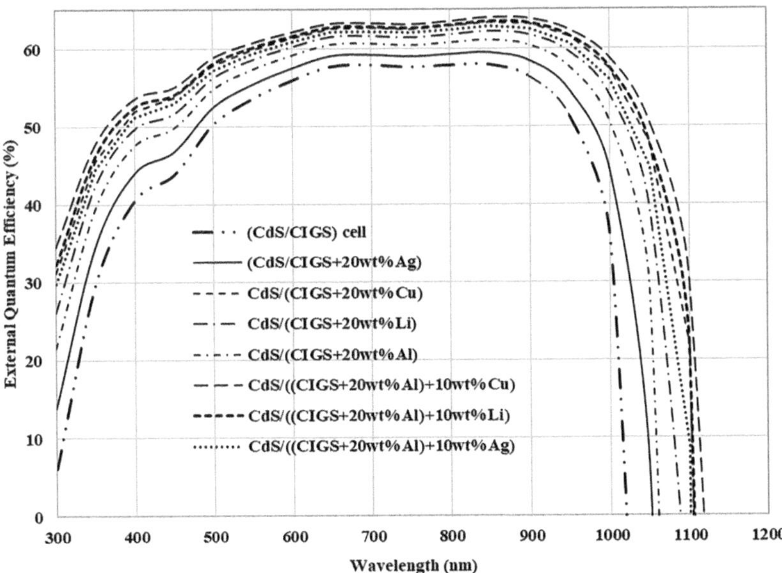

FIGURE 6.3 External quantum efficiency of ZnO/CdS/CIGS/Mo thin-film solar cells using individual and multimetal NPs.

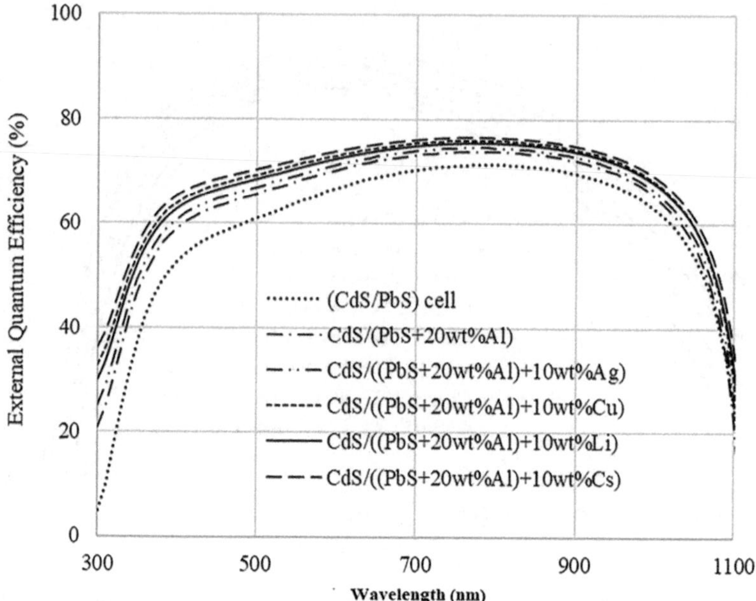

FIGURE 6.4 External quantum efficiency of CdS/(PbS + 20wt.%Al) nanocomposite thin-film solar cell.

TABLE 6.4
Temperature Effect on Efficiencies of ITO/CdS/PbS/Al and SnO_2/CdS/CdTe/Cu Thin-Film Solar Cells

	Efficiency % SnO2/CdS/CdTe/Cu cells				
Solar Cells	**300k**	**330k**	**350k**	**370k**	**400k**
CdS/CdTe	6.15	5.26	4.69	4.13	3.32
CdS/(CdTe+20wt.%Al)	11.86	9.79	8.89	8.02	6.76
CdS/(CdTe+20wt.%Al) +10wt.%Ag)	13.86	12.24	11.19	10.17	8.71
CdS/(CdTe+20wt.%Al) +10wt.%Cu	15.07	13.31	12.19	11.09	9.53
CdS/(CdTe+20wt.%Al) +10wt.%Li	14.49	12.80	11.71	10.66	9.14
	Efficiency % ITO/CdS/PbS/Al cells				
CdS/PbS	2.01	1.74	1.58	1.42	1.21
CdS/(PbS+20wt.%Al)	3.37	2.98	2.74	2.50	2.16
CdS/(PbS+20wt.% Al)+10wt.%Ag	4.16	3.73	3.45	3.17	2.78
CdS/(PbS+20wt.% Al)+10wt.%Cu	4.39	3.92	3.61	3.31	2.89
CdS/(PbS+20wt.% Al)+10wt.%Li	4.27	3.82	3.53	3.25	2.84

increased EQE, in contrast with the effects of adding 20wt.%Al + 10wt.%Ag to the base matrix. That is, when added to Al to form NPs, Cs had a more pronounced effect on the thin film's performance in the solar cell than did Ag.

Thermal condition is an effective parameter in design, production, and operation selection. Table 6.4 shows the effects of temperature on the efficiency of ITO/CdS/PbS/Al and SnO_2/CdS/CdTe/Cu nanocomposite thin-film solar cells using individual and multimetal NPs. Notably, increasing the temperature decreased the efficiency of CdS/CdTe and CdS/PbS cells. Adding 20wt.%Al individual NP filler to the absorbing layer, however, improved the performance of the selected thin-film solar cells at high temperatures. Finally, all results depict that adding multiple NPs improved the efficiency of CdS/(PbS + 20wt.%Al), CdS/(CdTe + 20wt.%Al) nanocomposite thin-film solar cells at high temperature. Moreover, adding 20wt.%Al and 10 wt.%Cu as composite NPs into PbS or CdTe base matrix absorb layers increased EQE the most. Cs NPs were limited at high temperature due to their low melting points.

6.3.3 Wind Energy

An anti-icing, self-cleaning, and drag-reduction super-hydrophobic coating is one of the most promising surface functional materials that can be developed for wind turbine blades. Embedding modified ZnO and SiO_2 NPs in PDMS resulted in a multiscale hydrophobic coating with mechanical flexibility.

Researchers modified ZnO and SiO_2 NPs by embedding a multiscale ZnO/SiO_2 micro-nano layer directly onto a substrate to form an extremely flexible superhydrophobic, easy-to-prepare anti-icing coating. Using two types of SiO_2 as a coating additive improved the coating's mechanical properties and its resistance to heat and weather, as well as its corrosion resistance. It has been also used PDMS to construct a multiscale micro-nano superhydrophobic coating that had the advantage of being easy to handle and able to construct micro-nano structures with low surface energy. In addition to its excellent mechanical stability, this coating is super-hydrophobic as well as having been modified in a way that does not affect the two types of particles. It is also possible to use this coating to self-clean, anti-freeze, and de-ice the blades of the wind turbine, which will shorten the time it takes to generate electricity from the turbine blades in icy conditions [96, 97].

FRPs (fiber reinforced polymers) with NPs in them also improved the mechanical properties of wind turbine blades and strengthened their structures. Al NPs (Al_2O_3) and graphene nanoplatelets (GNPs) were used to reinforce the FRP blades for fracture toughness before use. Microstructural analysis revealed excellent scattering in the FRP matrix of the reinforced NPs of Al_2O_3 and GNPs in a FRP matrix. Defect-free samples with exceptional performance in terms of strength performance and fracture toughness are obtained by dispersing NPs of GNPs and Al_2O_3 within the matrix of FRPs. In recent days, FRPs can be improved mechanically, tribologically, and fracture toughly by adding Al_2O_3, GNPs, and CNTs [98–101].

A teamwork in one group synthesized glass and carbon to make hybrid fiber-reinforced plastics by mixing nanocomposites using high-frequency sonication. Different NPs have different morphological, microstructural, and mechanical characteristics related to how well they disperse in a matrix, such as their content of Al_2O_3 and GNPs in woven glass fiber, carbon fiber, and hybrid glass fibers. Sonication at high frequencies dispersed the nanocomposite NPs efficiently on the woven hybrid glass and carbon fibers. A weight fraction of 1.5% for GNPs and 3% for Al_2O_3 led to the maximum hardness, while 3% achieved the maximum ultimate tensile strength. Microstructure evaluations show excellent scattering as well as greater distributions of both glass NPs and AlO in hybrid glass–carbon fibers. When the amount of GNPs and Al_2O_3 dispersed in the substrate increased, the hardness increased [101].

Conversely, as the amount of GNP and Al_2O_3 decreased, the mechanical properties declined as a result of the progressive influence of the GNP and Al_2O_3 grain growth mechanism. The teamwork observed that at a weight fraction of 3% of Al_2O_3 reinforced nanocomposite at initial flexural strength of 2.5%, the flexural strength of the composites reached its maximum and gradually decreased at weight fractions of 1.5% of GNPs and 1.5% of Al_2O_3 at a high concentration. The microstructure and mechanical properties of the hybrid glass and carbon fibers improved considerably when a weight fraction of 1.5% of GNPs and Al_2O_3 was added to the glass and carbon fibers [101].

6.4 NANOMATERIALS AND LIFE MODELS

In recent years, the development of green energy has taken a variety of forms; however, among the various forms of green, sustainable and safe energy, hydrogen is considered the most efficient approach thanks to its low price and the fact that it does not harm the environment. Furthermore, sunlight and water are important resources for the generation of hydrogen fuel.

Clean technologies are the result of nanotechnology, which aims to reduce environmental risks and improve human health. Nanotechnology products and manufacturing processes are at the center of this field. Using green nanotechnology to synthesize NPs with improved properties is a great way to produce NPs that can be put to use in various applications; NPs are created in order to increase sustainability and make them more environmentally friendly. In a perfect NP, light absorbs, causing electrons in the valence band to be excited, leading to holes in the valence band. NPs are excited by spontaneous electron and hole migration to their surfaces, where they oxidize and reduce water, respectively, to oxygen and hydrogen [102].

NPs are an ideal candidate for splitting water under sunlight under certain conditions:

1. Ultraviolet rays and sunlight should not damage the NPs.
2. An effective photocatalyst NP requires a large surface area and a high surface energy; the small particles allow charge carriers (electron and hole) to reach more efficiently.
3. Optimal NP crystallinity is required for efficient photoinduced electron–hole separation in NPs and carrier mobility to avoid the loss of energy during recombination and migration.
4. A conduction band must be more negative than hydrogen's and oxygen's reduction potentialthe s, allowing efficient transportation of electrons and holes for hydrogen and oxygen generation.
5. The band gap of NPs should match the visible energy (2.04 eV); the band gap needed to split a water molecule is 1.23 eV.

There is no doubt that in the new era of high efficiency supercapacitors, one of the most exciting areas of research both in academia and industry has turned out to be one of the most promising research fields: The use of carbon nanomaterials as electrodes in supercapacitors has been the subject of a wide range of research work. Supercapacitors work in a similar way to other capacitors; they are also made up of a separator and an electrolyte that is in contact with two solid and porous electrodes. Supercapacitors store charges at the interface between the electrolyte solution and the electrode, while electrolytic capacitors accumulate charges between the two conductors separated by a dielectric; however, electrolytic capacitors accumulate charges between the leads by accumulating voltage on the dielectric, and the porous dielectric materials can prevent the transport of charges between two electrodes [103].

A supercapacitor is typically classified as either a pseudo-capacitor or an electrical double-layer capacitor (ELDC) [104]. Pseudo-capacitors and ELDCs are fundamentally different in their charge transfer and storage characteristics: A pseudo-capacitor obeys the faradaic charge-transfer laws, while ELDCs obey the non-faradaic charge-transfer laws. The Helmholtz layer forms when there is no transfer of electrons between the storage medium and field, in which case the ELDC stores electrical energy electrostatically, and there is no electron transfer going on [105]. When carbon electrodes with high specific surface areas have redox-active moieties present on their surfaces, there is resulting pseudo-capacitance [106].

ELDCs work by transferring charge between the electrodes and the electrolytes via an electrochemical redox reaction; this means that the capacitance of EDLCs is higher than the that of pseudo-capacitors. The use of rechargeable metal-ion batteries is widespread. In fact, rechargeable metal-ion (e.g., Li^+, Na^+, K^+) batteries are going to be crucial for renewable energy sources to be efficiently stored and delivered as the essential intermediaries to provide the intermittent and fluctuating electricity generated by solar and wind power efficiently. A number of factors play a role in determining the charge storage process, including surface processes and transport kinetics. For example, the development of a nanoscale counterpart of bulk macroscale electrode materials could significantly improve the energy storage performance of rechargeable metal-ion batteries. By reducing diffusion paths for ionic transport and electronic conduction, nanomaterials can increase electrode electroactivity and increase electrode/electrolyte contact areas. Nanostructured electrode materials are usually found in powder form as NPs, nanowires, nanotubes, nanosheets, and nanoflakes, and they usually face the problem of agglomeration [107, 108].

Due to their high surface energies and extremely specific surface areas, powdery nanostructured electrodes tend to aggregate and agglomerate, making slurry casting a difficult method for evenly dispersing and mixing electrodes with conductive additives and polymeric binders. Supercapacitors are considered attractive alternatives to batteries for storing and delivering fast energy due to their high-power density, quick charge–discharge rate, and excellent cycle stability, but they have the shortcoming of limited energy density. A supercapacitor stores energy through one of two mechanisms: through ionic adsorption–desorption at the interface between the electrodes and electrolytes (electrochemical supercapacitors) or by rapidly and reversibly reacting between the electrodes and electrolytes (pseudo-capacitors). Consequently, one of the basic theories for improving the energy density of supercapacitors is increasing the surface area of active materials that are accessible to electrolytes by developing nanostructures in 3D. In addition to the advantages that 3D nanostructured materials can offer for rechargeable metal-ion batteries, most of them are also applicable to supercapacitors. Some of the disadvantages of supercapacitors have been observed as well (such as a large specific surface area to store more charge, a low charge transfer resistance to increase the rate capability, and a porous network that is interconnected for rapid diffusion of ions), while a few of the disadvantages are also occurring on supercapacitors. [109–113].

For life-model wind energy, it is well known that wind farms' economic benefits and operational reliability are seriously compromised due to the icing of wind turbine blades, which can result in a decrease in output power. Traditionally, deicing has been done with the help of costly equipment, and it consumes extensive amounts of energy. A passive deicing technique with low interfacial toughness and low energy dissipation could be a very effective method of deicing a plane. There is still a concern regarding the durability of polydimethylsiloxane (PDMS) coatings; therefore, researchers developed a physical blending technique for preparing low-interfacial toughness PDMS coatings. As wind turbine blade icing becomes more prevalent in cold regions, PDMS coatings can become crucial to improving wind turbine blade performance. In addition, it has been determined adding NPs and plasticizers to the surface of PDMS coatings increased the interfacial toughness to attain large-scale deicing of PDMS coatings [114]. The solvent evaporation-induced crosslinking of PDMS coatings allows for large-scale deicing operations to be performed using this coating [115].

The prepared PDMS coating exhibits a very strong ice adhesion strength that appears to primarily be affected by the size of the iced area and does not appear to be affected by the coating thickness in the critical length of the coating. Through the controlled addition of plasticizers and SiO_2 to the PDMS coating, a coating with low interfacial toughness was successfully produced. The coatings showed extremely low adhesion to ice, with ice adhesion strength of 12.63 kPa, and demonstrated outstanding deicing durability under four different simulations of operating conditions. The PDMS coatings maintain excellent ice-phobic and deicing properties despite some damage from salt spray corrosion. As well as having high effectiveness for large-scale deicing, the PDMS coating prepared has unparalleled durability [116].

Hydrogen is considered the most significant energy carrier of the future; in addition, water electrolysis is a clean and simple method for producing hydrogen that produces very clean gases. While this method has many benefits, it is characterized by slow kinetics, high energy consumption and a common electrocatalyst based on Pt, a rare and expensive metal. Researchers showed that a new structure of available electrocatalyst materials that increased the efficiency of electrolyzers reduced the overpotentials of HER. Scientists made a highly effective electrocatalyst for HER in H_2SO_4 solution by reducing graphene oxide (GO) and displaying a hyperbranched gold nanostructure. To obtain a high surface area of hyper-branched gold nanostructures, a variety of electrodeposition conditions were optimized and controlled. It has been also used graphite sheets (GS) as electrode substrates since they were a relatively cheaper alternative to metals. GS is an excellent electrocatalytic substrate due to its outstanding electrical conductivity; it is ideally suited to electrocatalysis. It is important to factor in the acidic nature of the electrolyte as well as the presence of GO when forming nanostructures. The HER process was more efficient on the bare

GS electrodes, the GS/GO electrodes, and the GS/Au oxide electrodes due to the higher current density and the higher positive onset potential. Moreover, this unique hybrid structure and its high chemical stability caused remarkable catalytic stability throughout the duration of the stability test [117, 118].

NPs have a quantum dimensional effect that causes them to have a wider band gap and a blue shift in their absorption bands because the particles are so small. The band gaps of the thin films obtained from nanocomposite layers were observed to be conventional and near the optimum. In these thin-film solar cells, CdS and PbS were suitable as absorbing layers, while CdS and CdTe were suitable as active layers, which means that they can be used instead of PbS, CIGS, or CdTe. As the dielectric constant and refractive index of the nanocomposites absorb layers decreased with the increased absorption energy band gap, the electron–hole generation rate also increased, and reverse saturation current density decreased. As the volume fractions of the CdS/PbS, CdS/CIGS, and CdS/CdTe nanocomposites increased, the EQE and energy conversion efficiency increased as well [72–76].

To improve the absorption of absorbing layers and to obtain more convenient energy band gaps near to the optimal value, researchers investigated using composite NPs to improve the absorption of absorbing layers, cell performance, and efficiency. In the course of improving the optical characteristics of base matrix semiconductor materials and layers, future metal/semiconductor nanocomposite absorbing layers are expected to consist of composite NPs. A significant factor that improves the optical properties of a nanocomposite is the arrangement of NPs within a base matrix for a heterogeneous array of NPs [72–76].

In solar power module stations, the main impact of heating the solar modules (nanocomposite CdS/(metal/CIGS), CdS/(metal/CdTe), and CdS/(metal/PbS)) from 25°C to 127°C is the decrease in open circuit voltage caused by the increase in temperature. Although the maximum power point and energy conversion efficiency dropped at constant solar irradiance 1000 watts per square meter, there was a slight increase in short circuit current with an increase in module temperature. In contrast, the irradiance of the sun increased the open circuit voltage of the modules at constant temperature, the short circuit current, and the efficiency of their energy conversion. When solar cells are operated at a high temperature, the efficiency drops to zero depending on the materials that make up the cells [72–76].

Scientists introduced metal coatings made with CdTe, metal/CIGS, and metal/CIGS nanocomposites to achieve high thermal stability in the cells by using individual and multiple absorbing layers. Efficiency decreased between 307°C and 487°C, in contrast to 273°C when using traditional crystalline NP. With a new technique of multiple NPs, the teamwork designed a solar power substation with efficiencies of 15.2%–18.45% based on CdS/(CIGS + 20wt.%Al) + Ag), CdS/(CIGS + 20wt.%Al) + Cu), CdS/(CIGS + 20wt.%Al) + Li), CdS/(CdTe + 20wt.%Al) + Cu) nanocomposite thin-film solar cells [72–76].

Using a facile and cost-effective fabrication method, researchers synthesized modern Al/PbS and Cu/PbS nanocomposites that contained volume fractions of Cu and Al with 10wt.%–20wt.% for solar substations with a capacity of 64,516.8 KWP; this process lowered the total number of modules that is required for the design process. Researchers applied a series of thin films via heated evaporation deposition to a row of prepared materials under a vacuum of 10^{-6} torr and then analyzed the powders and films to determine the crystal size, optical properties, depression characteristics, and dielectric properties. Nanocomposite powders and films have cubic natures and nanostructures, with crystal sizes of approximately 8.10–29.4 nm and 18.18–36.06 nm, respectively. PbS nanostructures absorbed lighter when Al and Cu fillers were added as individual components and as multiple nanocomposites [72–76].

Accordingly, the teamwork compared pure PbS film and PbS films doped with Cu or Al that showed a shift in absorption edge toward longer wavelengths depending on the concentration of dopant used, in contrast with a pure nanostructure sample with a constant absorption edge. For powders and films, it is possible to obtain an optical band gap ranging between 3.7–5.1 ev and 1.55–1.85 ev, respectively. In addition, the thin films had the optical characteristics of high absorption, low local conductivity, and low transmittance. It has been also measured a range of other parameters including reflectance, electrical conductivity, volume energy loss, surface energy loss, and refractive index. Al/PbS, Cu/PbS and Cu/(PbS + 20wt%Al) thin films showed very high optical characteristics, and their energy band gaps were close to optimal. Incorporating these layers as absorbent layers instead of PbS improved the performance of CdS/PbS thin-film solar cells [72–76].

REFERENCES

1. Suryanarayana C, "Nanocrystalline materials" *Inter. Mater. Rev.* 40, 41–64, 1995.
2. Gleiter H, "Nanostructured materials: Basic concepts and microstructure" *Acta Mater.* 48, 1–29, 2000.
3. El-Sherik AM and Erb U, "Synthesis of bulk nanocrystalline nickel by pulsed electrodeposition" *J. Mater. Sci.* 30, 5743–5749, 1995.
4. Birringer R, "Nanocrystalline materials" *Mat. Sci. Eng. A* 117, 33–43, 1989.
5. Lu K, "Nanocrystalline metals crystallized from amorphous solids: Nanocrystallization, structure, and properties" Mat. Sci. Eng. R: Rep. 16, 161–221, 1996.
6. Valiev RZ, Islamgaliev RK and Alexandrov IV, "Bulk nanostructured materials from severe plastic deformation" *Prog. Mat. Sci.* 45, 103–189, 2000.
7. Sun XK, Cong HT, Yang MC, et al., "Preparation and mechanical properties of highly densified nanocrystalline Al" *Metall. Mater. Trans. A* 31, 1017–1024, 2000.
8. Tellkamp VL, Lavernia EJ and Melmed A, "Mechanical behavior and microstructure of a thermally stable bulk nanostructured Al alloy" *Metall. Mater. Trans. A* 32, 2335–2343, 2001.
9. Bang JH and Suslick KS, "Applications of ultrasound to the synthesis of nanostructured materials" *Adv. Mater.* 22, 1039–1059, 2010.
10. Balaya P, "Size effects and nanostructured materials for energy applications" *Energy Environ. Sci.* 1, 645–654, 2008.
11. Li KC, Chu HC, Lin Y, Tuan HY and Hu YC, "PEGylated copper nanowires as a novel photothermal therapy agent" *ACS Appl. Mater. Interfaces* 8, 12082–12090, 2016.
12. Fedyanin DY, Yakubovsky DI, Kirtaev RV and Volkov VS, "Ultralow-loss CMOS copper plasmonic waveguides" *Nano Lett.* 16, 362–366, 2016.
13. Amendola V, Pilot R, Frasconi M, Marago OM and Iati MA, "Surface plasmon resonance in gold nanoparticles: A review" *J. Phys. Condens. Matter*, 29, 203002, 2017.
14. Sun L, Zhang C, Wang CY, Su PH, Zhang M, Gwo S, Shih CK, Li X and Wu Y, "Enhancement of plasmonic performance in epitaxial silver at low temperature" *Sci. Rep.* 7, 8917, 2017.

15. Ashcroft NW, Mermin ND and Rodriguez S, "Solid state physics" *Am. J. Phys.* 46, 116–117, 1978.
16. Linic S, Christopher P and Ingram DB, "Plasmonic-metal nanostructures for efficient conversion of solar to chemical energy" *Nat. Mater.* 10, 911–921, 2011.
17. Mascaretti LA, Dutta A, pán Kment Š, Shalaev VM, Boltasseva A, Zborˇil R and Naldoni A, "Plasmon-enhanced photoelectrochemical water splitting for efficient renewable energy storage" *Adv. Mater.* 1805513, 2019.
18. Baburaj EG, Froes FH, Shutthanandam V and Thevuthasan S, *Low Cost Synthesis of Nanocrystalline Titanium Aluminides Interfacial; Chemistry and Engineering, Annual Report*, Oak Ridge, TN, USA, 2000, pp. 1–4.
19. Schulz R, Boily S, Zaluski L, Zaluka A, Tessier P and Strom-Olsen JO, "Nanocrystalline materials for hydrogen storage" *Innov. Met. Mater.* 529, 1995.
20. Varin RA, Czujko T and Wronski ZS, *Nanomaterials for Solid State Hydrogen Storage*, Springer Publications, Berlin, Germany, 2009, p. 338, ISBN 978-0-387-77711-5.
21. Srinivasan S, Demirocak DE, Kaushik A, Sharma M, Chaudhary GR, Hickman N and Stefanakos E, "Reversible hydrogen storage using nanocomposites" *MDPI, Appl. Sci.* 10, 4618, 2020, doi:10.3390/app10134618.
22. Monsef R, Ghiyasiyan-Araniand M, Salavati-Niasari M, "Designof magnetically recyclable ternary Fe2O3/EuVO4/g-C3N4 nanocomposites for photocatalytic and electrochemical hydrogen storage" *ACS Appl. Energy Mater.* 4, 680–695, 2021.
23. Liang H, Chen DD, Thiry D, Li WJ, Chen MF and Snyders R, "Efficient hydrogen storage with the combination of metal Mg and porous nanostructured material" *Int. J. Hydrogen Energy* 44(31), 16824e32, 2019.
24. Liang H, Chen D, Chen M, Li W, Snyders R, "Studyofthesynthesisof PMMA-Mg nanocomposite for hydrogen storage application" *Int. J. Hydrogen Energy* 45(7), 4743–4753, 2020.
25. Su YS and Manthiram A, "Lithium-sulphur batteries with a microporous carbon paper as a bifunctional interlayer" *Nat. Commun.* 3, 1166–1172, 2012.
26. Eshetu GG, Armand M, Scrosati B and Passerini S, "Energy storage materials synthesized from ionic liquids" *Angew. Chem. Int. Ed.* 53, 13342–1359, 2014.
27. Li Y, Yang J, Song J, "Design structure model and renewable energy technology for rechargeable battery towards greener and more sustainable electric vehicle" *Renew. Sustain. Energy Rev.* 74, 19–25, 2017.
28. Hui P, Elisabeth R and Vehse M, "Cost-effective nanostructured thin-film solar cell with enhanced absorption" *Appl. Phys. Lett.* 7(105), 1–8, 2014.
29. Krishnakumar V, Barati A, Schip H, Klein A and Jaegermann W, "A possible way to reduce absorber layer thickness in thin film CdTe solar cells" *Thin Solid Films* 4(3), 233–236, 2013.
30. Bai Z, Yang J, DePba S and Wang D, "Thin film CdTe solar cells with an absorber layer thickness in micro- and sub-micrometer scale" *Appl. Phys. Lett.* 4(14), 1–4, 2011.
31. Belghachi A and Limam N, "Effect of the absorber layer band-gap on CIGS solar cell" *Chinese J. Phys.* 6(4), 1127–1134, 2017.
32. Mohamed HA, "Theoretical study of the efficiency of CdS/PbS thin film solar cells" *Solar Energy* 10(2), 360–369, 2015.
33. Ramanujam J and Singh U, "Copper indium gallium selenide based solar cells—a review" *Energy Environ. Sci.* 11(9), 1306–1316, 2017.
34. Al-Hamadani N, Mustafa F and Al-Kabi J, "Effect of thickness and annealing temperature on structural properties of (CIGS) thin films" *Int. J. Advan. Eng. Technol. Manage. Appl. Sci.* 5(1), 24–33, 2017.
35. Mannan M, Anjan M and Kabir M, "Modeling of current–voltage characteristics of thin film solar cells" *Solid State Electron.* 6(1), 49–54, 2011.

36. Prakash R and Singh S, "Designing and modelling of solar photovoltaic cell and array" *IOSR J. Elec. Electron. Eng.* 11(2), 35–40, 2016.
37. Puspita S, Hoque S and Shamim M, "Theoretical efficiency and cell parameters of AlAs/GaAs/Ge based new multijunction solar cell" *IEEE* 6, 1–6, 2017.
38. Tawsif E, Anjum A and Tanzidul M, "Parametric analysis of CdTe/CdS thin film solar cell" *Int. J. Advan. Res. Comput. Commun. Eng.* 5(6), 401–404, 2016.
39. Singh P and Ravindra N, "Temperature dependence of solar cell performance–an analysis" *Sol. Energy Mater Sol. Cells* 101(4), 36–45, 2012.
40. Derkacs D, Lim S, Matheu P and Mar W, "Improved performance of amorphous silicon solar cells via scattering from surface plasmon polaritons in nearby metallic nanoparticles" *Appl. Phys. Lett.* 89(9), 1–9, 2006.
41. Kord M, Hedayati K and Farhadi M, "Green synthesis and characterization of flower like PbS and metal- doped nanostructures via hydrothermal method" *Main. Group Metal Chem.* 575(40), 35–40, 2017.
42. Shkira M, Khanb MT, Khan A, Mohamed A, Aldalbahi A and AlFaify S, "Facilely synthesized Cu:PbS nanoparticles and their structural, morphological, optical, dielectric and electrical studies for optoelectronic applications" *Mater. Sci. Semicond. Process.* 96(3), 16–23, 2019.
43. Mohanty B, Chattopadhyay A and Nayak J, "Band gap engineering and enhancement of electrical conductivity in hydrothermally synthesized CeO2-PbS nanocomposites for solar cell applications" *J. Alloys Compd.* 850(5), 1–14, 2021.
44. Guo T, Yao M, Lin Y and Nan C, "A comprehensive review on synthesis methods for transition-metal oxide nanostructures" *R. Soc. Chem.* 17(19), 3551–3585, 2015.
45. Wang A and Kong X, " Review of recent progress of plasmonic materials and nanostructures for surface- enhanced Raman scattering" *Mater. Basel* 8(6), 3024–2052, 2015.
46. Sagadevan S, Pal K, Hoque E, Zaman Z, "A chemical synthesized Al-doped PbS nanoparticles hybrid composite for optical and electrical response" *J. Mater. Sci. Mater. Electron.* 15(28), 10902–10908, 2017.
47. Touati B, Gassoumi A and Kamoun N, "Structural, optical and electrical properties of Ag doped PbS thin films: Role of Ag concentration" *J. Mater. Sci. Mater. Electron.* 8(28), 18387–18395, 2017.
48. Soetedjo H, Siswanto B, Aziz I and Jatmoko S, "Deposition of Cu-doped PbS thin films with low resistivity using DC sputtering" *Results Phys.* 6(2), 903–907, 2018.
49. Kar A, Sain S, Rossouw D, Knappett R, Kumer S and Wheatley A, "Facile synthesis of SnO2–PbS nanocomposites with controlled structure for applications in photocatalysis" *R. Soc. Chem. Nanoscale* 8(5), 2727–2739, 2016.
50. Kumar S and Jeevan P, " Synthesis of PbS—Al2O3 nanocomposites by sol-gel process and studies on their optical properties" *Opt. Mater.* 7(46), 209–215, 2015.
51. Derkacs D, Lim S, Matheu P and Mar W, "Improved performance of amorphous silicon solar cells via scattering from surface plasmon polaritons in nearby metallic nanoparticles" *Appl. Phys. Lett.* 9(89), 1–9, 2006.
52. Piralaee M and Asgari A, " Modeling of optimum light absorption in random plasmonic solar cell using effective medium theory" *Opt. Mater.* 62(4), 399–402, 2016.
53. Rosario S, Kulandaisamy I, Deva K, Arulanantham A, Valanarasu S, Youssef M and Awad S, "Deposition of p-type Al doped PbS thin films for heterostructure solar cell device using feasible nebulizer spray pyrolysis technique" *Physica B Phys. Conden. Matter.* 8(3), 1–8, 2019.
54. Braun A, Katz A and Cordon J, " Basic aspects of the temperature coefficient of concentrator solar cell performance parameters" *Prog. Photovolt.* 21(3), 1087–1094, 2013.
55. Rau U and Kirchartz T, "Charge carrier collection and contact selectivity in solar cells" *Adv. Mater. Interfaces* 11, 2019.

56. Muhammad F, Waleed M, Khan S and Ahmed A, "Low efficiency of the photovoltaic cells: Causes and impacts" *Int. J. Sci. Eng. Res.* 8(11), 1201–1207, 2017.
57. Ferekides T, Morel C, Stefanakos D, Ullal E and Upadhyaya H, " Solar photovoltaic electricity: Current status and future prospects" *Sol. Energy* 28(8), 1580–1608, 2015.
58. Srinivas B, Balaji S, Nagendra M and Reddy Y, "Review on present and advance materials for solar cells" *Int. J. Eng. Res.* 5(3), 178–182, 2015.
59. Thabet A, Mobarak Y, Salem N and El-Noby A, "Performance comparison of selection nanoparticles for insulation of three core belted power cables" *Int. J. Electr. Comput. Eng.* 10(3), 2779–2786, 2020, doi:10.11591/ijece.v10i3.pp2779-2786.
60. Todd M and Shi F, "Molecular basis of the interphase dielectric properties of microelectronic and optoelectronic packaging materials" *IEEE Trans. Compon. Packaging Tech.* 26(3), 667–672, 2003, doi:10.1109/TCAPT.2003.817862.
61. Thabet A, *Design and Investment of High Voltage NanoDielectrics*, IGI Global, Publisher of Timely Knowledge, p. 363, August 2020, ISBN13: 9781799838296, ISBN10: 1799838293, EISBN13: 9781799838302, doi:10.4018/978-1-7998-3829-6.
62. Karkkainen KK, Sihvola AH and Nikoskinen KI, "Effective permittivity of mixtures: Numerical validation by the FDTD method" *IEEE Trans. Geosci. Remote Sensing* 38, 1303–1308, 2000, doi:10.1109/36.843023.
63. Polizos G, Tuncer E, Sauers I and More KL, "Properties of a nanodielectric cryogenic resin" *Appl. Phys. Lett.* 96(15), 152903, 2010, doi:10.1063/1.3394011.
64. Tagami N, Hyuga M, Ohki Y, Tanaka T, Imai T, Harada M, Ochi M, "Comparison of dielectric properties between epoxy composites with nanosized clay fillers modified by primary amine and tertiary amine" *IEEE Dielectr. Elect. Insul. Trans.* 17(1), 214–220, 2010, doi:10.1109/TDEI.2010.5412020.
65. Todd M and Shi F, " Characterizing the interphase dielectric constant of polymer composite materials: Effect of chemical coupling agents" *J. Appl. Phys.* 94, 4551–4557, 2003, doi:10.1063/1.1604961.
66. Yi Y and Sastry M, " Analytical approximation of the two-dimensional percolation threshold for fields of overlapping ellipses" *Phys. Rev. E* 66, 2002, doi:10.1103/PhysRevE.66.066130.
67. Thabet A and Salem N, "Experimental investigation on dielectric losses and electric field distribution inside nanocomposites insulation of three-core belted power cables" *Adv. Ind. Eng. Polym. Res.* 4(1), 1857–1864, 2021, doi:10.1016/j.aiepr.2020.11.002.
68. Thabet A, *Emerging Nanotechnology Applications in Electrical Engineering,* IGI Global, Publisher of Timely Knowledge, August 2021, ISBN13: 9781799885368, ISBN10: 1799885364, EISBN13: 9781799885382, ISBN13 Softcover: 9781799885375, doi:10.4018/978-1-7998-8536-8.
69. Dance S, et al., "A comparison of copper indium sulfide-polymer nanocomposite solar cells in inverted and regular device architecture" *Synth. Met.* 222(1), 115–123, 2016.
70. Thabet A, Abdelhady S, Ebnalwaled AA and Ibrahim AA, " Improvement optical and electrical characteristics of thin film solar cells using nanotechnology techniques" *Int. J. Electron. Telecommun.* 65(4), 625–634, October 2019.
71. Bellos E and Tzivanidis C, "Parametric analysis and optimization of an organic rankine cycle with nanofluid based solar parabolic trough collectors" *Renew. Energy* 114, 1376–1393, 2017.
72. Thabet A, Abdelhady S, Ebnalwaled AA and Ibrahim AA, " Innovative industrial Cu(In,Ga) Se2 thin film solar cell with high characterization using nanoparticles structure" *Indo. J. Elect. Eng. Informat.* 7(2), 382–392, June 2019.
73. Thabet A, Mobarak Y, Salem N and El-Noby A, "Performance comparison of selection nanoparticles for insulation of three core belted power cables" *Int. J. Electr. Comput. Eng.* 10(3), 2779–2786, 2020.

74. Thabet A, Abdelhady S and Mobarak Y, " Design modern structure for heterojunction quantum dot solar cells" *Int. J. Electr. Comput. Eng.* 10(3), 2915–2922, 2020.
75. Thabet A, Abdelhady S and Mobarak Y, " Ultra-optical characterization of thin film solar cells materials using core/shell absorber layer" *Int. J. Electr. Comput. Eng.* 12(2), 1377–1384, 2022.
76. Thabet A and Abdelhady S, "Strategy trends of core/multiple shell for quantum dotbased heterojunction thin film solar cells" *Aust. J. Electr. Electron. Eng.* 19(3), 203–218, 2022.
77. Mrazova M, "Advanced composite materials of the future in aerospace industry" *INCAS Bull.* 5(3), 139, 2013.
78. Appadurai M and Raj EFI, "Epoxy/silicon carbide (sic) nanocomposites based small scale wind turbines for urban applications" *Int J Energy Environ Eng.* 13, 191–206, 2022.
79. Rostamzadeh H and Rostami S, "Performance enhancement of waste heat extraction from generator of a wind turbine for freshwater production via employing various nanofluids" *Desalination* 478, 114244, 2020.
80. Ismael M, "A review and recent advances in solar-to-hydrogen energy conversion based on photocatalytic water splitting over doped-TiO2 nanoparticles" *Sol. Energy* 211, 522–546, 2020.
81. Han C, Gao Y, Liu S, Ge L, Xiao N, Dai D, Xu B and Chen C, "Facile synthesis of AuPd/g-C3N4 nanocomposite: An effective strategy to enhance photocatalytic hydrogen evolution activity" *Int. J. Hydrogen Energy* 42, 22765e75, 2017.
82. Prasad C, Tang H, Liu Q, Bahadur I, Karlapudi S and Jiang Y, "A latest overview on photocatalytic application of g-C3N4 based nanostructured materials for hydrogen production" *Int. J. Hydrogen Energy* 45(1), 337–379, 2020.
83. Cai J, Cao J, Tao H, Li R and Huang M, "Three-dimensional ZnO@TiO2 core-shell nanostructures decorated with plasmonic Au nanoparticles for promoting photoelectrochemical water splitting" *Int. J. Hydrogen Energy* 46(73), 36201–36209, 2021.
84. Luna AL, Dragoe D, Wang K, Beaunier P, Kowalska E, Ohtani B, Bahena Uribe D, Valenzuela MA, Remita H and Colbeau-Justin C, "Photocatalytic hydrogen evolution using Ni–Pd/TiO2: Correlation of light absorption, charge-carrier dynamics, and quantum efficiency" *J. Phys. Chem. C,* 121, 14302–14311, 2017.
85. Luna AL, Novoseltceva E, Louarn E, Beaunier P, Kowalska E, Ohtani B, Valenzuela MA, Remita H and Colbeau-Justin C, "Synergetic effect of Ni and Au nanoparticles synthesized on titania particles for efficient photocatalytic hydrogen production" *Appl. Catal. B* 191, 18–28, 2016.
86. Yuan X, Dragoe D, Beaunier P, Bahena D Uribe, Ramos L, Guadalupe M, Medrano M and Remita H, "Polypyrrole nanostructures modified with monoand bimetallic nanoparticles for photocatalytic H2 generation" *J. Mater. Chem. A* 8, 268–277, 2020.
87. Sadhasivam T, Kim HT, Jung S, Roh SH, Park JH and Jung HY, "Dimensional effects of nanostructured Mg/MgH2 for hydrogen storage applications: A review" *Renew. Sustain. Energy Rev.* 72, 523–534, 2017.
88. Ghobadi N, "Band gap determination using absorption spectrum fitting procedure" *Int. Nano Lett.* 4(2), 1–4, 2013.
89. Liao Y, Yikuo S, Tyng W, Lai F and Hsieh D, "Observation of unusual optical transitions in thin-film Cu(In,Ga)Se2 solar cells" *Opt. Express* 20(23), 1–8, 2012.
90. Safa R, Newaz A, Asaduzzaman M, Maksudur M and Ahmed K, "Numerical dataset for analyzing the performance of a highly efficient ultrathin film CdTe solar cell" *Data Brief* 5(3), 336–340, 2017.
91. Whitaker J, *The Electronics Handbook* (Chapter 9), IEEE Press, pp. 124–143, 1996.
92. Handbush G, *Mo Molybdenum: Physical Properties, Part 2* (Electrochemistry 8th edition), Springer, pp. 62–69, 2013.

93. Duley W, UV Lasers: Effects and Applications in Materials Science Book (Chapter 2), Cambridge University Press, pp. 40–56, 2005.
94. Micheal F, *Radiative Heat Transfer* (3rd edition), Elsevier, pp. 77–98, 2013.
95. Yanchuk B, *Laser Cleaning: Optical Physics Applied Physics and Materials Science*, World Scientific Press, pp. 136–145, 2002.
96. Peng C, Chen Z and Tiwari MK, "All-organic superhydrophobic coatings with mechanochemical robustness and liquid impalement resistance" *Nat. Mater.* 17(4), 355–360, 2018, doi:10.1038/s41563-018-0044-2.
97. Bao J, He J, Chen B, Yang K, Jie J, Wang R and Zhang S, "Multi-scale superhydrophobic anti-icing coating for wind turbine blades" *Energy Eng. J.* 118(4), 2021.
98. Abyzov A, "Aluminum oxide and alumina ceramics (review). Part 2. Foreign manufacturers of alumina ceramics. Technologies and research in the field of alumina ceramics 1" *Refract. Ind. Ceram.* 60, 33e42, 2019.
99. Sun J, Gao L and Jin X, "Reinforcement of alumina matrix with multi-walled carbon nanotubes" *Ceram. Int.* 31, 893e6, 2005.
100. Puchy V, Hvizdos P, Dusza J, Kovac F, Inam F and Reece M, "Wear resistance of Al2O3eCNT ceramic nanocomposites at room and high temperatures" *Ceram. Int.* 39, 5821e6, 2013.
101. Abu-Okail M, Alsaleh NA, Farouk WM, Elsheikh A, Abu-Oqail A, Abdelraouf YA and Abdel Ghafaar M, "Effect of dispersion of alumina nanoparticles and graphene nanoplatelets on microstructural and mechanical characteristics of hybrid carbon/glass fibers reinforced polymer composite" *J. Mater. Res. Technol.* 14, 2624–2637, 2021.
102. Arsh Basheer A and Ali I, "Water photo splitting for green hydrogen energy by green nanoparticles" *Int. J. Hydrogen Energy* 44, 11564e11573, 2019.
103. Banerjee J, Dutta K and Rana D, "Carbon nanomaterials in renewable energy production and storage applications, in: Rajendran, S., Naushad, M., Raju, K., Boukherroub, R. (eds) *Emerging Nanostructured Materials for Energy and Environmental Science. Environmental Chemistry for a Sustainable World* (Vol. 23), Springer, Cham, 2019.
104. Pan H, Li J and Feng YP, "Carbon nanotubes for supercapacitor" *Nanoscale Res. Lett.* 5, 654–668, 2010.
105. Emmenegger C, Mauron P, Sudan P, Wenger P, Hermann V, Gallay R and Züttel A, "Investigation of electrochemical double-layer (ECDL) capacitors electrodes based on carbon nanotubes and activated carbon materials" *J. Power Sources* 124, 321–329, 2003.
106. Gong K, Du F, Xia Z, Durstock M and Dai L, "Nitrogen-doped carbon nanotube arrays with high electrocatalytic activity for oxygen reduction" *Science* 323, 760–764, 2009.
107. Pomerantseva E, Bonaccorso F, Feng X, Cui Y and Gogotsi Y, "Energy storage: The future enabled by nanomaterials" *J. Sci.* 366, eaan8285, 2019.
108. Guo YG, *Nanostructures and Nanomaterials for Batteries: Principles and Applications*, Springer, Singapore, 2019.
109. Zhao H, Liu L, Vellacheri R and Lei Y, "Recent advances in designing and fabricating self-supported nanoelectrodes for supercapacitors" *Adv. Sci.* 4, 1700188, 2017.
110. Yu Z, Tetard L, Zhai L and Thomas J, "Supercapacitor electrode materials: nanostructures from 0 to 3 dimensions" *Energy Environ. Sci.* 8, 702–730, 2015.
111. Wang S, Zhang L, Sun C, Shao Y, Wu Y, Lv J and Hao X, "Gallium nitride crystals: Novel supercapacitor electrode materials" *Adv. Mater.* 28, 3768, 2016.
112. Wang S, Li L, Shao Y, Zhang L, Li Y, Wu Y and Hao X, "Transition-metal oxynitride: A facile strategy for improving electrochemical capacitor storage" *Adv. Mater.* 31, 1806088, 2019.
113. Lei Z, Liu L, Zhao H, Liang F, Chang S, Li L, Zhang Y, Lin Z, Kröger J and Lei Y, "Nanoelectrode design from microminiaturized honeycomb monolith with ultrathin and stiff nanoscaffold for high-energy micro-supercapacitors" *Nat. Commun.* 11, 299, 2020.

114. Yu YD, Chen L, Weng D, Hou YC, Pang ZB, Zhan ZW and Wang JD, "Effect of doping SiO2 nanoparticles and phenylmethyl silicone oil on the large-scale deicing property of PDMS coatings" *ACS Appl. Mater. Inter.* 14, 48250–48261, 2022.
115. Yu YD, Chen L, Weng D, Hou YC, Pang ZB, Zeng YS and Wang JD, "Solvent volatilization-induced cross-linking of PDMS coatings for large-scale deicing applications" *ACS Appl. Polym. Mater.* 5, 57–66, 2022.
116. Zhu T, Yuan Y, Song L, Wei X, Xiang H, Dai X, Hua X and Liao R, "A PDMS coating with excellent durability for large-scale deicing" *J. Mater. Res. Technol.* 29, 4526–4536, 2024.
117. Zhao H and Lei Y, "3D nanostructures for the next generation of high-performance nanodevices for electrochemical energy conversion and storage" *Adv. Energy Mater.* 2001460, 2020.
118. Yasaman S, Jaberi S and Ghaffarinejad A, "A graphite sheet modified with reduced graphene oxide-hyper-branched gold nanostructure as a highly efficient electrocatalyst for hydrogen evolution reaction" *Int. J. Hydrogen Energy* 44, 29922–29932, 2019.

7 Synthesis of Nanotech Green Energy

Nanotechnology allows for synthesizing materials with unique and functional properties for energy conversion and storage applications, such as green energy materials. There has been a growing trend lately of using low-cost nanomaterials to create efficient and inexpensive energy conversion and storage systems, regardless of the type of energy that is being used, to develop efficient and inexpensive energy production and storage systems. A new generation of nanomaterials with well-controlled sizes, shapes, porosities, and crystalline phases is essential for the development of sustainable energy technologies. In this context, it is imperative to synthesize and characterize new and novel nanomaterials. Hence, it is necessary to develop a suite of synthesis and characterization techniques for novel nanomaterials that need to be both inexpensive and highly efficient to meet the demands of industry. There is no doubt that the fabrication of thin films of nanomaterials and nano-systems plays a critical role in modern industry, environmental monitoring, biotechnology, nanotechnology, and even personal applications.

7.1 INTRODUCTION

Nanotechnology encompasses the production of nanomaterials, their use in fuel cells, the electrochemical reactions in hydrogen fuel cells, alkaline and acidic environments, and the use of various fuels. This leads to the development of nanomaterial fuel cells operating on small organic molecules rather than polymer electrolyte membrane fuel cells [1].

Nanosized catalysts on high surfaces provided long-term stability for reducing Pt in the cathode. Some of the industry directions followed over the last decade include reducing Pt loading or replacing Pt with low-cost nonnoble metal catalysts, improving performance and forming monolayers of catalysts. The catalysts can be prepared with a high index plane, with a large density of low-coordinated atoms arranged on steps and kinks, and with alloys being used to form core/shell structures and nanoporous and hollow catalysts [2, 3]. Several studies have shown that Pt-based catalysts with 1D structures such as nanorods, nanotubes, and nanowires were more advantageous alternatives to 0D Pt NPs with a large number of beneficial properties; this is because the former suffers less dissolution and Ostwald ripening/aggregation [4].

Bimetal and composite NPs based on Pt, Pd, and Pt-based alloys were synthesized and characterized for use in nano-catalysis, energy conversion, and fuel cells. Through a combination of chemistry and physics, it is possible to synthesize Pt and Pd NPs with modified nanostructures capable of serving as electro-catalysts.

DOI: 10.1201/9781003512486-9

Developing new catalysts at low cost can be achieved by determining the relationships between crystal structure, size, morphology, shape, and composition of cheap base metal catalysts. NPs are shaped in many different ways at the nanoscale, including polyhedra, cubes, octahedra, tetrahedra, bars, and rods, particularly those made of noble and inexpensive metals. In the modern fuel cell industries, it is essential that uniform Pt-based nano-systems are produced with an inner structure, an external structure, a shape, and morphology on a nanometer scale. Recently, the modification of Pt and Pd-based catalysts in alloy, core/shell, and mixture structures has proved to be very effective, durable, and stable in the field of nano-catalysis, energy conversion, and fuel cells, especially for large-scale commercialization [5].

Nanomaterials are divided into 0D, 1D, 2D, or 3D. In chemistry, pharmaceutics, electronic engineering, and engineering chemistry, 1D nanomaterials have a wide range of applications [6]. Monolayers or thin films refer to a wide variety of nanomaterials that are useful in electronics, storage systems, and the fabrication of LEDs, and their sizes can range from 1 to 100 nm [7]. Currently, 2D nanomaterials are being used to construct nanodevices, surpassing the limitations earlier nanometric materials because of their 2D size range [8]. Nanomaterials with two dimensions are useful in the construction of nanocontainers, photocatalysts, nanoreactors, and templates for 2D structures. It should be noted that CNTs are classified among the NPs that can be found in two dimensions. It is important to consider that nanomaterials behave differently based on their shapes, sizes, and morphologies, which determine their potential applications and performance as nanostructures [9].

Nanocomposites are becoming increasingly popular among researchers and industries alike. In recent years, nanocomposites have found a wide range of applications including the production of wind turbine blades. Due to their excellent properties, glass fiber-reinforced epoxy matrix composites are becoming increasingly popular with wind turbine blade manufacturers [10]. Wind energy harvesting's efficiency and the economic viability of wind power can both be attributed to the life span of wind turbines. Blades are an important part of wind turbines, and materials for them must be selected with great care [11].

Composites are often joined using polymer adhesives, which are highly efficient and economical. This solution offers many advantages, including a high strength-to-weight ratio, better fatigue and environmental resistance, lower stress concentrations, fewer processes to adhere to, as well as relatively low cost. A potential route to progress in this area is as nanofiller-toughened adhesives, which include CNTs [12].

Moreover, CNTs have the potential to provide significant benefits when applied to wind turbine blades. CNT-based components of complex composites and hybrid materials typically perform better mechanically than CNT-based components of similar composites. In hierarchical composites, the arrangement and adhesion of CNTs against and adhering to the co-filler is critical, as are filler density, distribution, and homogeneity. There is a prominent trade-off between properties and economy in the manufacture of CNT-based final materials for wind turbine blades at an industrial scale. In the global wind energy industry, the high production cost and the limited uniformity of production batches make an important economic argument for using new CNT-based nanomaterials in blades [13].

7.2 FABRICATION METHODOLOGY

Essentially, nanotechnology revolves around NPs, which are incredibly small particles with a radius less than 100 nm that can be composed of carbon, metal, metal oxides, or organic substances [14]. The size and shape of NPs play a significant role in their modeling and characterization; a wide range of applications can be achieved with NPs due to their ability to behave in different ways, such as electronics, energy, and medicine; it is essential to synthesize NPs that have an appropriate size, structure, dispersity, and morphology for these applications to be successful. NPs and nanomaterials can be safely and reliably synthesized with nanotechnology [15].

7.2.1 NANOTECH HYDROGEN AND FUEL CELLS

Sustainability and environmental responsibility are two of the most important challenges facing the world right now. The most effective strategy is to use very efficient energy together with renewable and sustainable (carbon-neutral) sources. A fuel cell is a device that stores energy by converting chemical energy into electrical energy and vice versa. In most cases, hydrogen is used as the energy storage medium and oxygen as the oxidizer, as shown in Equations 7.1–7.3, and the only byproduct that is emitted in the process is water [16, 17]:

$$Anode: H_{2(g)} \rightarrow 2H^{+}_{aq} + 2e^{-} \tag{7.1}$$

$$Cathode: \frac{1}{2}O_{2(g)} + 2H^{+}_{aq} + 2e^{-} \rightarrow +H_2O_{aq} \tag{7.2}$$

$$Net\ Equation: \ H_{2(g)} + \frac{1}{2}O_{2(g)} \rightarrow H_2O_{aq} \tag{7.3}$$

Sunlight is the most promising renewable source for supporting the clean energy requirements of the future. Hydrogen can be produced from water by photoelectrolysis, making it a renewable energy source. For mobility and transport applications, large-scale hydrogen storage is an important issue. To measure the hydrogen storage capacity in many compounds that contain hydrogen, the following reaction is used: Aqueous HCl solution is reacted with Zn metal; solid Mg hydride is hydrolyzed in the presence of acetic acid by an ideal gas law apparatus for measuring hydrogen yield and using a fuel cell to consume hydrogen. Metal hydrides are an excellent choice for hydrogen storage in solid compounds since they contain high volumes of hydrogen within a small mass. There are various types of resistance that can occur during water vapor transfer through a membrane as depicted in [18, 19].

Membrane resistance is the most prominent resistance that must be overcome for vapor transfer through membranes to be effective in the majority of applications; the membrane diffusion coefficient is used to describe this phenomenon. In most membranes, the selective layer is composed of a porous structure with a support structure surrounded by porous layers. Concentration polarization at the boundary layer of a membrane usually occurs on both sides of the membrane because of resistance between them. Consequently, the most common method for measuring water vapor transfer is a time-lag method referred to as constant-volume, variable-pressure, with a single gas. The second method, which measures membrane resistance only, is a mixed-gas steady state method [20, 21]. An economy based on clean and sustainable technologies can benefit greatly from hydrogen as an energy carrier. In recent years, several automotive manufacturers have begun commercializing fuel cell electric vehicles (FCEVs) that convert hydrogen chemical energy into electricity using polymer electrolyte fuel cells [22].

FCEVs remain in the niche for some time until further technological advances are made, and nano-catalysts in the electrodes are one of the most significant challenges. In practice, Pt-based catalysts, primarily spherical NPs supported on carbon supports with high surface areas, have been extensively studied and are currently in use in the form of spherical NPs. Industry researchers have also made parallel efforts to develop nano-catalysts, such as shape-controlled NPs, which improve cell efficiency and reduce the use of Pt [23].

However, catalyst degradation has demonstrated itself to be a significant problem, especially shape-controlled catalysts, as high activities depend heavily on the fine surface structures of nano-catalysts and the functionality of the alloying base metals. There are also challenges associated with durability given that it is necessary to increase the catalyst's specific surface area to improve power density; this requires a higher degree of stability. The origins of these issues have been clarified through research efforts relating to fuel cell technologies; it is widely believed that the trade-off problem between activity, durability, and power density will soon be solved using a variety of material and structure design approaches, on both the catalyst and the ionomer sides of the system [23].

7.2.2 Nanotech Solar Cells

Today, NP solar cells are the dominant technology in PV technology; about 80% of the PV modules installed are NP mono- or multicrystalline in nature [24]. However, there is still a chance that alternative materials for solar cells are developed since NP is a material with an indirect band gap and low absorption coefficient that inhibits the development of alternative materials. Since polycrystalline NP absorbs little light, polycrystalline NP solar cells require thick structures in the range of 200 m, which increases their cost. Solar cell applications can be achieved using a variety of non-NP intermediate band gap materials [25–28].

Inorganic nanocomposites in thin-film organic solar cells provide for harvesting more solar energy and assist in charge transport charges during the charging process. An important characteristic of a nanocomposite is its optical absorption and conductivity, attributes that determine whether it is suitable for use in solar energy applications [29–32]. Promising materials for semiconductors for solar cells must have a direct band gap with an appropriate band gap value for deposited films. The deposited film should always absorb as much sunlight as possible with this direct band gap. In the spectral range of the visible and the ultraviolet, PbS has a direct and narrow band gap of 0.41 eV at room temperature, making it an attractive material for opto-electrical devices such as solar cells because it has a high absorption in the UV–Vis range [33–35].

The structure and electron transport properties of thin films doped with metal atoms in photovoltaics and other devices are influenced by the doping of metal atoms with PbS thin films. This phenomenon makes it an ideal material for use in photovoltaics and other applications [36–38]. There has been considerable attention on the efficiency of impurity atoms sprinkled into nanocrystal thin films with IV–VI doping; it has demonstrated remarkable effects and has made it possible to introduce these impure atoms into films of nanosized thicknesses. Consequently, the semiconductor lattice can be sharpened to offer atomic levels that contribute to the electrons and holes being held in place by their lattice. Nanostructured semiconductors have been developed using a variety of different dopants, whose properties and concentration play important roles in their properties and efficiency in the semiconductor industry. Doping PbS structures with metal ions such as Au, Cu, Zn, and Al can alter their optical properties [39, 40].

A new chemical route has been developed and applied for structural, morphological, optical, and electrical studies of PbS NPs and hybrid composites made from pure and Al-doped PbS [41]. This strategy was aimed at tailoring the physical properties of PbS films by doping them with tratoal elements Zn, Cu, Cd, and Al to tailor the physical properties of PbS [42–44]. A very common phenomenon associated with metal ions is that they can be incorporated into PbS semiconductors in order to alter their band structure and extend the visible range of photo response [45].

Researchers placed purified organic material in a Mo or tungsten boat to undergo thermal sublimation in vacuum, which is one of the most common methods of depositing small molecules. In the process of passing current through the boat or crucible, the temperature of the material is raised above its sublimation point, resulting in the material evaporating and depositing everywhere on the chamber walls and on the target substrate. During the process of preparing thin films, it is very important to clean the substrate because contaminants such as oil, dust, and fingerprints can severely affect the properties of thin films. Following the cleaning of the gadgets in the vacuum chamber with acetone, the filament holder of the vacuum chamber was filled with the evaporation source boat [46–51].

The Mo boat used in the experiment was a rotary and diffusion pump that was used in combination with evacuation of the Mo chamber at a pressure better than 10^{-5}–10^{-6} torr using a mixture of PbS powder prepared at different reaction times. This automatically switched on the heater, which allowed for heating the multi-nanocomposite PbS powder sample to a much higher temperature than its melting point. Following this procedure, the PbS powders were evaporated in the glass. The researchers fabricated modern films 500 nm thick using the powder of multiple nanocomposites of PbS [46–51].

7.2.3 Nanotech Wind Energy

In wind turbines, the kinetic energy of the wind is converted into mechanical energy that is then converted to electricity in a subsequent process. During the design phase of a wind turbine, it is critical to pay attention to the blades' structural health and how they will endure the life span of the turbine, in addition to their shape and geometrical dimensions. That is, the performance of a wind turbine is directly dependent upon the design of the blades. A blade's structural integrity is compromised by poor erosion resistance, resulting in suboptimal blade performance. Blade erosion patterns contain four phases: upper core breach, lower core breach, leading edge core breach, and advanced erosion [52, 53].

It is essential to develop tools for predicting erosion lifetime of coatings for wind turbine blades, as well as for identifying the optimal combination of coatings and composite substrates for these coatings. There is a significant need to protect the lower edge of wind turbine blades against rain erosion if the researchers are to achieve competitive energy costs for wind turbines. Sol-gel-based coatings typically have a maximum thickness of 10 mm, which has negligible effects on the blade weight compared with a coating derived from sol particles. Sol-gel coatings increase rain erosion resistance, increase the hydrolysis water content, hardness, and flexibility, and decrease the organic/inorganic molar ratio as long as the coating is adhered directly to the substrate. CNPs, including CNTs and graphene, are frequently added to coatings to increase their erosion resistance; however, adding CNPs to coatings can also reduce the erosion resistance. To achieve effective dispersion of CNPs, it is necessary to properly functionalize the particles so that they can disperse efficiently and positively influence the coating properties [54].

Researchers accelerated rain erosion to test the durability of the coatings and demonstrated that that their coatings could withstand rain erosion. The mechanical tests they conducted were measuring the pull-off force, peeling and adhesion of the coating, and nanoindentation tests. They developed numeric models of rain droplet impact simulations to characterize erosion processes and understand the failure mechanism in order to understand the details of erosion processes [54].

There is also the issue of blade icing, which results in significant aerodynamic impairments and an adverse effect on the safety of wind turbines operating in cold regions. An effective method of preventing wind turbine blades from becoming iced up during the winter months is by applying a coating to them. A major difference between active and passive anti-icing and deicing technologies for wind turbine blades can be found in the way they collect and remove moisture [55, 56].

A number of methods are used for active deicing including mechanical, thermal, ultrasonic vibration, and so on. However, these methods are often extremely expensive, energy-intensive, less efficient, and short-lived, as well as requiring specialized equipment, all of which can make them not very practical in engineering settings [57–60]. Therefore, there has been a gradual shift in the attention of scholars from active deicing to passive anti-icing [61]. Using the passive method, water droplets are reduced in adhesion to the coating surface due to the hydrophobic properties of certain materials. Consequently, under natural environment conditions, the superhydrophobic coating greatly reduces the accumulation of liquid water on the blade's surface [62, 63].

Recently, it has been undertaken using both the hydrothermal method and the liquid phase method to synthesize ultra-superhydrophobic materials with nano-rough surfaces. Upon contact with the surface, a micro-nano rough structure was created that was used to create a superhydrophobic surface. A molybdenum disulphide (MoS_2)/zinc oxide (ZnO)/PDMS coating surface expressed in the form of a porous and rough surface was formed by a layer of air cushion in contact with liquid, so the adhesion between liquids and the coating surface reduced; therefore, the adhesion between the coating surface and ice decreased as well. In the icing wind tunnel test, the blades coated with MoS_2/ZnO/PDMS nanomaterials produced significantly less icing under the same conditions as those coated without nanomaterials. There was a reduction of 21.4% in icing thickness along the leading edge, as well as a reduction of around 28.3% in icing surface area [64].

There were some early applications of nanotechnology in the electrical power industry that can be traced to the development of coronavirus-resistant enamel wire used in inverter-fed motors. Hydro generators and turbines that generate high currents were designed with high-voltage insulation to improve their electrical properties, but inorganic particles of 10 to 50 nm loaded into the generator insulation showed erosion and treeing; more efficient ground wall insulation with NPs was developed for large generators. Adding spherical SiO_2 NPs to a well-approved epoxy-mica ground wall insulation system significantly improved the performance: The electrical resistance of the component increased considerably, resulting in a much longer lifetime of the component before it is damaged by partial discharge erosion or electrical treeing. The nanocomposites also showed improved mechanical and thermal properties, which are important for the windings of the stator of a large generator. In addition, electrical breakdown tests were conducted on stator bars with epoxy–mica insulation systems impregnated with vacuum, and these showed optimum dielectric strength with mica content. A long and narrow treeing path should be used at the junction between resin and mica to achieve the highest electrical strength; therefore, the main wall insulation was impregnated and sealed as heavily as possible with mica and as many tape layers as possible to slow tree propagation [65]. In a follow-up, that study's teamwork published their findings for the observed erosion degradation as well as specific erosion degradation parameters including surface roughness, depth of erosion, eroded volume, and PD energy [66, 67].

There was a strong correlation between the amount of filler present and the efficacy of erosion inhibition in nanocomposite structures that consist of epoxy resin containing differing amounts of nanosized silica filler. An electrical lifetime significantly greater than the original generator stator bars was observed in voltage endurance tests using SiO_2 NPs for well-approved insulation, in contrast with a NP-based high-voltage insulation system. There is a reasonable possibility that the researchers will be able to increase the wattage of the generators by 10 compared to the approved reference [68].

The new insulation system facilitated a more efficient generator stator winding design and more sustainable power plants in the near future. As well as offering a more effective stator winding design, the nanocomposite insulation system enables less thickness of insulation, and more copper material to be incorporated into the stator, providing a better heat transfer as a result of reduced thickness and filler content. In order to increase the power output and retrofit the generators for industrial applications, nanotechnology provides us with various generator design options [68].

7.3 THIN FILM MORPHOLOGY

7.3.1 Nanotech Hydrogen and Fuel Cells

A thin liquid/gas diffusion layer (LGDL) produces hydrogen with a high efficiency by splitting water and using it to split molecules. As a consequence of corrosion and carbon consumption, LGDL's conductivity and interfacial contacts are decreased, decreasing the performance and efficiency of the proton exchange membrane electrolysis cell (PEMEC). Efforts have been made to develop materials that are coated with a high level of corrosion resistance as a means of counteracting impact. Ti is one of the most desirable materials to be used as the LGDL in a PEMEC because of its desirable properties; its high corrosion resistance makes it an excellent structural/functional material for aerospace, marine, nuclear, electronics, and medical instruments. In addition, it features excellent mechanical properties as well as high thermal and electrical conductivity [69, 70].

Today, Ti meshes, felts, and foams are the most used LGDLs at the PEMEC anode site due to their high electrical conductivity and fluid resistance. This material comes in 150 mm thick with a significant electrically conductive path and a high level of fluidic resistance. Complex pore morphology causes unusual interfacial contact resistance, and their random structures make water, electron, and thermal distribution impossible. Although Ti has many advantages, such as corrosion resistance, excellent electrical conductivity, and good mechanical properties, it is difficult to fabricate the material itself using conventional methods even though it is the ideal raw material for tiny LGDLs [71].

For large-scale commercial applications, additive manufacturing is still costly and time consuming, even though it does allow fabricating LGDLs with relatively planar surfaces in prototypes. Manufacturing micro and submicron features in these products poses significant challenges. Photochemical machining is a technique available for mass producing Ti thin/well-tunable LGDLs that demonstrates high precision, high repeatability, and less raw material consumption. In nanofabrication facilities, Ti thin and well-tunable LGDLs are made using small foils patterned with lithographic images and chemical wet etching procedures to separate them from the foils [71].

Photomasks are the first and most important step in controlling the size, shape, porosity, and other characteristics of the pores in a Ti thin LGDL. The PEMEC is one of the most important components of the system, and it is sandwiched between the catalytic layer, the current distributor field, and the current gas diffusion layer, all of which are important parts of the system. This layer helps with mixing liquids and gases and is one of the key components [72].

Researchers examined the morphological characteristics of the LGDLs used in testing using field-emission SEM, and the SEM images of the newly designed Ti thin LGDLs reflected varied surface structures and pore morphologies, in contrast with the felt LGDL, which displayed random pore shapes and distributions. The pore size, shape, and distribution were well controlled in the new thin LGDL made from a thin Ti foil. SWCNTs were used to support the membrane electrode assembly, which was flat; its porosity and pore size are measured by micrometers. To develop a carbon fiber electrode utilizing simple electrophoretic deposition, scientists sequentially deposited layers of SWCNTs and Pt on the electrode. Nanotubes and Pt remained morphologically intact on the carbon fiber surface on SEM and Raman spectroscopy. The charge–transfer reaction resistance measured by electrochemical impedance spectroscopy for the HER was orders of magnitude lower on the CNT-coated electrodes than on the carbon black electrodes [73].

Proton MEAs allow for seeing if SWCNTs can reduce charge transfer resistance and thereby improve the fuel cell's performance. Researchers constructed a proton exchange membrane assembly using SWCNT/Pt as the catalyst for a H_2/O_2 fuel cell and evaluated its performance in a fuel cell under normal operating conditions. Carbon materials that possess a high surface area are becoming very popular as new electrode materials, including activated carbon, carbon nanofibers, and CNTs, which have shown significant potential for improving fuel cell performance in recent years [74–77].

Using CNTs as electrode materials in fuel cells for both oxidation and reduction reactions gives improved conductivity and charge transfer at electrode interfaces. This reduces the amounts of precious metal catalysts that are required for fuel cells, which will increase their commercial viability. To construct a robust electrode–catalyst assembly with low resistivity, CNTs must be anchored on the conducting surface. A polymer binder and CNTs directly grown on carbon paper or cloth were used as nanostructured carbon supports for proton exchange membrane fuel cells [78–81].

Bindings are convenient when casting SWCNT films, and they significantly increase resistivity as well. One research group proposed anodic electrocatalysts based on Ru and Fe metal nanostructures and employed graphene as a carbon base as well as SWCNTs and MWCNTs. They conducted physical and electrochemical tests to assess the morphology and activity of the synthesized electrocatalysts and identified PtRuFe/SWCNT as the most suitable. A comparative analysis of PtRuFe/SWCNT electrocatalysts demonstrates that their electrochemical activity, mass activity, and mass transfer were higher than the commercial Pt/C catalyst and the other synthesized catalysts with regard to methanol oxidation, and their electrochemical active surface area was the highest. The electrocatalysts were tested in a micro-direct methanol fuel cell, and PtRuFe/SWCNT showed significantly better electrical properties of zero-voltage and maximum power than did commercial Pt/C [82].

7.3.2 Nanotech Solar Cells

It is possible to analyze individual and multiple NPs within the nanostructure by analyzing their morphology in the powder characterization of metal/PbS nanocomposites. Nanocomposites have been recently attracting attention due to their high properties as well as their optical characteristics; the morphology of the Cu/(PbS + 20wt.%Al) nanocomposite films prepared was examined by SEM with a 10 kV accelerated voltage (JEOL SEM model JSM-5500, Japan; Figure 7.1). XRD was used to characterize the PbS nanocomposites using a Cu target and a graphite monochromator (λ = 1.54056A). The researchers also examined the absorption and transmittance of the nanocomposite films in the wavelength range of 300–1100 nm. The different percentages of Cu nanofiller (10wt.%, 15wt.%, 20wt.%) were deposited on glass substrates with nanocomposite Cu/(PbS + 20 wt.%Al). The images confirm that a PbS nanocomposite layer formed on the substrate.

A conventional electromagnetic spectrometer consists of two basic units, an excitation source and a spectrometer/detector. Researchers applied an accelerated voltage of 200 kV to metal/PbS nanocomposites to analyze their morphological traits using a high-resolution transmission electron microscope (HRTEM), (JEOL, JEM-2100,

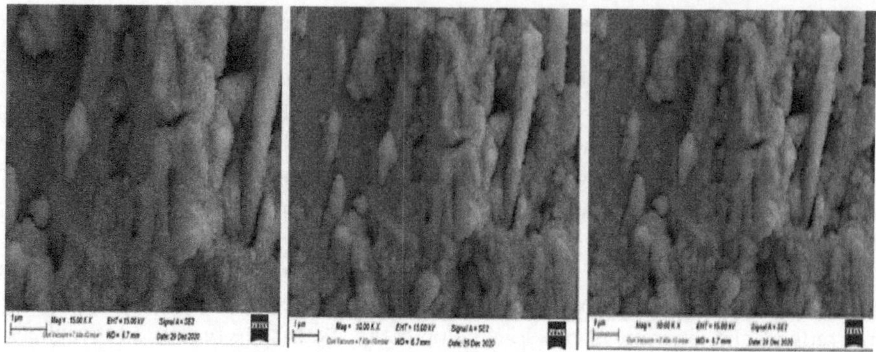

FIGURE 7.1 SEM images for Cu/(PbS + 20wt.%Al) films: left: **S1** 10wt.%Cu/(PbS + 20wt.%Al); middle: **S2** 15wt.%Cu/(PbS + 20wt.%Al); right: **S3** 20wt.%Cu/(PbS + 20wt.%Al).

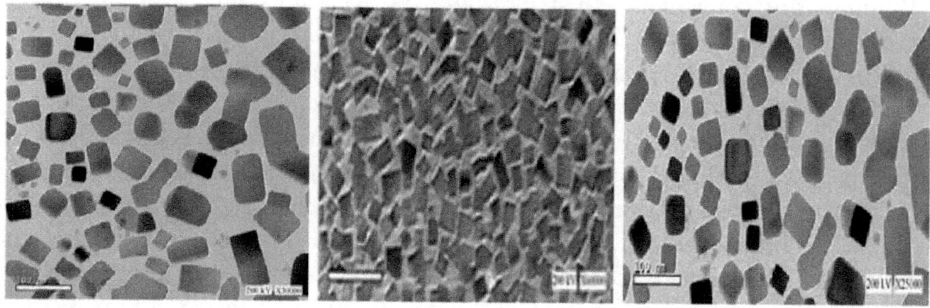

FIGURE 7.2 HRTEM images of the prepared PbS nanostructure: left: pure PbS; middle: PbS + 15wt.%Al; right: PbS + 15wt.%Cu.

Japan) to obtain direct information regarding the shape and size of the nanostructures formed by producing pure PbS NPs. Figure 7.2 presents typical HRTEM images of a pure nanostructure PbS sample prepared using a hydrothermal reaction time of 24 hr.

Table 7.1 shows the optical parameters for the prepared Cu/(PbS + 20wt.%Al) nanocomposites films. It summarizes the refractive indices (n) and extinction coefficients for all prepared films at wavelength 500 nm. For the prepared films, the energy gap changed according to the volume fraction of Cu as the second filler in the prepared PbS samples due to the increased density of localized states in the energy gap. The table shows that the films with higher energy band gaps had lower refractive indices, extinction coefficients, infinity dielectric constants, and lattice dielectric constants. Sample 2, the 15wt.%Cu/(PbS + 20wt.%Al) film, had the highest energy band gap and the lowest refractive index and extinction coefficient. The increased energy band gap resulted from the decrease in the size of the Cu/(PbS + 20wt.%Al) film.

Fourier transform IR (FTIR) spectroscopy is a sensitive technique for examining intractable samples. IR spectra in a chemical sample are passed through a sample, and the amount of incident radiation that is absorbed at a particular energy level is

TABLE 7.1
Optical Parameters for Prepared Cu/(PbS + 20wt.%Al) Nanocomposite Films

Sample	E_d (ev)	E_o (ev)	ε_l	N/m (gm-1.m-3)	ε_∞	n	k	E_g (ev)
S1	5.39	1.38	24	3E+41	4.9	2.20	0.67	2.15
S2	3.47	1.5	22	2E+41	3.3	2.15	0.57	2.25
S3	5.25	1.16	28.5	3E+41	5.5	2.53	0.786	1.9

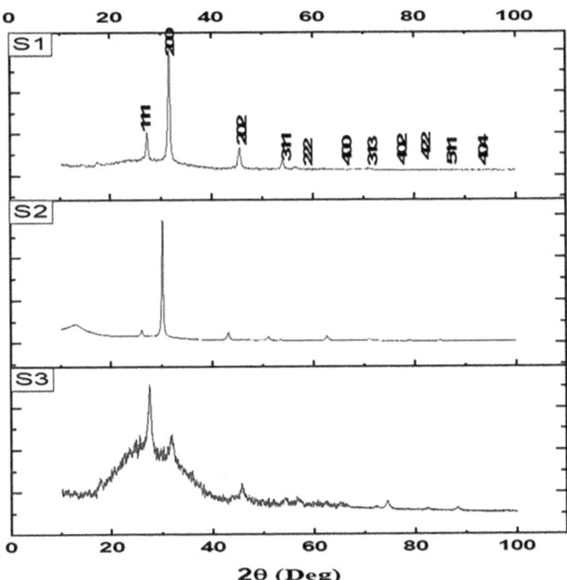

FIGURE 7.3 X-ray diffraction patterns for Cu/(PbS + 20wt.%Al) nanocomposite films.

measured using the vibrations of atoms. Molecular vibration frequencies are represented as group frequencies by absorption spectra that correspond to the energy peak. There tends to be a variety of species-specific group frequencies in the mid-IR region of the IR spectrum, and in the near-IR and far-IR regions, unique molecular motions are responsible for the characteristics of the IR bands [83].

The purity of metal/PbS nanocomposites particles was determined by FTIR spectrometry. A KBr pellet was placed in an evacuated die, followed by PbS samples, to prepare each sample for the spectroscopy. To evaluate the functional groups in the samples, the researchers measured the spectra of all the prepared samples using an IR spectrometer (Jasco Model 4100, Japan) with a resolution of 4.00 cm^{-1} at room temperature that covers the wave number range of 4000–400 cm^{-1} [84].

Figure 7.3 displays the XRD analysis results of the synthesized PbS + 20wt.%Al + Cu nanocomposite films. The existence of narrow, sharp, and well-defined peaks clearly reflects the crystallinity of the prepared multiple nanocomposites films. For PbS nanocomposite films grown directly on glass substrate, peaks were clearly observed at (111), (200), (220), (311), (400), (313), (222), (422), (511), and (404) XRD peaks corresponding to the cubic PbS phase.

The fundamental electron excitation spectrum of the Cu/(PbS + 20wt.%Al) films was described using the frequency dependence of the complex electronic dielectric constant. The real part of the dielectric constant (ε_1) relates to electronic polarizability and the local field within the material when an electric field is applied. The imaginary part of the dielectric constant (ε_2) is responsible for attenuating the local field inside the material, causing dielectric loss. These parts are related to the refraction index and extinction coefficient. Figures 7.4 and 7.5 show the spectra of the real and imaginary parts of the dielectric constant for the Cu/(PbS + 20wt.%Al) nanocomposites film at 10wt.%, 15wt.%, and 20wt.% of Cu as a second filler. The real parts of the constants decreased with increasing photon energy and remained constant at high photon energy. On the other side, the imaginary parts of the dielectric constants for the films increased with increasing photon energy.

Figures 7.6–7.8 show the variations in tanδ (the loss tangent) and, consequently, the surface and volume energy loss functions as a function of photon energy for the (PbS + 20wt.%Al) + Cu nanocomposite films at various percentages of Cu nanofiller (10wt.%–20wt.%). The volume energy loss function and surface energy loss function depend on the complex dielectric constants of the films. The figures show that tanδ increased with increasing photon energy but then decreased at high photon energy.

Figures 7.9 and 7.10 show the optical and electrical conductivities of the Cu/(PbS + 20wt.%Al) nanocomposite film with different concentrations of Cu as a second filler at various percentages (10wt.%–20wt.%). Optical and electrical conductivity

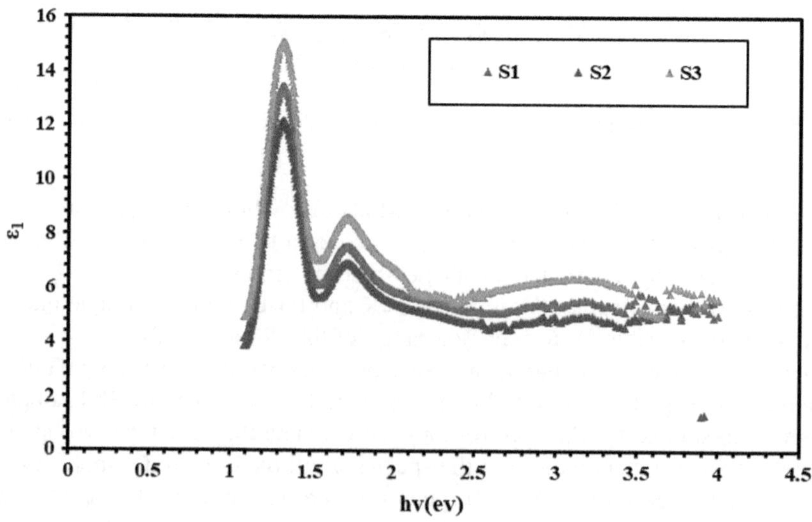

FIGURE 7.4 The spectra of the real parts of the dielectric constant for Cu/(PbS + 20wt.%Al) nanocomposite films.

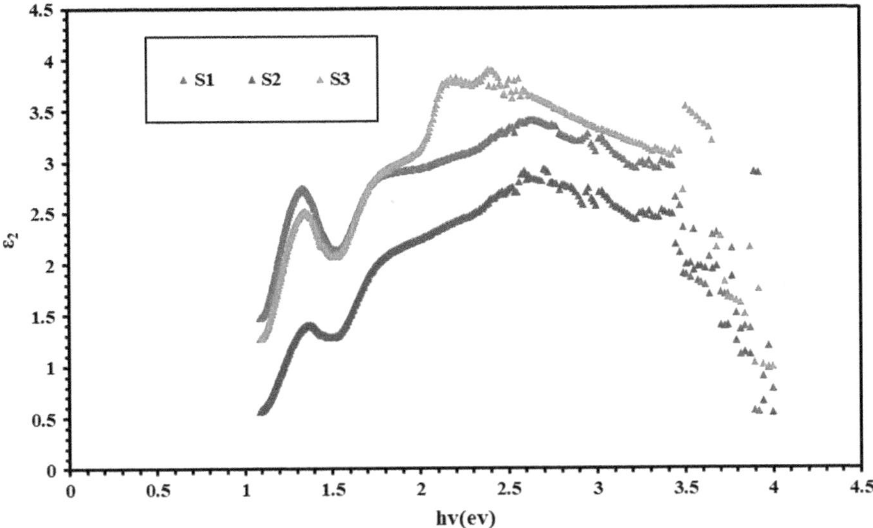

FIGURE 7.5 The spectra of the imaginary parts of the dielectric constant for Cu/(PbS + 20wt.%Al) nanocomposite films.

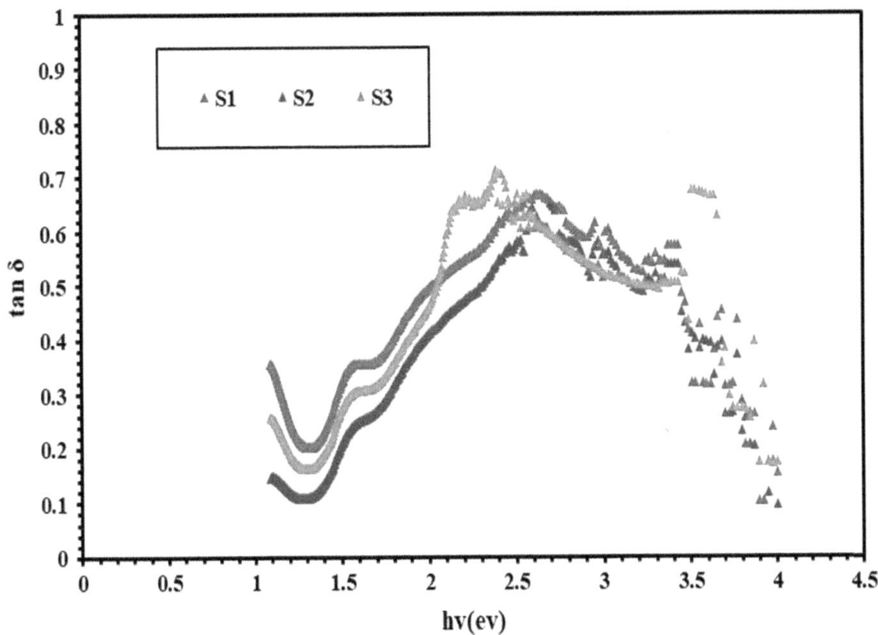

FIGURE 7.6 Tanδ as a function of photon energy for Cu/(PbS + 20wt.%Al) nanocomposite films.

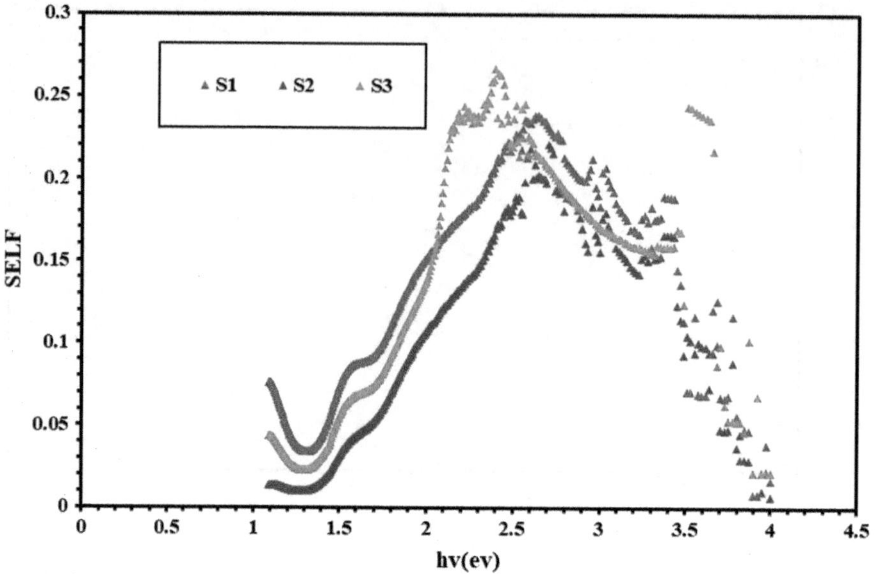

FIGURE 7.7 Surface energy loss as a function of photon energy for Cu/(PbS + 20wt.%Al) nanocomposite films.

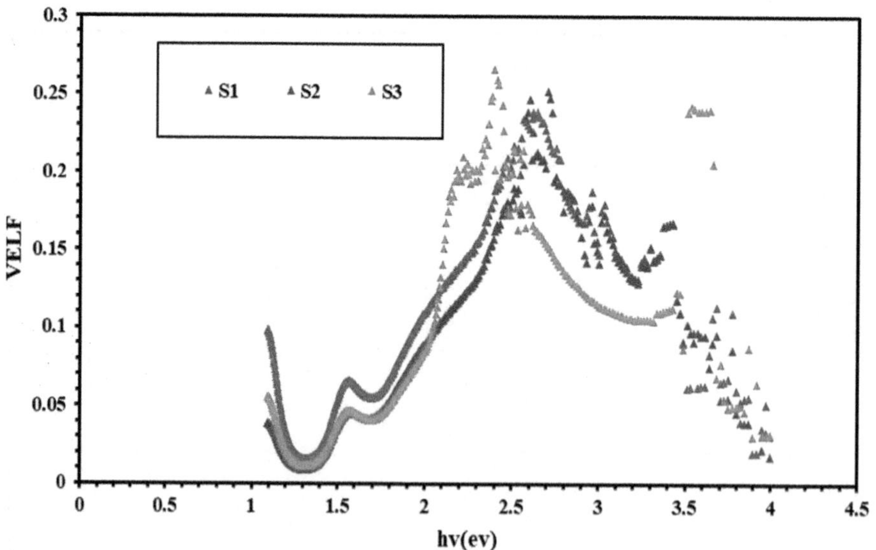

FIGURE 7.8 Volume energy loss as a function of photon energy for Cu/(PbS + 20wt.%Al) nanocomposite films.

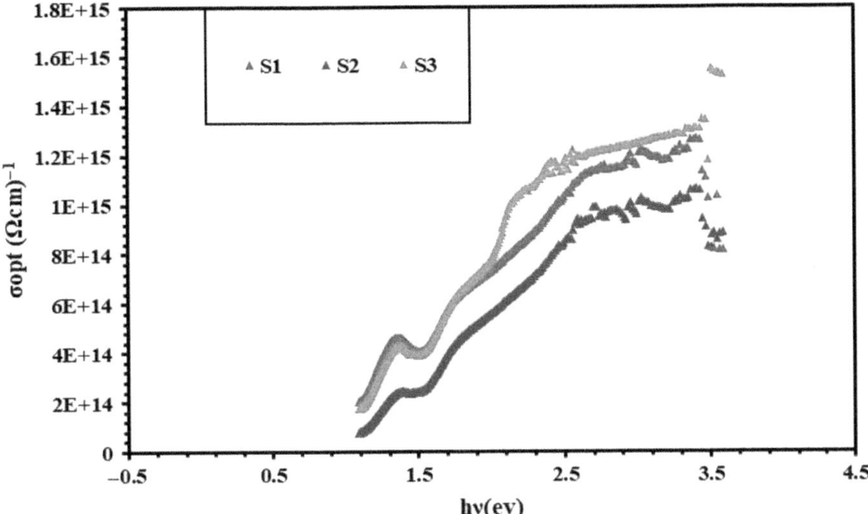

FIGURE 7.9 Optical conductivity as a function of photon energy for Cu/(PbS + 20wt.%Al) nanocomposite films.

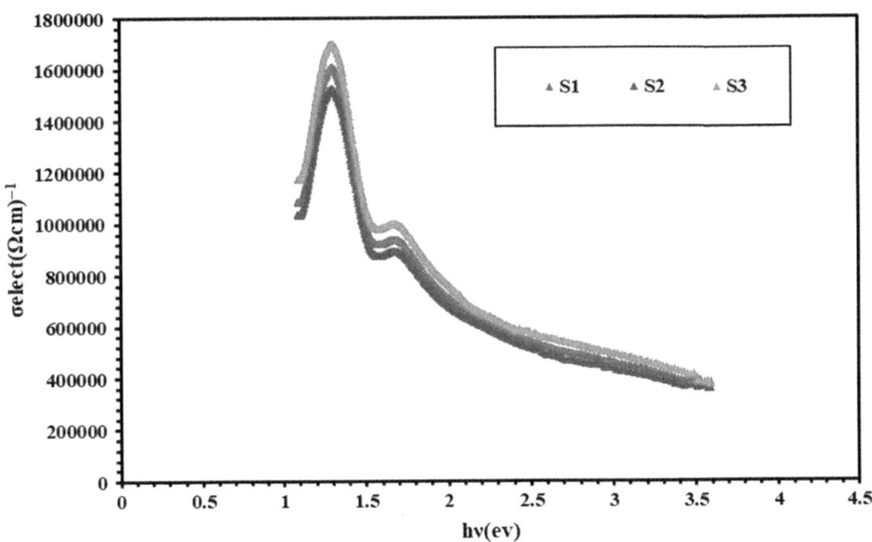

FIGURE 7.10 Electrical conductivity as a function of photon energy for Cu/(PbS + 20wt.% Al) nanocomposite films.

can be calculated using the absorption coefficient, and the optical conductivity of the prepared films increased with increasing photon energy. The improved optical conductivity of the prepared nanocomposite films at high photon energies was related to the high absorbance of the films, which resulted in electron excitation by photon energy. S3 had the highest optical and electrical conductivities, followed by S1 and S2.

7.3.3 Nanotech Wind Energy

Researchers tested nanomaterials in a wind turbine to examine the effects of a polymer matrix. Cracks are one of the leading causes of turbine failure, and nanomaterials prevent this by replacing or blocking nanopores; the loaded stresses are directed or distributed to NPs, thereby preventing catastrophic failures. In this study, the nanomaterials have high mechanical properties and provided greater fracture toughness and better mechanical properties the than polymer composite [85]. Among all the nanomaterials that could be used in composites, it is clear that CNTs have the best mechanical properties. CNT-based nanocomposites are also able to produce multifunctional composites due to the high thermal and electric conductivities of CNTs; this makes them an excellent candidate for wind turbine applications owing to their promising properties [86].

There has been a great deal of research on passive anti-icing coatings, especially superhydrophobic nano-coatings, to prevent ice formation. Active ice-protection systems consume a considerable amount of energy and are neither simple nor inexpensive to develop and to maintain [86]. Passive methods of anti-icing (such as coatings), on the other hand, prevent ice accumulation and reduce maintenance costs on wind turbine equipment because they do not require external energy technologies and reduce the amount of water required for ice to accumulate [87].

Advances in nanotechnology and material engineering in recent years have made biomass nano-coatings a practical, intelligent, efficient, adaptive, and durable hydrophobic coating [88]. A variety of nanocarbon materials, most of which are superhydrophobic and super-ice phobic, demonstrate good anti-icing properties. Because of the high synthesis costs associated with these carbon materials, their applications have been limited. Biomass carbon has been reported as a promising hydrophobic material since it is abundant, reproducible, avirulent, and more cost-effective than the others [89].

Researchers prepared hydrophobic carbon coatings on wind turbine blades using rice straw biogas residues. Nanocarbon exhibits a significant anti-icing function that results in dynamic freezing adhesion, ice resistance, and changes in the mechanical characteristics of the adhesive force. These effects were observed when ice layer pull-off failure occurred and the hydrophobic corn-straw-biogas-residue coating was applied to an icing blade in a wind tunnel. The superhydrophobic nanocarbon biomass residue coating showed excellent hydrophobic properties and an outstanding anti-icing effect. The thin film of commercial-grade polyvinylidene difluoride (PVDF) was made from a liquid Li battery factory in Harbin, China, and required no processing. The wind turbine blade was made in a laboratory using a laboratory-made blade airfoil with a chord length of about 100 mm [90].

7.4 NANOMATERIAL CHARACTERIZATION

7.4.1 Nanotech Hydrogen and Fuel Cells

Hydrogen is a renewable energy source that can be converted to hydrocarbons with high energy conversion efficiency, which is why new clean and sustainable economies need hydrogen energy; solid oxide fuel cells (SOFCs) can be used to convert hydrogen into hydrocarbons [91–93].

Researchers used chemical vapor deposition to reduce the thickness of yttria-stabilized zirconia (YSZ) electrolytes to tens of nanometers using atomic layer deposition (ALD). Using a wafer-scale self-assembly nanotexturing technique, the manufacturing method produced corrugated electrolyte membranes with reduced manufacturing complexity; nanosphere lithography was applied to create 3D patterns on NP substrates by patterning ALD and YSZ electrolytes on wafer-scale surfaces [94–96].

Nanospheres can form close-packed patterns because of their ability to self-assemble; this avoids the need to employ conventional lithography, which involves mask design, fabrication and photoresist application, development, and removal. As a result, the dry etching conditions of the process determine the geometry of this nanostructured pattern. The patterned substrate was coated with a layer of YSZ electrolyte using ALD to replicate this 3D nanostructured geometry with increased surface area. Electrolytes must be uniform throughout the active fuel cell area to achieve pinhole-free performance, this ALD coating consisted of zirconia doped with 7%–8% mol percent of yttria. As part of its unmatched functionality, it is a unique property of ALD to deposit ultrathin conformal coatings over corrugated geometries without leaving pinholes behind. This is not possible with physical vapor deposition [97, 98].

Reducing the density of the current in a fuel cell reduces the loss of polarization, although it also lowers the output power of the fuel cell. A significant amount of power can be obtained by increasing the electrolyte membrane area by a factor of five. Researchers developed a method for fabricating 3D corrugated thin-film electrolytes with a surface area five times greater than noncorrugated thin-film electrolytes via photolithography steps. Their method was less complex more cost-effective than the complex fabrication process, but it remains difficult and expensive [99].

Currently, Pt-based catalysts are used as microparticles supported on carbon supports with high surface areas. However, it is challenging to ensure the durability of the catalyst because of the requirement to increase its specific surface area to increase its power density. Researchers have taken a number of approaches to material and structure design that aim to achieve the trade-off between activity, durability, and power density on both the catalyst and ionomer sides. The performance of a cell made of nonprecious metal catalysts was comparable with that of a cell made of a Pt catalyst. While further improvements in the performance and durability of these catalysts for FCEV applications are still needed, ongoing research on their industrialization is expected to help them achieve increased industrial success [100].

7.4.2 Nanotech Solar Cells

Multiple oxide films such as Cu/PbS, Al/PbS and Cu/(PbS + 20wt.%Al) materials with volume fraction are classified as p-type materials with high absorption coefficients and an appropriate energy band gap range. Band gaps were wider for nanocomposite films containing Cu, Zn, and Al at different volume fractions ranging from 10% to 20%, demonstrating quantum confinement as a result of the minimum grain size of the crystalline films in the range of 18.18–36.06 nm. These findings agreed with the theoretical maximum band gap for an effective absorber layer (1.5 eV) [101–104].

In solar cells, visible light absorption is essential for reducing the band gap; a nanocomposite's band gap could be reduced to 1.55 eV with increasing the volume fraction. In nanostructured PbS, Cu/PbS, Al/PbS and Cu/(PbS + 20wt.%Al) composite films incorporated into thin-film solar cells, s p+ layer increased the short-circuit current to increase efficiency. The reason for the high absorption coefficient of absorbing layers is that their absorption edges are shifted to the visible region by metal NPs, which enhances the films' ability to separate and collect the charge carriers generated in an individual PbS semiconductor. It is crucial to achieve high short circuit currents in solar cells to increase the efficiency of the cell's conversion process. This requires efficient charge carrier separation and absorbance. The modern composites of CuPB, AlPb, and Cu-Cu/(PbS + 20wt.%Al) have the potential to be used as absorption layers in solar cells at a variety of concentrations. Furthermore, an average reflectance of less than 45% of the incident light was obtained for the PbS thin films [105, 106]. However, these thin films were also specularly reflective and were therefore capable of reflecting the majority of the incident light. Optical devices requiring low transmittance and reflectance can use films with low transmittance and reflectance as radiation-absorbing layers. Thin films with high absorption coefficients can be used to prepare high-efficiency thin film cells, reducing the thickness of the absorbing layer [105, 106].

This allowed the fabrication of highly stable and efficient devices with absorb layers as thin as 0.5 mm. The prepared PbS + 20wt.%Al nanocomposite films were well suited for use as second fillers in the absorb layer in NP thin-film cells because of their good optical conductivity as well as their low refractive index for efficient energy conversion. The energy conversion efficiency of CdS/PbS was 2.16% at 0.5 microns in the absorbing layer and increased to 4.13% at 2 microns [105, 106].

In recent years, QDs containing energy band structures of different semiconductors with varying size-dependent energy band structures have generated considerable interest in fabrication of functional materials with adjustable optical and electrical properties. A growing body of scientific knowledge continues to improve the performance of NP solar cells. PbS QDs are one of several attempts to synthesize these NPs, offering a band gap that can be adjusted from near-IR to visible wavelengths. A variety of applications have been proposed for them, including photodetectors, optical switches, and solar cells [107–112].

Based on experimental and theoretical results, multiple nanocomposites of Al/PbS, Cu/PbS, and Cu/(PbS + 20wt.%Al) had better optical properties than those with individual NPs. Researchers observed high absorption capacity with a high energy band gap for the film, close to the maximum energy band gap for effective absorbing layers. There was good agreement in the absorption edge of the multiple nanocomposite films, in contrast with the bulk PbS absorption edge. For powder materials, recent studies indicate a high blue shift in the onset of absorption edge (200–400 nm). It is evident, however, that the onset of the absorption edge (600–900 nm) shows a noticeable blue shift. Despite the fact that particle size decreases as a result of the phenomenon of blue shift, the absorption edge continues to shift from a wavelength region within the visible spectrum to a region within the IR spectrum [107–112].

The blue shift is primarily caused by the quantum dimensional effect of NPs, which increases the band gap and blue shifts the absorption band as the particle size decreases. Additionally, NPs are subjected to the surface effect of large surface forces, resulting in crystal lattice aberration and low crystal constant caused by the large surface force. This reduction in refractive index occurs because the metal NPs in the base matrix of PbS are able to reduce the reflectance of PbS, resulting in a decrease in its refractive index. There is theoretical evidence that metal/PbS nanocomposites with higher volume fractions have enhanced optical properties; however, experimental data reveals that the enhancement varies according to the type of metal and the concentration of metal in the composites. Experimentally, 15wt.%Cu/20wt.%Al + PbS film showed the best results for enhancing the optical properties of PbS when used as the absorbing layer, but theoretically, the 20wt.%Cu/(20wt.%Al + PbS) was most effective [107–112].

PbS thin films were synthesized by hydrothermal growth and deposited by vacuum thermal evaporation; XRD was used to analyze the QD PbS thin film and confirm the formation of PbS. SEM analysis of the surface morphology and composition together with an EDAX study were used to determine the optical properties and band gap energy, as well as the optical depression and dielectric characterization for the QD film. Theoretical modeling produced predictable data sheets that showed the current trends for QD PbS film as well as the I-V and P-V characteristics curves for the design of HJQD solar power substations with total installed power capacity of 64,518.6 KW. It has been designed the cells using the same QD window layers and QD PbS absorbing layers with varying radii and materials [107–112].

Decreasing the size of the NPs increases the Eg as a result of the confinement effect. It is this increased absorption of light from the solar spectrum that is responsible for increasing the intensity of light absorbed by the cell, as well as the increase in the driving force for electron and hole injection into surrounding phases, which has a direct effect on the efficiency of the solar cell [113]. Various types of solar cells use QDs. Some of the most well-known are Schottky solar cells [114], depleted heterojunction SCs [115], hybrid organic/inorganic solar cells [116, 117], and QD-sensitized SCs [118, 119].

In the QD solar cell world, there are three major types of QD solar cells: quantum funnels (QF), quantum junctions (QJ), and heterojunctions (HJ). Researchers built a quantum funnel solar cell (QFSC) with a PCE of 2.7% on a graded device using a layer-by-layer technique; they selected a sequence of layers consisting of QDs of different sizes and thus different electron affinities. First, a TiO_2 layer was coated with colloidal PbS QDs with larger diameters; then, layers of colloidal QDs of smaller diameter were deposited on top to form the upper layer. From the point of generation of the photoelectrons, the quantum funnels are able to drive the photoelectrons toward the electron acceptor layer. QJSC layers are made of the same semiconductor material but with different levels of doping; they are made of both p-type and n-type PbS QDs. PbS layers can be deposited by a variety of techniques such as electrodeposition [120], pulsed laser deposition [121], spray pyrolysis [122], and thermal evaporation and chemical bath deposition (CBD) [123]. Optical properties of QD PbS deposited by vacuum thermal evaporation were studied after hydrothermal growth.

7.4.3 Nanotech Wind Energy

Composite materials are widely used in the manufacturing of wind turbine blades. Most often, composite materials consist of synthetic polymers such as glass, Kevlar, and carbon-reinforced matrix. A composite made from Kevlar and epoxy is durable, lightweight, and dimensionally stable; it is lightning resistant due to Kevlar's dielectric strength. In many cases, composite materials can provide higher stiffness than metals, resulting in less weight added to finished parts. To meet design requirements, composite materials require unique manufacturing techniques. In thermoplastics, impregnation is difficult because they are more viscous than thermosets; these unique techniques can be applied to thermosetting polymers and thermoplastic polymers. Generally, increasing the temperature decreases viscosity, but the decomposition of some polymers can occur before they reach very low viscosities due to their inherent properties. Most wind turbine blades are manufactured using epoxy or vinyl ester as the matrices. The mechanical and physical properties of epoxy can be improved by using functionalized CNTs including amino and carboxylate and functionalized MWCNTs [124, 125].

Applications for Kevlar, a synthetic, high-modulus fiber based on p-phenylene terephtalamide include its use in polymer fiber reinforcements. Fibers were produced in a solution extruder that allowed for controlling strength; the extruder produced two samples of Kevlar fibers with a super molecular structure. In one, a hydrogen bond formed in a molecule arranged in a sheet along the crystal plane (100) of the molecule; these sheets were then arranged around the fiber axis, with the crystal direction (010) pointing radially. In the other, cylindrical surfaces were sequentially arranged in a coaxial direction to construct the fiber structure [126].

Wind turbine blades can experience delamination damage caused by deflections of the blade, which can lead to delamination damage and, eventually, lead to blade failure within a given period of time. These factors can influence failure mechanisms and produce failure modes based on materials, structural defects, and manufacturing processes. Researchers mixed MWCNTs with epoxy resin employing sonication to improve CNT dispersion and determine if the coating could reduce the maximum deflection of Kevlar-reinforced epoxy (0.4wt.%). The sonication processes produced a satisfactory dispersibility of the F-MWCNTs (0.5wt.%). The interfacial strength of the matrix-fiber composite was examined using SEM, and the Kevlar/epoxy composites performed poorly under high CNT loadings (0.5wt.%) of unmodified MWCNTs [127].

The scientists also studied the morphologies of the fracture surfaces using vacuum infusion and measured the ultimate tensile strength of the CNTs. The nanotube bundles were pretreated with acetone in ultrasonic baths at room temperature for 30 minutes before dispersion to guarantee a uniform distribution. To remove the acetone from the blends, the samples were placed in a vacuum oven with a temperature of 70°C for 1 hour; they increased the temperature by 50°C in epoxy resin for 2 hours to improve the dispersibility of the MWCNTs in the epoxy resin to reduce the viscosity of the resin. A Ti alloy probe was immersed in the epoxy/MWCNT mixture to achieve strong adhesion between the epoxy and the MWCNT mixture. Both unmodified and modified MWCNTs were added at 0.1% and 0.5% using ultrasonics [127].

7.5 TRENDS AND RECOMMENDATIONS

In this chapter, it has been discussed the development of thin Ti liquid-gas diffusion layers (LGDLs) with flat interfacial surfaces is presented and their application to a PEMEC is demonstrated in this paper. They demonstrate superior multifunctional performance over conventional LGDLs having thin and well-tunable Ti surfaces. A further investigation was conducted by ex situ and in situ characterization experiments for better understanding of the involved mechanisms; thin and well-tunable LGDL with flat surface features displayed a remarkable reduction of its overall resistances, as well as producing significant improvements in PEMEC efficiency and performance. Furthermore, LGDL thickness can be greatly reduced in comparison to conventional LGDLs, and PEMEC stacks can be greatly reduced, resulting in new approaches for developing low-cost, high-performing PEMECs. For developing PEMEC models and validating simulations of PEMECs, its well-tunable features will come in very handy to optimize and repeatable performance of its pore size, shape, distribution, and porosity. Pore sizes, shapes, distributions, and porosities, along with their associated modeling, are being investigated in depth as a part of the thin LGDL optimization of pore morphologies. A simple and versatile way of designing membrane electrode assemblies (MEAs) for fuel cells based on SWCNTs has been discovered through electrophoretic deposition of SWCNTs on carbon fiber electrodes. A SWCNT is well suited to anchor Pt catalysts to an electrochemical oxidation, and reduction reaction mixture just as a CB is well suited to anchor commercially available Pt catalysts to an electrochemical reaction mixture. H_2-based PEM fuel cells can be improved by SWCNT/Pt nanostructures under higher pressure by significantly improving performance. The MEA based on SWCNTs is an attractive candidate for developing next generation fuel cell devices because of its decreased charge transfer resistance and greater hydrogen adsorption capabilities. Fuel cells operating at higher anodic back pressures are showing promising results in overcoming the activation energy barrier encountered in SWCNT-based electrodes.

7.5.1 Trends in PbS Nanocomposite Films

It was discovered that in the visible and near-infrared region of the visible spectrum of Cu/PbS and Al/PbS nanocomposites films, with various ratios of Cu and Al fillers, the transmission was below 50%. Since thin films are manufactured with low reflectance and transmittance, they do a good job of absorbing radiation and can, therefore, be used effectively as absorbing layers in NP cells. With the preparation of PbS nanocomposites films that contain varying ratios up to 20% of Al and Cu fillers, these nanocomposites have good optical conductivity, high energy band gaps, and a low refractive index. Thus, they can be used as absorbing layers in NP cells as they are very suitable for achieving high performance, decreasing the reverse saturation current, and maximizing the energy conversion efficiency of NP cells with fewer materials. Multiple layers of Al:PbS can be produced when the Cu/20wt.% is used, and it is used for novel PbS photovoltaics. With an excellent optical property derived from its solution-processability, lead sulfide is an effective material for making Cu:PbS and Al:PbS nanocomposites as well as Cu/20wt.% Al:PbS multiple nanocomposites. Using a facile and cost-effective fabrication growth method, Cu:PbS and Al:PbS nanocomposites are synthesized, and deposition of thin films is carried out on glass by means of thermal evaporation deposition under vacuum of 10^{-6} torr.

Crystal size and optical properties are influenced by the concentration of Al and Cu fillers; PbS multiple nanocomposites can be produced conveniently and economically using hydrothermal growth, as evidenced by yield, crystallinity, and crystal size characteristics. As individual fillers in PbS or as multiple NPs in 20wt.% Al:PbS, Al or Cu NPs increase the absorption and optical band gap. Using the alpha-pbs/pbs mixtures and Cu:Al alloys nanocomposites materials, as well as Cu/20wt.% Al:PbS multiple nanocomposites materials, it has been detected that there are high optical band gaps which reveal that these powders have a significant role to play in achieving high optical characteristics in electronic devices. In addition to the optical properties of the films obtained from this technique, the electron energy band gap values for the nanocomposite films in Cu:PbS, Al:PbS and Cu/20wt.% Al:PbS with various volume fractions indicated that the obtained films showed high absorption. When pure or Cu or Al doped films were compared with original samples, the absorption edge of pure films shifted toward longer wavelengths depending on the concentration of dopant. Adding Al and Cu fillers individually and in multiple nanocomposites increased the absorption of pure PbS nanostructures; achieving optimum absorption and conventional energy band gap characteristics of the thin film absorbing layer of Al:PbS, Cu:PbS and Cu/20wt.%. As an alternative to PbS, Al:PbS thin films can serve as absorbers for CdS/PbS thin film solar cells. On glass substrates, different Cu filler ratios have been deposited in multiple nanocomposites films containing Cu/(PbS + 20wt.%Al) binary compounds with various x, y and z fillers. In XRD analysis, the films containing Cu/(PbS + 20wt.%Al) binary compounds exhibit cubic nanostructures. Multiple nanocomposites film as-prepared Cu/(PbS + 20wt.%Al) was recorded with its lowest refractive index. A drastic increase in photon energy ranges is observed in optical conductivity when photon energy increases; this corresponds to the optical energy gap region and the optical absorption region. Several nanocomposites' films prepared from Cu/(PbS + 20wt.%Al) showed excellent optical characteristics, a low dielectric constant and refractive index, enabling them to be used in both infrared and visible light spectrums.

7.5.2 Trends in Multiple NPs on Thin-Film Solar Cells

Thin-film solar cell research has been one of the most important topics in the last decade, with efforts being made to further reduce the thickness of the absorbers, to reduce the cost and time of processing, and to increase the voltage of the cells.

The quantum confinement effect in multiple nanocomposite films with different volume fractions of Cu:PbS, Al:PbS, and Cu/20 wt.%Al:PbS can be induced by the grain size of the polycrystalline films ranging from 18.18 to 36.06 nm, which increases the band gap. Enlarging the electron transition from the valence band into the conduction band enhances the electrical properties of solar cells. When the band gap increases, the solar cell efficiency will typically increase [5, 48].

A new group of nanocomposites films with the characteristics of Cu:PbS, Al:PbS, and Cu/20wt.% Al:PbS at different volume fractions was identified as p-type materials. They have high absorption coefficients and band gaps in the vicinity of the 1.5 eV optimum band gap of the absorbent layer in thin-film solar cells. This change in E_g from 0.41 eV (bulk PbS) to 1.55–2.83eV had positive effects on solar cells, and this result was in good agreement with optimum theoretical band gaps for good solar energy conversion [6].

A new group of Cu:PbS, Al:PbS and Cu/20wt.%Al:PbS multiple nanocomposites films have been developed with concentration. In recent studies, CdS/PbS thin film solar cells, increasing the thickness of the absorbing layer of CdS/PbS up to 0.5 mm, resulted in an increase of energy conversion efficiency from 2.16% to 4.13% at thickness of 2 μm [49]. Adding nanostructures of PbS, Cu:PbS, Al:PbS, and Cu/20wt.% Al:PbS to solar cells increased the short-circuit current by enhancing the efficiency of thin-film cells. Since the absorption edge of the absorbing layer is shifted into the visible region by metal NPs due to their high absorption coefficients, there is a short-circuit current in thin film cells. The obtained films improve the transport of charge carriers in PbS semiconductors. Creating effective short-circuit currents in solar cells based on charge carrier separation and absorbance can improve cell conversion efficiencies by improving the efficiency of the short circuit currents [50].

Researchers proposed using selected metal NPs (Al, Cu, Li, Ag) to increase the absorption coefficient and energy band gap of a PbS absorbing layer. Increasing the energy band gap to near the optimum value decreased both the space charge region and the reverse saturation current. Decreasing the absorbing layer thickness in CdS/PbS thin-film solar cells increased the electron–hole generation rate and recombination rate, increasing the performance and energy conversion efficiency and saving material [5–9, 18, 23, 25, 50, 51].

7.5.3 Trends in PbS Multiple-Nanocomposite Films

Several nanocomposites containing a mixture of Cu/(PbS + 20wt.%) fillers exhibited transmittance and reflectance below 45% as measured based on the ratio of the fillers in the composition; a thin film with high reflectance and transmittance is highly absorbent. Using nanocomposites in NP cell layers with low transmittances and low reflectance resulted in improved radiation absorption, and devices with low reflectance and transmittances below 45% had the same characteristics [31].

7.5.4 Trends in QD PbS Film on PV Power Stations

IR detection and NP cells can be achieved with Al/PbS and Cu/PbS thin films, which are effective materials for absorbing IR and visible light; the synthesized films were characterized by a distinct sharp, narrow structure. The films were effective radiation absorbers and can be used as layers in optoelectronic devices due to their high optical absorbance and band gap energy.

The Al/PbS nanocomposite film with 15wt.% Al/PbS displayed the highest band gap energy and the lowest refractive index and extinction coefficient at 2.3 eV. The type and ratio of nanofillers affected the PV nanocomposite thin films; in particular, Al/PbS and Cu/PbS films had controlled optical properties according to the metal content and type. As optical components and devices, NP thin films from nanocomposites can be tailored and modeled using these experimental results. PbS nanocomposites with increased Al and Cu filler ratios have a lower refractive index, which makes them suitable as absorbing layers for solar cells. Due to the good optical conductivity, absorbance, low reflectivity, and low refractive index of Al/PbS and Cu/PbS nanocomposites, they are suitable for use as absorbing layers in NP cells.

As a good absorbent of both visible and IR light, QDPbS thin films are highly suited to IR detection. The crystallinity, crystal size, morphology, and optical properties of QDPbS thin films were studied using thermal evaporation. A crystal structure of excellent quality was observed in the obtained films, with sharp, narrow, and distinct peaks; these films were effective radiation absorbers with energy band gaps ranging from 350 to 900 nm in the visible range. QD PbS is ideal for use as an absorbing layer in NP cells because of its excellent optical conductivity, absorbance, and low reflectivity. Utilizing QD absorbing layers and window layers with radii 1 nm instead of 1.5 nm, 2 nm and 3 nm reduced the numbers of modules and strings.

REFERENCES

1. Dhathathreyan KS, Rajalakshmi N, Balaji R, *Nanomaterials for Fuel Cell Technology* (Chapter 24; B. Raj, M. Van de Voorde, Y. Mahajan, Eds.), Wiley-VCH Verlag GmbH & Co., KGaA, 2017, Print ISBN:9783527340149 | Online ISBN:9783527696109.
2. Chen A, Holt-Hindle P, "Platinum Based Nanostructured Materials: Synthesis, Properties, and Applications", *Chemical Reviews*, 110(6), 3767–3804, 2010.
3. Long NV, Yang Y, Thi CM, Minh NV, Cao Y, Nogami M, "The Development of Mixture, Alloy, and Core-Shell Nano Catalysts with Nanomaterials Up Ports for Energy Conversion in Low-Temperature Fuel Cells", *NanoEnergy*, 2(5), 636–676, 2013.
4. Guo S, Li D, Zhu H, Zhang S, Markovic NM, Stamenkovic VR, Sun S, "Fe Pt and Co Pt Nanowires as Efficient Catalysts for the Oxygen Reduction Reaction", *Angewandte Chemie International Edition*, 52, 3465–3468, 2013.
5. Viet Long N, Thi CM, Yong Y, Nogami M, Ohtaki M, "Platinum and Palladium Nano-Structured Catalysts for Polymer Electrolyte Fuel Cells and Direct Methanol Fuel Cells", *Journal of Nanoscience and Nanotechnology*, 13, 4799–4824, 2013.
6. Gopi S, Amalraj A, Thomas S, "Effective Drug Delivery System of Biopolymers Based on Nanomaterials and Hydrogels—a Review", *Drug Design*, 5(129), 2169, 2016.
7. Arroyo CR, et al., "Reliable Tools for Quantifying the Morphological Properties at the Nanoscale", *Biology and Medicine,* 8(3), 1, 2016.
8. Jibowu T, "The Formation of Doxorubicin Loaded Targeted Nanoparticles Using Nanoprecipitation, Double Emulsion and Single Emulsion for Cancer Treatment", *Journal of Nanomedicine & Nanotechnology,* 7(379), 2, 2016.
9. Sreelakshmy V, Deepa MK, Mridula P, "Green Synthesis of Silver Nanoparticles From Glycyrrhiza Glabra Root Extract for the Treatment of Gastric Ulcer", *Journal of Developing Drugs,* 5(152), 2, 2016.
10. Ijaz I, Gilani E, Nazir A, Bukhari A, "Detail Review on Chemical, Physical and Green Synthesis, Classification, Characterizations and Applications of Nanoparticles", *Green Chemistry Letters and Reviews*, 13(3), 59–81, 2020.

11. Muhammed KA, Kannan CR, Stalin B, "Performance Analysis of Wind Turbine Blade Materials Using Nanocomposites", *Proceedings*, 33(Part 7), 4353–4361, 2020.
12. Yadollahi M, "Effect of Nanofillers on Adhesion Strength of Steel Joints Bonded with Acrylic Adhesives", *Science and Technology of Welding and Joining,* 20, 443–450, 2015.
13. Boncel S, Kolanowska A, Kuziel AW, Krzyżewska I, "Carbon Nanotube Wind Turbine Blades—How Far Are We Today from Laboratory Tests to Industrial Implementation?", *ACS Applied Nano Materials,* 1(12), 6542–6555, 2018.
14. Hasan S, "A Review on Nanoparticles: Their Synthesis and Types", *Research Journal of Recent Sciences,* 2277, 2502, 2015.
15. Khan Y, et al., "Classification, Synthetic, and Characterization Approaches to Nanoparticles, and Their Applications in Various Fields of Nanotechnology: A Review", *MDPI, Catalysts*, 12, 1386, 2022.
16. Abruna HD, "Energy in the Age of Sustainability", *Journal of Chemical Education,* 90, 1411–1413, 2013.
17. Press RJ, Santhanam KSV, Miri MJ, Bailey AV, Takacs GA, *Introduction to Hydrogen Technology*, John Wiley & Sons, Hoboken, NJ, 2008.
18. Yazici MS, "Hydrogen and Fuel Cell Activities at UNIDO-ICHET", *International Journal of Hydrogen Energy,* 35, 2754–2761, 2010.
19. Bailey A, Andrews L, Khot A, Rubin L, Young J, Allston TD, Takacs GA, "Hydrogen Storage Experiments for an Undergraduate Laboratory Course Clean Energy: Hydrogen/Fuel Cells", *Journal of Chemical Education,* 92(4), 688–692, 2015.
20. Koester S, Lolsberg J, Lutz L, Marten D, Wessling M, "On Individual Resistances of Selective Skin, Porous Support and Diffusion Boundary Layer in Water Vapor Permeation", *Journal of Membrane Science,* 507, 179, 2016.
21. Cahalan T, Rehfeldt S, Bauer M, Becker M, Klein H, 'Experimental Set-Up for Analysis of Membranes Used in External Membrane Humidification of PEM Fuel Cells", *International Journal of Hydrogen Energy*, 41(31), 13666–13677, 2016.
22. Yoshida T, Kojima K, Toyota MIRAI Fuel Cell Vehicle and Progress Toward a Future Hydrogen Society, *Interface*, 24, 45–49, 2015.
23. Kodama K, Nagai T, Kuwaki A, Jinnouchi R, Morimoto Y, "Challenges in Applying Highly Active Pt-Based Nanostructured Catalysts for Oxygen Reduction Reactions to Fuel Cell Vehicles", *Nature Nanotechnology*, 16, 140–147, February 2021.
24. Mola G, Mbuyise X, Oseni O, Dlamini M, Tonui P, Arbab E, Kaviyarasu K, Maaza M, "Nano Composite for Solar Energy Application", *Nano Hybrids and Composites*, 20, 90–107, 2018.
25. Rasukkannu M, Velauthapillai D, Ponniah V, "A Promising High-Efficiency Photovoltaic Alternative Non-Silicon Material: A First-Principle Investigation", *Scripta Materialia*, 156, 134–137, 2018.
26. Blachowicz T, Ehrmann A, "Recent Developments of Solar Cells from PbS Colloidal Quantum Dots", *Applied Science*, 16, 2020.
27. Rasukkannu M, Velauthapillai D, Bianchini F, Vajeeston P, "Properties of Novel Non-Silicon Materials for Photovoltaic Applications: A First-Principle Insight", *Materials Journal*, 17, 2018.
28. Rex Rosario S, Kulandaisamy I, Deva K, Arulanantham AMS, Valanarasu S, Youssef MA, Awwad S, "Deposition of p-Type Al Doped PbS Thin Films for Heterostructure Solar Cell Device Using Feasible Nebulizer Spray Pyrolysis Technique", *Physica B: Physics of Condensed Matter*, 8, 2019.
29. Touati B, Gassoumi A, Kamoun N, "Structural, Optical and Electrical Properties of Ag Doped PbS Thin Films: Role of Ag Concentration", *Journal of Materials Science: Materials in Electronics*, 28, 18387–18395, 2017.

30. Shkir M, Yahia I, AlFaify S, "A Facilely One Pot Low Temperature Synthesis of Novel Pt Doped PbS Nanopowders and Their Characterizations for Optoelectronic Applications", *Journal of Molecular Structure*, 1192, 68–75, 2019.
31. Sagadevan S, Pal K, Hoque E, Zaman Z, "A Chemical Synthesized Al-Doped PbS Nanoparticles Hybrid Composite for Optical and Electrical Response", *Journal Material Science: Material Electron*, 28(15), 10902–10908, 2017.
32. Soetedjoa H, Siswanto B, Aziz I, Sudjatmoko S "Deposition of Cu-Doped PbS Thin Films with Low Resistivity Using DC Sputtering", *Results in Physics*, 8, 903–907, 2018.
33. Ravishankar S, Balu AR, Balamurugan S, et al., "TG–DTA analysis, structural, optical and magnetic properties of PbS thin films doped with Co2+ ions", *Journal of Mateialr Science: Materials in Electronics*, 29, 6051–6058, 2018.
34. Mola G, Mbuyise XG, Oseni O, Dlamini M, Tonui P, Arbab EAA, Kaviyarasu K, Maaza M, "Nano Composite for Solar Energy Application", *Nano Hybrids and Composites*, 20, 90–107, 2018.
35. Kumar S, Jeevanandam P, "Synthesis of PbS–A12O3 Nanocomposites by sol-gel Process and Studies on Their Optical Properties", *Optical Materials*, 46, 209–215, 2015.
36. Saeed Hassanien A, El Radaf IM, Akl AA, "Physical and Optical Studies of the Novel Non-Crystalline CuxGe20-xSe40Te40 Bulk Glasses and Thin Films", *Journal of Alloys and Compounds*, 17, 2020.
37. El Radaf IM, Al-Zahrani HYS, Hassanien AS, "Novel Synthesis, Structural, Linear and Nonlinear Optical Properties of pType Kesterite Nanosized Cu2MnGeS4 Thin Films", *Journal of Materials Science: Materials in Electronics*, 31, 8336–8348, 2020.
38. Ahmed Aklab A, El Radafcd M, Hassanien AS, "Intensive Comparative Study Using X-Ray Diffraction for Investigating Microstructural Parameters and Crystal Defects of the Novel Nanostructural ZnGa2S4 Thin Films", *Superlattices and Microstructure Journal*, 143, 106544, 2020.
39. Shkir M, Chandekar VK, Alshahrani T, Kumar A, AlFaify S, "A Novel Terbium Doping Effect on Physical Properties of Lead Sulfide Nanostructures: A Facile Synthesis and Characterization", *Journal of Material Research*, 12, 2020.
40. Kar A, Sain S, Rossouw D, Knappett R, Kumer S, Wheatley A, "Facile Synthesis of SnO2–PbS Nanocomposites with Controlled Structure for Applications in Photocatalysis", *The Royal Society of Chemistry Nanoscale*, 8, 2727–2739, 2016.
41. Kumar Yadav S, Jeevanandam P, "Synthesis of PbS—A12O3 Nanocomposites by solgel Process and Studies on Their Optical Properties", *Optical Materials*, 7, 2015.
42. Chaudhuri TK, Kothari AJ, Tiwari D, Ray A, "Photoconducting Nanocomposites Films of PbS Nanocrystals in Insulating Polystyrene", *Physica Status Solidi A,* 210, 356–360, 2013.
43. Ebnalwaled AA, Essai MH, Hasaneen BM Mansour HE, "Facile and Surfactant-Free Hydrothermal Synthesis of PbS Nanoparticles: The Role of Hydrothermal Reaction Time", *Journal of Materials Science: Materials in Electronics,* 28, 1958–1965, 2017.
44. Rohom AB, Londhe PU, Jadhav PR, Bhand R, Chaure NB, "Studies on Chemically Synthesized PbS Thin Films for IR Detector Application", *Journal of Materials Science: Materials in Electronics*, 28(8), 17107–17113, 2017.
45. Kord M, Hedayati K, Farhadi M, "Green Synthesis and Characterization of Flower Like PbS and Metal-Doped Nanostructures via Hydrothermal Method", *Main Group Metal Chemistry,* 40, 35–40, 2017.
46. Thabet A, Abdelhady S, Ebnalwaled AA, Ibrahim AA, "Innovative Industrial Cu(In,Ga) Se2 Thin Film Solar Cell with High Characterization Using Nanoparticles Structure", *Indonesian Journal of Electrical Engineering and Informatics (IJEEI)*, 7(2), 382–392, June 2019.

47. Thabet A, Abdelhady S, Ebnalwaled AA, Ibrahim AA, "Improvement Optical and Electrical Characteristics of Thin Film Solar Cells Using Nanotechnology Techniques", *International Journal of Electronics and Telecommunications (IJET)*, 65(4), 625–634, October 2019.
48. Thabet A, Mobarak Y, Salem N, El-Noby A, "Performance Comparison of Selection Nanoparticles for Insulation of Three Core Belted Power Cables", *International Journal of Electrical and Computer Engineering (IJECE)*, 10(3), 2779–2786, 2020.
49. Thabet A, Abdelhady S, Mobarak Y, "Design Modern Structure for Heterojunction Quantum Dot Solar Cells", *International Journal of Electrical and Computer Engineering (IJECE)*, 10(3), 2915–2922, 2020.
50. Thabet A, Abdelhady S, Mobarak Y, "Ultra-Optical Characterization of Thin Film Solar Cells Materials Using Core/Shell Absorber Layer", *International Journal of Electrical and Computer Engineering (IJECE)*, 12(2), 1377–1384, 2022.
51. Thabet A, Abdelhady S, "Strategy Trends of Core/Multiple Shell for Quantum Dot-Based Heterojunction Thin Film Solar Cells", *Australian Journal of Electrical and Electronics Engineering*, 19(3), 203–218, 2022.
52. Li D, et al., "A Review of Damage Detection Methods for Wind Turbine Blades", *Smart Materials and Structures*, 24(3), 033001, 2015.
53. Fiore G, Selig M, *Simulation of Damage Progression on Wind Turbine Blades*, American Institute of Aeronautics and Astronautics, Reston, 2016.
54. Dashtkar A, Hadavinia H, Sahinkaya MN, Williams NA, Vahid S, Ismail F, Turner M, "Rain Erosion-Resistant Coatings for Wind Turbine Blades: A Review", *Polymers and Polymer Composites*, 1–33, 2019.
55. Chen J, Luo Z, An R, Marklund P, Bjorling M, Shi Y, "Novel Intrinsic Self-Healing Poly-Silicone-Urea with Super-Low Ice Adhesion Strength", *Small*, 18, 2200532, 2022.
56. Liu Z, Feng F, Li Y, Sun Y, Tagawa K, "A Corncob Biochar-Based Superhydrophobic Photothermal Coating with Micro-Nanoporous Rough-Structure for Ice-Phobic Properties", *Surface and Coatings Technology*, 457, 129299, 2023.
57. Liao C, Yi Y, Chen T, Cai C, Deng Z, Song X, Lv C, "Detecting Broken Strands in Transmission Lines Based on Pulsed Eddy Current", *Metals*, 12, 1014, 2022.
58. Mu Z, Li Y, Guo W, Shen H, Tagawa K, "An Experimental Study on Adhesion Strength of Offshore Atmospheric Icing on a Wind Turbine Blade Airfoil", *Coatings*, 13, 164, 2023.
59. Li L, Liu Y, Tian L, Hu H, Hu H, Liu X, Hogate I, Kohli A, "An Experimental Study on a Hot-Air-Based Anti-/De-Icing System for Aero-Engine Inlet Guide Vanes", *Applied Thermal Engineering*, 167, 114778, 2020.
60. Li Y, Shen H, Guo W, "Effect of Ultrasonic Vibration on the Surface Adhesive Characteristic of Iced Aluminum Alloy Plate", *Applied Sciences,* 12, 2357, 2022.
61. He Z, Xie H, Jamil MI, Li T, Zhang Q, "Electro-/Photo-Thermal Promoted Anti-Icing Materials: A New Strategy Combined with Passive Anti-Icing and Active De-Icing", *Advanced Materials Interfaces*, 9, 2200275, 2022.
62. Liu Y, Zhao Z, Shao Y, Wang Y, Liu B, "Preparation of a Superhydrophobic Coating Based on Polysiloxane Modified SiO2 and Study on Its Anti-Icing Performance", *Surface and Coatings Technology*, 437, 128359, 2022.
63. Rivero PJ, et al., "Evaluation of Functionalized Coatings for the Prevention of Ice Accretion by Using Icing Wind Tunnel Tests", *Coatings*, 10, 636, 2020.
64. Liu B, Liu Z, Li Y, Feng F, "Wind Tunnel Test of the Anti-Icing Properties of MoS2/ZnO Hydrophobic Nano-Coatings for Wind Turbine Blades", *MDPI, Coatings*, 13, 686, 2023.
65. Baumann T, Hillmer T, Klee P, Stoll D, *Optimized, Emission Free Production Process for Main Wall Insulation with Enhanced Lifetime Properties*, EIC, Annapolis, USA, 2011.

66. Brockschmidt M, Pohlmann F, Kempen S, Groeppel P, *Testing of Nano-Insulation Materials*, EIC, Annapolis, 2011.
67. Hoffmann C, Brockschmidt M, Peier D, *Unter-suchungen zur TE-Resistenz von EP-Harzen mit nanoskaligen SiO2-Füllstoffen*, ETG-Fachtagung "Grenzflachen in Isoliersystemen", Wurzburg, 2008.
68. Weidner JR, Pohlmann F, Groppel P, Hildinger T, *Nanotechnology in High Voltage Insulation Systems for Turbine Generators—First Results*, International Symposium on High Voltage Engineering, Hannover, Germany, August 22–26, 2011.
69. Toops TJ, et al., "Evaluation of Nitrided Titanium Separator Plates for Proton Exchange Membrane Electrolyzer Cells", *Journal of Power Sources*, 272, 954–960, 2014.
70. Mo J, Steen SM, Zhang FY, Toops TJ, Brady MP, Green JB, "Electrochemical Investigation of Stainless-Steel Corrosion in a Proton Exchange Membrane Electrolyzer Cell", *International Journal of Hydrogen Energy*, 40, 12506–12511, 2015.
71. Mo J, Steen SM, Retterer S, Cullen DA, Terekhov A, Zhang FY, "Mask-Patterned Wet Etching of Thin Titanium Liquid/Gas Diffusion Layers for a PEMEC", *ECS Transactions*, 66, 3–10, 2015.
72. Mo J, Kang Z, Yang G, Retterer ST, Cullen DA, Toops TJ, Green JB Jr., Zhang FY, "Thin Liquid/Gas Diffusion Layers for High-Efficiency Hydrogen Production from Water Splitting", *Applied Energy*, 177, 817–822, 2016.
73. Girishkumar G, Rettker M, Underhile R, Binz D, Vinodgopal K, McGinn P, Kamat P, "Single-Wall Carbon Nanotube-Based Proton Exchange Membrane Assembly for Hydrogen Fuel Cells", *Langmuir*, 21, 8487–8494, 2005.
74. Serp P, Corrias M, Kalck P, "Carbon Nanotubes and Nanofibers in Catalysis", *Applied Catalysis A*, 253, 337–358, 2003.
75. Li WZ, Liang CH, Zhou WJ, Qiu JS, Zhou ZH, Sun GQ, Xin Q, "Preparation and Characterization of Multiwalled Carbon Nanotube-Supported Platinum for Cathode Catalysts of Direct Methanol Fuel Cells", *The Journal of Physical Chemistry B*, 107, 6292–6299, 2003.
76. Liu ZL, Lin XH, Lee JY, Zhang W, Han M, Gan LM, *Langmuir*, "Preparation and Characterization of Platinum-Based Electrocatalysts on Multiwalled Carbon Nanotubes for Proton Exchange Membrane Fuel Cells", *Langmuir,* 18, 4054–4060, 2002.
77. Steigerwalt ES, Deluga GA, Lukehart CM, "Pt–Ru/Carbon Fiber Nanocomposites: Synthesis, Characterization, and Performance as Anode Catalysts of Direct Methanol Fuel Cells. A Search for Exceptional Performance", *The Journal of Physical Chemistry B*, 106, 760–766, 2002.
78. Sun X, Li R, Stansfield B, Dodelet JP, Desilets S, "3D Carbon Nanotube Network Based on a Hierarchical Structure Grown on Carbon Paper Backing", *Chemical Physics Letters*, 394, 266–270, 2004.
79. Sun X, Li R, Villers D, Dodelet JP, Desilets S, "Composite Electrodes Made of Pt Nanoparticles Deposited on Carbon Nanotubes Grown on Fuel Cell Backings", *Chemical Physics Letters*, 379, 99–104, 2003.
80. Sun X, Stansfield B, Dodelet JP, Desilets S, "Growth of Carbon Nanotubes on Carbon Paper by Ohmically Heating Silane-Dispersed Catalytic Sites", *Chemical Physics Letters*, 363, 415–421, 2002.
81. Choi WB, Chu JU, Pak CH, Chang H, "Method of Fabrication of Carbon Nanotubes for Fuel Cells", in *U.S. Patent Application Published* (Vol. 1), Samsung SDI Co., Ltd., South Korea, p. A1, 2004.
82. Mehrpooya M, Valizadeh F, Askarimoghadam R, Sadeghi S, Pourfayaz F, Ali Mousavi S, "Fabrication of Nano-Platinum Alloy Electrocatalysts and Their Performance in a Micro-Direct Methanol Fuel Cell", *The European Physical Journal Plus*, 135, 589, 2020.

83. Cotton FA, Wilkinson G, Murillo CA, Bochmann M, *Advanced Inorganic Chemistry*, John Wiley & Sons, pp. 53–65, 1999.
84. John C, Russ S, *Fundamentals of Energy Dispersive X-Ray Analysis*, Springer, pp. 69–75, 2013.
85. Buyuknalcaci FN, Polat Y, Negawo TA, Doner E, Alam MS, Hamouda T, Kilic A, "Carbon Nanotube-Based Nanocomposites for Wind Turbine Applications", *Polymer-Based Nanocomposites for Energy and Environmental Applications*, 635–661, 2018.
86. Wei KX, Yang Y, Zuo HY, Zhong DQ, "A Review on Ice Detection Technology and Ice Elimination Technology for Wind Turbine", *Wind Energy*, 23, 433–457, 2020.
87. Li C, Li XH, Tao C, Ren LX, Zhao YH, Bai S, Yuan X, "Amphiphilic Antifogging/Anti-Icing Coatings Containing POSS-PDMAEMA-b-PSBMA", *ACS Applied Materials & Interfaces*, 9, 22959–22969, 2017.
88. Nguyen-Tri P, Tran HN, Plamondon CO, Tuduri L, Vo DVN, Nanda S, Mishra A, Chao HP, Bajpai AK, Recent Progress in the Preparation, Properties and Applications of Superhydrophobic Nano-Based Coatings and Surfaces: A Review", *Progress in Organic Coatings*, 132, 235–256, 2019.
89. Tong G, Li Y, Tagawa K, Feng F, "Effects of Blade Airfoil Chord Length and Rotor Diameter on Aerodynamic Performance of Straight-Bladed Vertical Axis Wind Turbines by Numerical Simulation", *Energy*, 265, 126325, 2023.
90. Feng F, Wang R, Yuan W, Li Y, "Study on Anti-Icing Performance of Biogas-Residue Nano-Carbon Coating for Wind-Turbine Blade", *Coatings*, 13, 814, 2023.
91. Steele BCH, Heinzel A, "Materials for Fuel-Cell Technologies", *Nature*, 414, 345–352, 2001.
92. Yamamoto O, "Solid Oxide Fuel Cells: Fundamental Aspects and Prospects", *Electrochimica Acta*, 45, 2423–2435, 2000.
93. Minh NQ, "Solid Oxide Fuel Cell Technology, Features and Applications", *Solid State Ionics*, 174, 271–277, 2004.
94. Deckman HW, Dunsmuir JH, "Natural Lithography", *Applied Physics Letters*, 41, 377–379, 1982.
95. Haynes CL, Van Duyne RP, "Nanosphere Lithography: A Versatile Nanofabrication Tool for Studies of Size-Dependent Nanoparticle Optics", *The Journal of Physical Chemistry B*, 105, 5599–5611, 2001.
96. Hsu CM, Connor ST, Tang MX, Cui Y, "Wafer-Scale Silicon Nanopillars and Nanocones by Langmuir_Blodgett Assembly and Etching", *Applied Physics Letters*, 93, 133109–133111, 2008.
97. Ritala M, Leskela M, "Atomic Layer Epitaxy—a Valuable Tool for Nanotechnology?", *Nanotechnology*, 10, 19–24, 1999.
98. George SM, "Atomic Layer Deposition: An Overview", *Chemical Reviews*, 110, 111–131, 2010.
99. Chao CC, Hsu CM, Cui Y, Prinz FB, "Improved Solid Oxide Fuel Cell Performance with Nanostructured Electrolytes", *ACS NANO Journal*, 5(7), 5692–5696, 2011.
100. Kodama K, Nagai T, Kuwaki A, Jinnouchi R, Morimoto Y, "Challenges in Applying Highly Active Pt-Based Nanostructured Catalysts for Oxygen Reduction Reactions to Fuel Cell Vehicles", *Nature Nanotechnology*, 16, 140–147, February 2021.
101. Premkumar M, Sowmya R, "An Effective Maximum Power Point Tracker for Partially Shaded Solar Photovoltaic Systems", *Energy Reports*, 5, 1445–1462, 2019.
102. Srimath A, Innocenzo D, Petrozza G, Enrico D, Filippo D, Henry S, Alison W, "Chapter 4. Photo Physics of Hybrid Perovskites Unconventional Thin Film Photovoltaics", *The Royal Society of Chemistry*, 107–140, 2016.
103. Wurfel U, Cueva A, Wurfel P, "Charge Carrier Separation in Solar Cells," *IEEE Journal of Photovoltaics*, 5(1), 461–469, 2014.

104. Abdel Rafea M, Roushdy N, "Study of Optical Properties of Nanostructured PbS Films", *Philosophical Magazine Letters*, 90(2), 113–120, 2010.
105. Talapin DV, Murray CB, "PbSe Nanocrystal Solids for n- and p- Channel Thin Film Field-Effect Transistors", *Science*, 310, 86–89, 2005.
106. Kowshik M, Vogel W, Urban J, Kulkarni SK, Paknikar KM, "Microbial Synthesis of Semiconductor PbS Nanocrystallites", *Advanced Materials*, 14, 815–818, 2002.
107. Kane RS, Cohen RE, Silbey R, "Theoretical Study of the Electronic Structure of PbS Nanoclusters", *Journal of Physical Chemistry*, 100, 7928–7932, 1996.
108. Pan AC, Canizo C, Canovas E, Santos NM, Leitao JP, Luque A, "Enhancement of Up-Conversion Efficiency by Combining Rare Earth-Doped Phosphors with PbS Quantum Dots", *Solar Energy Materials and Solar Cells*, 94, 1923–1926, 2010.
109. Choudhury N, Sarma BK, "Structural Analysis of Chemically Deposited Nanocrystalline PbS Films", *Thin Solid Films*, 519, 2132–2134, 2011.
110. Jiao Y, Gao X, Lu J, Chen Y, Zhou J, Li X, "A Novel Method for PbS Quantum Dot Synthesis", *Materials Letters*, 72, 116–118, 2012.
111. Emin S, Loukanov A, Wakasa M, "Photostability of Water-Dispersible CdTe Quantum Dots: Capping Ligands and Oxygen", *Chemistry Letters*, 654–656, 2010.
112. Olson J, Rodriguez Y, Yang L, Alers GB, Carter SA, "CdTe Schottky Diodes from Colloidal Nanocrystals", *Applied Physics Letters*, 96(24), 2010.
113. Debnath R, Greiner M, Kramer I, Fischer A, Tang J, Barkhouse D, Wang X, Levina L, Lu Z, Sargent EH, "Depleted-Heterojunction Colloidal Quantum Dot Photovoltaics Employing Low-Cost Electrical Contacts", *Applied Physics Letters*, 97(2), 2010.
114. Liu CP, Wang HE, Ng TW, Chen ZH, Zhang WF, Yan C, Tang YB, "Hybrid Photovoltaic Cells Based on ZnO/Sb2S3/P3HT Heterojunctions", *Physica Status Solidi*, 627–633, 2012.
115. Chang JA, Rhee JH, Lee YH, Kim H, Seok S, Nazeeruddin M, Gratzel M, "High-Performance Nanostructured Inorganic–Organic Heterojunction Solar Cells", *Nano Letters*, 10(7), 2609–2612, 2010.
116. Huang X, Huang S, Zhang Q, Guo X, Li D, Luo Y, Shen Q, Toyodo T, Meng Q, "A Flexible Photoelectrode for CdS/CdSe Quantum Dot Sensitized Solar Cells (QDSSCs)", *Chemical Communications*, 2664–2666, 2011.
117. Kamat PV, "Boosting the Efficiency of Quantum Dot Sensitized Solar Cells Through Modulation of Interfacial Charge Transfer", *Accounts of Chemical Research*, 1906–1915, 2012.
118. Lee H, Leventis H, Moon S, Chen P, Haque S, Zakeeruddin S, Nazeeruddin M, "PbS and CdS Quantum Dot-Sensitized Solid-State Solar Cells: 'Old Concepts, New Results'", *Advanced Functional Materials*, 2735–2742, 2009.
119. Kramer I, Levina L, Debnath R, Zhitomirsky D, Sargent EH, "Solar Cells Using Quantum Funnels", *Nano Letters*, 3701–3706, 2011.
120. Perera S, et al., "Hydrothermal Synthesis of Graphene-TiO2 Nanotube Composites with Enhanced Photocatalytic Activity", *ACS Catalysis*, 2, 2012.
121. Mathewsa NR, Angeles-Chavez C, Cortes-Jacome MA, Toledo Antonio JA, "Physical Properties of Pulse Electrodeposited Lead Sulfide Thin Films", *Electrochimica Acta*, 99, 76–84, 2013.
122. Atwa DMM, Azzouz IM, Badr Y, "Studies on Chemically Synthesized PbS Thin Films for IR Detector Application", *Journal of Applied Physics B*, 103, 161–192, 2011.
123. Veena E, Bangera KV, Shivakumar GK, "Studieson Chemically Synthesized PbS Thin Films for IR Detector Application", *Applied Physics A*, 123, 366–378, 2017.
124. Mostovoy AS, Yakovlev AV, Tseluikin VN, Lopukhova M, "Epoxy Nanocomposites Reinforced with Functionalized Carbon Nanotubes", *Polymers*, 12, 1816, 2020.
125. Roy S, Petrova RS, Mitra S, "Effect of Carbon Nanotube (CNT) Functionalization in Epoxy-CNT Composites", *Nanotechnology Reviews*, 7, 475–485, 2018.

126. Beaumont PWR, Zweben CH, Gdutos E, Talreja R, Poursartip A, Clyne T, Ruggles-Wrenn MB, *Comprehensive Composite Materials* (2nd edition), Elsevier, Amsterdam, The Netherlands, 2018.
127. Elhenawy Y, Fouad Y, Marouani H, Bassyouni M, "Performance Analysis of Reinforced Epoxy Functionalized Carbon Nanotubes Composites for Vertical AxisWind Turbine Blade", *Polymers*, 13, 422, 2021.

8 Nanotech Modules for Green Energy Power Plants

Nanotechnology has enabled great advances in the field of hydrogen generation, photovoltaics, and turbines using organic and inorganic materials. Nanocarrier devices deliver materials safely to prevent severe defects caused by hollow tubes with spheres in nanowires, which are addressed by mechanical sections. As industry has progressed, carbon NPs have gained in demand as MWCNTs in addition to semiconductors such as silicon, ceramics, and other materials that are common in the world. A NP possesses excellent conductivity, and its thermal resistance allows it to be modified via laser tuning and multitarget precursors to achieve greater control of its properties. Nanotechnology modules in energy-related applications plays a crucial role, for instance, in solar PV, fuel cell, Li-ion batteries, wind turbine blades, and other applications.

8.1 INTRODUCTION

Energy is delivered to loads through power systems such that there is no barrier of conductivity to the transmission of electricity between loads. Energy production based on nanotechnology is much more efficient than traditional methods such as laser ablation and vapor deposition. Graphene is an extremely useful material for powering energy systems due to its physical properties, such as electron mobility, high conductivity, and broad range of applications. To create a significant impact in the generation of green energy, hydrogen, solar cells, and wind energy are used across a variety of components to convert the green energy and produce electricity in the form of fuel and NP modules. Global energy demand increased exponentially over the past few decades, which caused an increase in energy consumption and generation. In response, scientists and policy makers are investigating renewable energy as a solution for replacing traditional sources of energy [1–3].

Wind turbines and PV panels were analyzed for their environmental impact, air pollution, greenhouse gas emissions, wastes produced, fossil fuel consumption, wildlife, health benefits, etc. [4]. Three main factors can be cited as the reasons for the challenge of increasing renewable energy production: a) their cost compared with the cost of conventional power generation; b) whether they are capable of being integrated into the grid, which may require new controls; and c) if an ideal storage system is not available. Meeting the growing electricity demand in the future will require green energy systems that are configured to emit few greenhouse gases, rely less on external sources, and have lower cost [5]. Energy storage systems can help reduce

DOI: 10.1201/9781003512486-10

the uncertainty or unavailability of renewable energy sources, which can help wind turbines and solar systems combine to increase the reliability of a hybrid renewable energy system [6].

There is a strong seasonal cycle in irradiance related to rising temperatures caused by the rotation of the earth, which is approximately anticorrelated with wind speed, meaning that windy days are not highly likely to be cloudier [7]. Many renewable energy technologies face land occupation issues, particularly as their deployment increases; dual use of land for wind and solar PV is expanding. Wind–solar hybrid farms can save land and lead to lower utility costs; however, shading caused by wind turbines could cause energy loss as the PV panels are not exposed to sunlight. An optimal hybrid farm will avoid wind turbine shading to reduce the average percentage of land loss for PV panels [8, 9].

Nanotechnology enhances features such as mechanical properties, cost, weight, and electrical performance. This technology is therefore needed to improve the efficiency of hydrogen, wind–solar cell hybrids, and other green energy modules. As previously described, nanotechnology is sometimes referred to as 'third-generation photovoltaics' [10] and essentially refers to components of nanoscale size that are capable of modifying the energy band gap of solar cells, thus allowing them to perform better conversion rates [11]. Solar cells are among the most widely used nanotechnology devices to produce PV cells, nanotubes, QDs, and 'hot carriers' [12].

8.2 DESIGN MODULES

8.2.1 Hydrogen Energy Modules

Self-supported tubular membrane modules must be designed based on a) chemical, mechanical, and geometry considerations; b) hydrogen mass transfer mechanisms; and c) operating parameters (pressure, temperature, flow rate, etc.). Researchers developed an assessment and optimization tool for membrane modules consisting of dense metal tubes in single- or multitube configurations to determine the relationships among these main design criteria and parameters. The design process consists of establishing the relationship between the maximum hydrogen recovery, the operative pressure, and the dilution of the feed stream. Hydrogen recovery factor h can be calculated by taking the ratio between the flow rate Qperm quenched through the membrane by hydrogen permeating through the membrane and the flow rate Q_{H2} fed by hydrogen distributed through the membrane (mol s-1) [13]:

$$\eta = \frac{Q_{perm}}{Q_{H_2,feed}} \tag{8.1}$$

Metal tubes' mechanical strength and hydrogen permeability are both affected by the temperature of the membranes. The dependence of the membrane permeability from the temperature is given by an Arrhenius expression:

$$P_e = P_{e_o} . e^{\frac{-E_a}{R.T}} \tag{8.2}$$

where E_a is the activation energy of the permeation process [J mol^{-1}], R is the gas constant [J mol^{-1} K^{-1}], and Pe0 is the preexponential factor [mol s^{-1} m^{-1} Pa$^{-0.5}$].

Once the diameter of the tubes is fixed, their length has to be chosen. This choice determines the number of tubes required, given the permeation area A [m2]. Parameter l, the specific length, is defined as the ratio between the length and the diameter:

$$\lambda = \frac{L}{D} \tag{8.3}$$

whereas the area of single tube is $A_0 = \pi DL$, and the number of tubes required α can then be obtained by the ratio of A with A_0:

$$\alpha = \frac{A}{A_O} = \frac{Q_{perm,\varnothing}}{P_e(T).\Delta P.\pi.D.\lambda} \tag{8.4}$$

It is possible that the designer needs to perform the procedure iteratively; more tubes may be required to meet the demands of the application, or the diameter may need to be changed to meet those demands. As soon as the operative pressure is doubled, the permeation area of the membrane will be doubled, resulting in a higher flux across the membrane. As long as mechanical safety is assured, the tube wall thickness is the most important geometrical parameter that affects membrane module design; number of tubes is linearly affected by tube diameter and length. The S-I cycle has gained considerable attention in recent years since it is one of the only processes that splits water thermochemically to produce hydrogen efficiently and without producing CO_2 as a byproduct. There are three chemical reactions that can be used in this thermochemical S-I process to breakdown water into its elements of hydrogen and oxygen by means of I and S compounds [14]:

$$SO_2 + I_2 + 2H_2O \rightarrow H_2SO_4 + 2HI \tag{8.5}$$

Sulfuric acid decomposition (900°C)

$$H_2SO_4 \rightarrow H_2O + SO_2 + 0.5O_2 \tag{8.6}$$

Hydrogen iodide (HI) decomposition (500°C)

$$2HI \rightarrow H_2 + I_2 \tag{8.7}$$

In thermochemical S-I, hydrogen is produced in a silica membrane reactor as part of the decomposition of HI. Although the silica membrane reactor has conversion limitations as well as the highest conversion attainable, HI decomposition membrane reactors are also the most efficient; the relationship between HI conversion and the number of silica membrane tubes can be used to help determine the optimal design of the HI decomposition membrane reactor for thermochemical S-I hydrogen productions. The fabrication of a silica membrane for membrane reactors for the decomposition of HI has been extensively investigated [14–16].

Hydrogen and solid interaction and mechanisms of hydride formation in nanomaterials change significantly because of their structural differences. NPs with composite architectures, such as core/shell or composite NPs, as well as NPs on porous or grapheme-like supports, are among these NPs. Bulk nanomaterials processed by mechanochemical routes are also a part of this category, as are thin films and multilayers. Surface, interface, and grain boundaries exhibit a shift in equilibrium pressure as a result of hydrogen-induced changes in specific free energy, as shown experimentally in multilayers and composite NPs [17, 18].

A similar destabilization of the hydride phase can be caused by compressive elastic strains of microstructure or geometrical origin. This aspect is important for applications of NPs in hydrogen sensors requiring a small pressure hysteresis and good reproducibility. The design of materials should be based on knowledge in conjunction with a careful analysis of the relationships between microstructure, properties, and mechanisms. Nanoobjects can be realized with composite architectures in which different functionalities such as catalysis, hydrogen storage, oxidation protection, and catalysis are combined at the nanoscale. These advances have not only contributed to fundamental understanding of hydrogen in metals but also led to excellent hydrogen sorption materials, which has unraveled transformation mechanisms at the nanoscale. A great deal of emphasis is placed on developing in situ microscopy and spectroscopy techniques in this context, for they offer the unique opportunity to conduct high-resolution experiments that are also relevant to other branches of materials science [17, 18]. There are many influences on phase transformation, including mechanical stresses that produce local plasticity and nanocrystallization. Additionally, Mg thin films exhibit a destabilizing stress state at ambient temperatures and cause the hydride to become unstable [19, 20].

8.2.2 Solar Energy Modules

Currently, the design of thin-film NP solar cell modules combines a core/shell absorbing layer for HJQD PV solar cells, such as CdS/CdTe, CdS/CIGS, and CdS/PbS. These modules consist of a core/shell absorbing layer made up of QD metals and a QD window layer. Traditional HJQD thin-film solar cells were enhanced with a QD absorbing layer and a bulk window layer. Sub-micro-sized absorbing layers are used to achieve high efficiency with low cost and quick turnaround time, along with requiring less material [21–23].

Researchers determined that metal-semiconductor core/shell absorbing layers improve optical characteristics, including energy band gaps and absorption, thereby improving the performance of HJQD ITO/CdS/QDPbS/Au, SnO_2/CdS/CdTe/Cu, and ZnO/CdS/CIGS/Mo thin-film solar cells. Nanostructured cells are a wonderful technology that has been developed to produce high-efficiency modules that have high performance characteristics and efficiency at high temperatures. Due to the intense interaction between metal and semiconductor, metal-semiconductor heterostructures have different properties from their individual components. A hetero-structured semiconductor can modulate surface plasmon resonances in metal; by near-field interactions, energy can flow from the excited state of semiconductor to plasmon. However, plasmon–exciton interactions can also affect semiconductor properties (enhance or quench), as well as their radiative properties. It is possible to suppress direct carrier recombination by using appropriate band structures to enhance charge separation in semiconductors transferring to metal. Thus, solar cells convert photoelectric energy more efficiently [21–23].

In an explicit model for spherical inclusions in QDs, the Maxwell Garnett expression was used to obtain the dielectric constant of the core/shell QD [24]:

$$\varepsilon_{efic-s} = \varepsilon_{shell} \frac{\left(\varepsilon_{core1} + 2\varepsilon_{shell}\right) + 2\mathcal{F}_{core1}\left(\varepsilon_{core1} - \varepsilon_{shell}\right)}{\left(\varepsilon_{core1} + 2\varepsilon_{shell}\right) - \mathcal{F}_{core1}\left(\varepsilon_{core1} - \varepsilon_{shell}\right)} \tag{8.8}$$

where $\mathcal{F}_{core1}$ the core filling factor, and

$$\mathcal{F}_{core1} = \left(\frac{R_1}{R_2}\right)^3 \tag{8.9}$$

R_1 is the radius of the QD metal core material;

R_2 is the radius of the QD semiconductor shell material, and

ε_{shell} is the dielectric constant of QD shell semiconductor layer material described by the Drude model shown in [25–28] as follows:

$$\varepsilon_{shell} = \varepsilon_{shell\infty}\left(1 - \frac{\omega_b^2}{\omega^2 + i\omega\left(\gamma_{aQD}\right)}\right) \tag{8.10}$$

where ω_b is the plasma angular frequency of semiconductor layer material and y_{aQD} is the macroscopic damping constant.

The damping constant was adjusted to the size of the particle to reflect the finite size of the NPs:

$$y_{aQD} = \gamma_{mb} + \left(\frac{3v_{fb}}{4R}\right) \tag{8.11}$$

v_{fb} is the Fermi velocity of semiconductor layer material.

The infinity dielectric constant of QD semiconductor shell layer material can be calculated as follows:

$$\varepsilon_{shell\infty QD} = \varepsilon_o\left(1 + \frac{\omega_b{}^2}{\left(E_{g(Qdots)}\right)^2}\right) \tag{8.12}$$

$\varepsilon_{shell\infty QD}$ is less than the value of bulk material because the nano-confined dimensions have a lower refractive index than incident wavelengths. Furthermore, semiconductors have lower refractive indices with rising energy band gaps. QDs have a lower infinity dielectric constant than bulk materials due to their increased energy band gap, which gives QDs an advantage for use in thin films for solar cell applications; the films also have low refractive indices, which reduces reflections and in turn increases the absorption index [29, 30].

The energy band gap $E_{g(Qdots)}$ of QD layer material can be approximated by [29–31] as follows:

$$E_{g(Qdots)} = E_{gbulk} + \frac{\pi^2 h^2}{2R^2}\left(\frac{1}{m_{eb}} + \frac{1}{m_{hb}}\right) - \frac{1 \cdot 8Q^2}{4\pi\varepsilon_b\varepsilon_o R_2} \tag{8.13}$$

where
R_2 is radius of QD layer material;
m_{eb} is the electron effective mass for layer material;
m_{hb} is the hole effective mass for layer material;
h is Plank constant;
ε_o is the permittivity of free space;
Q is elementary charge.

Varshni's relation describes the temperature dependence of the band gap in semiconductors that are used in absorber or window layers as in [32].

ε_{corel} is the dielectric constant of core metal material that can be described by the Drude model [23–35] as follows:

$$\varepsilon_{corel} = \varepsilon_{INTRi} + 1 - \frac{\omega_{pi}^2}{\omega^2 + i\omega\left(\gamma_{mi} + \left(\frac{3v_{fi}}{4R_1}\right)\right)} \tag{8.14}$$

where γ_{mi} is the macroscopic damping constant and ω_{pi}, v_{fi} are the plasma angular frequency and Fermi velocity of the core metal material.

As mentioned in [33–36], the electron–hole generation rate in the core/shell absorb layer can be calculated based on the density distribution of individual metal NPs as follows:

$$G_{ics}(\lambda) = \frac{\alpha_{bics}(\lambda)e^{-\alpha_{wiQD}(d_w)}\left[1 - R(\lambda)\right]\lambda I_o(\lambda)}{hc} \tag{8.15}$$

where λ is wavelength, $I_o(\lambda)$ is the intensity of the solar spectral, c is the speed of light, h is Plank constant, and $\alpha_{bics}(\lambda)$ is the absorption of the core/shell absorber layer. α_{wiQD} is the absorption of the QD window layer in core/shell thin-film solar cells. $\alpha_{bics}(\lambda)\, and\, \alpha_{wiQD}$ can be calculated by Beer–Lambert's law based on refractive index of layers as in [30, 34, 35]. The width of the depletion layer changes because of changing in the dielectric constant and was calculated as in [30, 34, 35] as follows:

$$W_{ics} = \sqrt{\frac{2\varepsilon_0\varepsilon_{efic-s}(V_{bi} - V)}{Q(N_a - N_d)}} \tag{8.16}$$

where $J_{diodics}(V)$ is the forward diode current for HJQD thin-film solar cell based on core/shell absorber layer and QD window layer based on [30, 33–35]:

$$J_{diodics}(V) = J_{Oics}\left(\exp\left(\frac{Q\left(V + J(V)R_{ser}\right)}{nKT}\right) - 1\right) \tag{8.17}$$

J_{Oics} is the reverse saturation current density in the new design for HJQD thin-film solar cells based on metal-semiconductor core/shell absorber layer.

$$J_{Oics} = \frac{W_{ics} Q \sqrt{N_{con} N_{val} \exp\left(-\frac{E_{gbics}}{KT}\right)}}{\sqrt{t_e t_h}} \quad (8.18)$$

where V_{bi},V are the built-in potential and applied voltage, J_{Oics} is the reverse saturation current density in HJQD cells using metal-semiconductor core/shell absorber layer. N_{con}, N_{val} are the effective state densities in the conduction and valence bands. R_{ser} is the series resistance. n is diode ideality factor. T is absolute temperature. K is Boltzmann constant. t_e . t_h are the electron and hole lifetimes.

E_{gbics} is the energy band gap of core/shell absorber layer material that has been affected by the metal NPs. The energy band gap of a semiconductor is enhanced with decreasing refractive index according to various empirical rules and expressions of refractive index. The energy band gap of metal-semiconductor core/shell absorber layer material E_{gbi} can be calculated as shown in [30, 34–37], as follows:

$$E_{gbics} = \frac{36.3}{e^{n_{fic-s}}} \quad (8.19)$$

The total photo-generated current density $J_{phics}(V)$ using a QD metal-semiconductor core/shell absorber layer and a QD window layer, which was obtained by integrating overall incident photon wavelengths of the solar spectrum [34, 35].

$$J_{phics}(V) = \int_0^{\infty} J_{Tics}(\lambda.V) d\lambda \quad (8.20)$$

where $J_{Tics}(\lambda.V)$ is the resultant photocurrent density and
$y J_{indcs}(V)$ is the net external current density of a solar cell.

Next is the QD metal-semiconductor core/shell absorber layer, the QD window layer, and the open circuit voltage equation [34, 35]:

$$J_{indcs}(V) = J_{phics}(V) - J_{dics}(V) - \left(\frac{V + J(V) R_{ser}}{R_{sh}}\right) \quad (8.21)$$

where R_{sh} is the shunt resistance.

$$V_{ocics} = \left(\frac{KT}{Q}\right) \ln\left[\frac{J_{phics}}{J_O} + 1\right] \quad (8.22)$$

The output power density of the new HJQD solar cell is based on metal-semiconductor core-shell absorber layer structure and QD window layer:

$$P_{ics} = J_{phics} \times V \quad (8.23)$$

P_{maxics} is the maximum power point density of the proposals cells which estimated from P_{ics} – V curve.

The calculation of fill factor to an excellent accuracy and the final equation for the efficiency of the new HJQD solar cell based on metal-semiconductor core-shell absorber layer structure and QD window layer are indicated in [30, 34, 35] and as follows:

$$F.F_{icore-shell} = \frac{P_{maxics}}{V_{ocics} \times J_{phics}} \quad (8.24)$$

The final equation for the energy conversion efficiency of the solar cell is in [30, 34, 35]:

$$\eta_{icore-shell} = \frac{J_{phics} V_{ocics} F.F_{icore-shell}}{P_{in}} \times 100 \quad (8.25),$$

whereas the external quantum efficiency $EQE_{icore-shell}(\lambda)$ of the new HJQD solar cell based on metal-semiconductor core-shell absorber layer and QD window layer structure has been calculated with the following equation [30, 34, 35]:

$$EQE_{icore-shell}(\lambda) = \frac{J_{phics}(\lambda)}{Q\varnothing(\lambda)} \quad (8.26)$$

where $\varnothing(\lambda)$ is the spectral photon density.

A metal-semiconductor core/shell absorb layer is employed in selected HJQD models to achieve high efficiency with low absorbing layer thickness, resulting in reduced material cost by enhancing the efficiency and performance of traditional models based on submicron absorb layer thickness. SnO_2/CdS/CdTe/Cu, ZnO/CdS/CIGS/Mo, and ITO/CdS/PbS/Al thin-film solar cells showed better high efficiency and J-V characteristics than do traditional modules [39–41]. Based on a QD window layer and two layers of core-shell absorbers, the of the modified HJQD NP appeared solar cell as shown in [18]. Tables 8.1–8.3 show the results of comparing the main parameters of the materials used in HJQD thin film models with the metallic-semiconductor core/shell absorber layers. Consequently, the usage of materials in modules, using the radius of the QD window layer of 1 nm and the radius of the QD core/shell layer, and under AM1.5 solar irradiation allows for obtaining the NP performance.

TABLE 8.1
Parameters of Materials for Use in HJQD Cell Modules Based on Core/Shell Absorbing Layer and QD Window Layer [30, 34, 35, 39–41]

Parameters	CdTe	CIGS	PbS	CdS
Energy band gap (eV) at zero Kelvin	1.6077	1.25	0.543	2.58
Band gap parameter $G(eVK^{-1})\times10^{-4}$	3.1	1.02	5	4.202
Band gap parameter β (K)	108	272	0.40	147
Electron mobility (cm^2/Vs)	320	100	1000	350
Hole mobility (cm^2/Vs)	40	25	80	50
Effective mass of electron	0.11	0.09	0.1	0.2
Effective mass of holes	0.35	0.75	0.1	0.7
Plasma angular frequency ($\omega_p\times10^{16}$ rad/s)	0.052	0.039	0.014	0.082
Damping constant ($\gamma_m\times10^{13}$ s^{-1})	8.88	19.5	1.76	17.6
Fermi velocity ($v_f\times10^6$ m/s)	0.59	0.45	8	0.89

TABLE 8.2
Parameters of HJQD Cells Based on Core/Shell Absorber Layer and QD Window Layer [30, 34, 35, 39–41]

Parameters	CdS/CdTe	CdS/CIGS	CdS/PbS
Absorber layer thickness(nm)	500	360	500
Window layer thickness(nm)	100	40	100
Front layer thickness (nm)	100	450	100
Electron lifetime (s)	1×10^{-8}	16×10^{-7}	10^{-9}
Hole lifetime (s)	5×10^{-8}	$1\cdot6\times10^{-5}$	10^{-9}
N_a-N_d the concentration of uncompensated acceptors (cm^{-3})	2.44×10^{12}	7.56×10^{13}	0.19×10^{10}
Diode quality factor	1.6	1.5	1.4
Series resistance (Ω. Cm^2)	1.08	2.5	2.1
Shunt resistance (Ω. Cm^2)	103	320	204

8.2.3 Wind Energy Modules

Several multifunctional carbon nanofiber (CNF) papers were developed and optimized to work as platforms for composite materials such as polymer composites. These composite materials were then integrated into polymer composites utilizing the VARTM process, which is compatible with the fabrication of wind turbine blades

TABLE 8.3
Characteristics of Materials Used as a Substrate Layer or QD Core Materials in QD Shell Absorbing Layer for HJQD Thin-Film Modules [34, 35]

Materials	Plasma angular frequency (ω_p 10^{16}rad/s)	Damping constant (γ_m 10^{13} s^{-1})	Fermi velocity (v_f 10^6m/s)
CdTe	0.052	8.88	—
PbS	0.014	1.76	—
Cs	0.54	0.756	0.75
Li	1.225	1.85	1.29
Cu	1.03	5.26	1.57
Ag	1.40	2.80	1.39
Al	1.09	12.4	2.03
Mo	0.19	1.13	—

(WTBs). A CNF nanocomposite was integrated with other NPs to provide effective protection for wind turbines installed in desert regions. This nanocomposite was highly resistant to impact friction and could be used to mitigate sand erosion at wind turbines. Moreover, the vibrational test results of the nanocomposite laminates indicated significant improvement in the damping ratios of the laminates. This improvement is important for the reinforcement of the blade structure, dynamic response, position control, and durability of the WTBs [42].

Among the many options being considered for improving WTBs, CNTs appear to be the most promising. CNT-based components in complex composites and hybrid materials are typically mechanically superior to other components, although they have yet to be studied systematically. The mechanical performance of CNT-based components is usually determined by matrix physiochemistry, matrix-filler interactions, filler–co-filler interactions, CNT morphology, filler density, distribution, and homogenization. A trade-off occurs when the properties of the CNT-based final material are weighed against the economy of manufacturing WTBs on an industrial scale [43].

CNT-based WTBs have a great deal of value. One of the most obvious and promising features of CNT-based WTBs is its ability to de-ice in extremely cold climates, either permanently or transiently. Ice accumulation causes issues with performance, durability, and safety issues. Conductive CNT-based composites provide synergetic electrothermal heating of WTBs, repelling water droplets, maintaining high surface temperatures at low wetting of superhydrophobic WTBs could retard icing. There is no doubt that systems based on CNTs are the closest to being implemented on a large scale. There was a greater positive effect on anti-icing performance between surfaces with high hydrophobicity and surfaces with active de-icing. Surfaces with higher hydrophobicity had lighter accumulated ice and negligible ice adhesion as well. It is not fully possible to exploit the whole potential of CNTs due to their nonuniform, viscous dispersions that formed from entangled, defective CNTs, although some of the trials resulted in successful results, such as the production of horizontally oriented CNT films of a particular aspect ratio by drawing a continuous sheet of CNTs [44].

CNT layers were applied on a nonwoven substrate; they were covered by a GF layer (positioning of the CNT layer on a nonwoven substrate was proved to be successful) and characterized by negligible weight, uniform heating, efficient energy consumption, and rapid anti-icing/de-icing [45]. This anti-icing system applied to a piece of equipment completely removed ice layers of 3–4 mm within 25 minutes at –5°C as a photothermal effect of CNTs; it was [46]. The desired properties of advanced materials for the WTB industry include being light, reducing gravitational forces, being strong enough to withstand the winds and gravitational forces of the blade, being fatigue-resistant to withstand repeated cyclic loads, and being stiff enough to make sure the blade stays in the optimal shape. A composite reinforced with nanomaterials exhibited good mechanical properties; the composites were allotropes of carbon that have a very high aspect ratio because of their nanostructures.

Nanotubes are isotropes of carbon that have carbon as its constituent element. WTBs benefit from these cylindrical carbon molecules because of their special properties. A wide range of resins is available for the reinforcement of CNTs, allowing them to exhibit different properties. As the wind turbine industry is modernizing, new innovative blade designs with advanced material systems are being developed. Changing environmental factors have led to the development of thermoplastics and natural biocomposites for use in WTBs. A CNT-reinforced composite optimized the material system's strength and density. Composite materials with CNTs can be further developed to improve their properties by hybridizing with natural fibers to provide an ecofriendly, high-strength, low-weight material for use in WTBs [47, 48].

Hybrid composite materials combine multiple individual materials into one composite that typically contains properties that are not present in a single fiber. The bonding parameters have a big effect on the hybrid composite material. A nanocomposite with high strength and light weight makes it a desirable material system for generating WTBs. Nanocomposite with high strength and light weight makes it an attractive material system for generating WTBs [49–51].

8.3 SYNTHESIS OF NANOTECH HYDROGEN AND FUEL CELLS

A proton exchange membrane fuel cell (PEMFC) is an extremely efficient, environment-friendly, and noise-free energy conversion device that generates electricity through a combination of hydrogen and oxygen. Nevertheless, PEMFCs still face numerous practical applications challenges, including insufficient power density, high cost, and poor durability. The oxygen reduction reaction (ORR) on the cathode is slow and difficult because the catalyst is not stable and does not have sufficient catalytic activity. The development of advanced Pt-based catalysts is, therefore, very important in order to be able to produce low Pt loads in PEMFCs and enable long-term operation of membrane electrode assemblies (MEAs) to enhance the performance of PEMFCs in the future. The modern PEMFC focuses on two main areas: Pt-based catalysts and the structural design of the catalyst layers that allow for efficient fuel combustion. Recent work has focused on making controllable Pt-based ORR catalysts and designing catalytic layers [52].

PEMFCs were made using Pt-based oxygen reduction catalysts, and the scientists analyzed the influence of ligand, strain, and synergistic effects on the ORR. To optimize ORR and MEA performance, the researchers examined the following features: structural design, morphology control, composition regulation, and support optimization. The influence of gradient and ordered CL structures on MEA performance correlated with the design of Pt-based catalysts, the cost and challenges associated with Pt-based catalysts, together with the future directions of improvement [52].

The Pt-based catalyst has been the focus of much research for years to improve its performance. A wide range of sizes of Pt-based nanostructures with adjustable compositions have been designed as well as support systems to optimize catalysts. For instance, research was conducted to optimize a catalyst microstructure to maximize its intrinsic activity [53, 54]. To improve the stability and activity of Pt-based catalysts, it is possible to design their structures so that they can be generated using a special surface structure, such as low-dimensional nanostructures (e.g., 2D nanoplates and 1D nanowires). In addition to having high conductivities, these structures can also ensure that the nanocrystals are in full contact with the support, which inhibits Ostwald ripening with excellent stability and effectively enhances the conductivity [55–57]. Functional nanopatterned materials can be prepared through self-assembly of block copolymers (BCs) [58–61]. The use of block-selective solvents can facilitate the self-association of di-BCs containing two incompatible blocks into micelles [62], which can then be used to generate nanopatterns consisting of uniform surface features such as spherical, cylindrical, and lamellate nanodomains [63, 64].

Developing high-density nanopatterns is highly challenging, requiring both low polymerization and high segment-segment interaction between the BC templates. To achieve the intended goal of harnessing these materials, sequentially doubling or tripling BC layers is an approach that can offer higher density. The density of a monolayer Pt-NP array is great with multiple di-BC routes based on identical self-assemble and plasma cleaning processes. Multiplication is a well-defined method of controlling spacing precisely and synthesizing hexagonal arrays that have predictable densities derived from duplication and translation. Pt utilization of monolayer arrays is not negatively affected by density multiplication; it is possible to obtain ultra-low-Pt fuel cells with better performance by optimizing the cell modules. Multiplying the densities of hexagonal arrays is based on smaller sized Pt NPs, and the latter is expected to achieve far better performance than conventional low-Pt fuel cells. NPs arrays from noble metal such as Au or Ir can also be supported by density multiplication; there are several di-BC routes in catalysis that can be used for this purpose, as well as other areas such as photoelectricity [65].

The usage of a three-component ORR/OER catalyst is introduced and illustrated using a unitized reversible fuel cell (URFC) oxygen electrode, a modular, multicomponent catalyst. A single URFC using liquid electrolyte displayed unprecedented catalytic performance. Different components were prepared and optimized separately and physically mixed during electrode preparation, depending on whether they were active for ORR or OER. Recently, a modern modular URFC catalyst, Cu-α-MnO2/XC-72R/NiFe-LDH, combined a carbon-supported, Cu-stabilized α-MnO_2.

An ORR catalyst incorporated with a NiFe-LDH OER catalyst improved activity and stability under URFC cycling compared with Platinum Group Metal reference catalysts. Optimizing the carbon and the interlayer anions in the OER component led to a further improved derivative, Cu-α-MnO2/O-MWCNTs/NiFe-LDH-Cl^-. On the rotating disk electrode scale of the URFC catalyst, this catalyst performed better than all previous materials [66]. Early in the development of the Pt-based catalyst, the research focused on the composition of the catalyst, and by adding different transition metals to the Pt-based NPs, it was hoped that the performance of the catalyst could be improved [67]. Nanotechnology has caused to pay more attention to the preparation of Pt-based catalysts with controllable morphologies since the morphology of nanomaterials determines their properties. Pt nanocrystals with different morphologies are prepared primarily for the purpose of exposing more highly active sites and increasing the catalytic activity of the nanocrystals. Nanocrystals of various morphologies have also been observed to have Pt atoms distributed in different ways across their surfaces.

8.4 SYNTHESIS OF NANOTECH SOLAR CELLS

As a renewable electric power source, solar technologies play an important role in the generation of direct currents because they do so without having a negative impact on the environment. A NP system is a collection of NP cells that can be grouped together to form panels or modules. Many cells are connected in a module to generate electrical energy since each cell has a small power output [68]. A solar array consists of a series of NP modules that are connected to form the string; a PV array consists of a string of PV modules coupled together to form a row [69, 70]. The performance of NP modules has improved through numerous studies; there are now techniques available for improving the performance [71].

Researchers designed NP modules using single-diode and two-diode models and used the distributions as standards in the PV industry. They observed the characteristics of the PV cells by looking at the terminals of the PV modules to see how they performed at 25°C under standard test conditions (STC); practical modules consist of several PV cells connected in series or parallel. It is generally provided in the datasheets how PV cells, which is to say in AM1.5 spectrum [70].

Due to the strong dependency between the composition of the metal NPs and the surface plasmon resonance properties, metal-semiconductor hetero-structures have tunable optical properties. The core/shell structure layer has been used to synthesis solar power substation modules with modern and high efficiency structures. In the optoelectronics industry, core/shell nanostructures of metal and semiconductor are extremely important. In most cases, the metal core increases the charge separation in the semiconductor, thereby enhancing its light absorption when light is stimulated to the semiconductor, thus enhancing the semiconductor's photocatalytic activity and its energy conversion efficiency [72–77].

Many efforts have been made to control the electronic structures of NPs in a variety of ways, such as by controlling the size, shape, composition, and atomic structure of those particles; the atomic structure and composition of a nanomaterial have to do with its electronic, optical, and magnetic properties. A core/shell structure is an alternative method to modify the properties of NPs. There have been multiple advantages to using such nanoparticles due to the combination of different compositions with shelled structures combined with a core/shell structure. In recent years, hybrid core/shell structures based on metal and semiconductor materials have attracted considerable attention due to their bifunctional properties [72–77].

In addition, there is the concept of peak power, which describes how much electric power can be produced by a NP module when the module receives insulation of 1 kW/m$_2$ at a cell temperature of 25°C. [78]. According to [71, 79–81], the total output current of the modern modules used with HJQD Core/shell cells or multiple HJQD Core/shell cells in series resistance and shunt resistance at standard test conditions is calculated at STC:

$$I_{MiSTC}(V) = I_{phiSTC}(V) - I_{dMiSTC}(V) - \left(\frac{V + I_{SiTC}(V) R_{ser}}{R_{sh}} \right) \tag{8.27}$$

where I_{phiSTC} is the light-generated current of the proposed PV module at STC depends on the generation rate and absorbance characteristics of the window and absorbing layers. This also depends on the metal inclusion added to the absorber layer material, as evaluated in [35, 36].

I_{dMiSTC} is the forward diode current of the module, which is determined by the following equation:

$$I_{dMiSTC}(V) = I_{OiSTC} \left(\exp\left(\frac{(V + I(V) R_{ser})}{V_{tMSTC}} \right) - 1 \right) \tag{8.28}$$

V_{tMSTC} is the thermal voltage of the modules, I_{OiSTC} is the reverse saturation current of the proposed PV module at STC, which depends on the energy band gap and effective dielectric constant of the HJQD core/shell absorbing [35, 36].

Moreover, the modern model of HJQD Core/shell solar modules at nominal operating cell temperature (NOCT) has been synthesis based on high quality energy yield predictions. A high-quality energy yield prediction is necessary to ensure the growth of the NP industry; a variety of factors play a role in affecting the energy yield of a system. Solar cells are very sensitive to temperature fluctuations due to the incident solar irradiance being reflected onto the module plane, which is one of the main factors affecting their performance. PV modules experience a significant drop in open-circuit voltage with an increase in PV module temperature, whereas the short-circuit current increases slightly with increased temperature [82].

PV modules are highly dependent on irradiance and temperature during their operating conditions; repeated increase in temperature results in the degradation of the PV modules. Among the degrading processes that occur as a result of humidity and UV light are hydrolysis and photodegradation. It is imperative for a manufacturer of solar cells to measure the efficiency of energy conversion of the cells under standard test conditions to evaluate their performance [83]. The cells should indicate temperature dependence when a module is exposed to 800 W/m^2 irradiation at 20°C ambient temperature and wind speed of about 1 m/s as defined by IEC standards [84]. The resistance of single-diode HJQD core and shell modules are calculated using series and shunt resistances for nominal operating cell temperatures for core and shell modules, [80–85]:

$$I_{HM}(V)=I_{phMH}(V)-\ I_{dHM}(V)-\left(\frac{V+I_{HM}(V)R_{ser}}{R_{sh}}\right) \tag{8.29}$$

where

I_{dHM} is the forward-diode current of the proposed module at temperature higher than 25°C, which is determined by the following equation:

$$I_{dHM}(V)=I_{OHM}\left(\exp\left(\frac{(V+I(V)R_{ser})}{V_{tM}}\right)-1\right) \tag{8.30}$$

The output light-generated current of the proposed NP HJQD core/shell module depends on the solar irradiation and the temperature according to the following expression:

$$I_{phMH}(V)=(I_{phiSTC}(V)+\Upsilon_{ic}\Delta T)\frac{\varnothing_o}{\varnothing_{oSTC}} \tag{8.31}$$

The reverse saturation current of proposals HJQD core/shell models affected by variations in the module temperature:

$$I_{OHM}=\frac{I_{phiSTC}+\Upsilon_{ic}\Delta T}{\exp\left[\frac{(V_{ocSTC}+\Upsilon_{vc}\Delta T)}{nV_{tM}}\right]-1} \tag{8.32}$$

where

$\varnothing_o$ is the actual sun irradiation;

$\varnothing_{oSTC}$ is nominal sun irradiation (1000W/m^2);

ΔT is the difference between actual temperature and standard temperature 25°C; and

Υ_{ic} is the temperature coefficient of the short-circuit current (A/°C) of the HJQD core/shell solar PV module. The energy conversion efficiency of proposed NP cells depends on the module temperature, which is depicted by temperature coefficients for the current, the voltage, and the power. These temperature coefficients can be calculated using the following equation [84–86]:

$$\Upsilon_{ic}=\frac{\Delta I_{iSc}}{\Delta T}=\frac{I_{isc,T2}-I_{isc,Tstc}}{T_2-T_{STC}} \tag{8.33}$$

Υ_{vc} is temperature coefficient of the open circuit voltage (V/°C) of the HJQD Core/shell solar PV module.

$$\Upsilon_{vc} = \frac{\Delta V_{ioc}}{\Delta T} = \frac{V_{ioc,T2} - V_{ioc,T1}}{T_2 - T_{STC}} \tag{8.34}$$

where

$I_{isc,T1}, I_{isc,T2}$ is the short circuit current of HJQD Core/shell module at standard test condition and at temperature T_2;

$V_{ioc,T1}, V_{ioc,T2}$ is the open circuit voltage of nanocomposite module at standard test condition and at temperature T_2;

There are several standard test conditions that must be met for the solar cell efficiency of the HJQD core/shell modules. At temperatures above 25°C, the electrical efficiency of HJQD Core/Shell modules can be estimated using the following equation [20, 35, 36]:

$$n_{iT2} = n_{iSTC}\left[1 - C_{iref}\left(T_2 - T_{STC}\right)\right] \tag{8.35}$$

where n_{iT2} is the efficiency of the solar cell/module at temperature higher than the STC, n_{iSTC} is the efficiency of the cell/module at STC; and C_{iref} is the temperature coefficient of HJQD core/shell modules.

In addition to the PV material of the QD window layer and the absorbing layer of the HJQD core/shell for the cell, the actual temperature coefficient varies based on the STC temperature (25°C) T_{STC}. It is given by the following expression [87, 88]:

$$C_{iref} = \frac{1}{\left(T_o - T_{STC}\right)} \tag{8.36}$$

where T_0 is the (high) temperature at which the PV HJQD core/shell modules electrical efficiency drops to zero [20] for crystalline NP solar cells T_0 is 270 °C [88].

The total number of modules required to achieve modules is calculated using the following equation:

$$N_M = \frac{Total\,installed\,capacity\left(TIC\right)}{P_{imax}} \tag{8.37}$$

P_{imax} is the maximum output power for HJQD core/shell module which calculated using the following equation:

$$P_{imax} = P_{mmaxden} * A \tag{8.38}$$

where $P_{mmaxden}$ is the maximum power density for the HJQD core/shell module, which depends on the materials cells type:

$$N_S = \frac{N_M}{number\ of\ modules\ /\ string} \tag{8.39}$$

When comparing the power point of the modern module with the life-cycle NP solar substations that are required to power solar power substations, the area used for the modern module should be the same. In Figure 8.1, typical modules are connected in arrays and converted to AC by inverters. Inverters prepare DC power for use in a tracker system. An electric transformer steps up the power output of the site's generator to the voltage needed by the nearby grid; the electricity is then transmitted from the site's generator to a grid substation where it is distributed into the wider grid for distribution to the rest of the country. Tables 8.4 and 8.5 present the data sheets of the modules and modules allocation that have been used in the substation. A QD window layer and a metal or semiconductor absorbing layer are used in modern HJQD CdS/PbS, CdS/CdTe, and CdS/CIGS thin-film solar cells [87, 88].

Figures 8.2–8.4 show the module allocation and efficiency for design solar power substations with total installed capacity of 64,516.8 KWP that based on crystalline NP solar cells. They also predict the efficiency and module allocation for solar power substations based on multiple-nanocomposites thin-film solar cells. The figures show that using multiple NPs leads to a solar power substation with high efficiency more than 15% based on CdS/(CIGS + 20wt.%Al) + Ag), CdS/(CIGS + 20wt.%Al) + Cu), CdS/(CIGS + 20wt.%Al) + Li) multiple-nanocomposite thin-film solar cells. They can also decrease the total number of modules for design solar power substations with total installed capacity of 64,516.8 KWP.

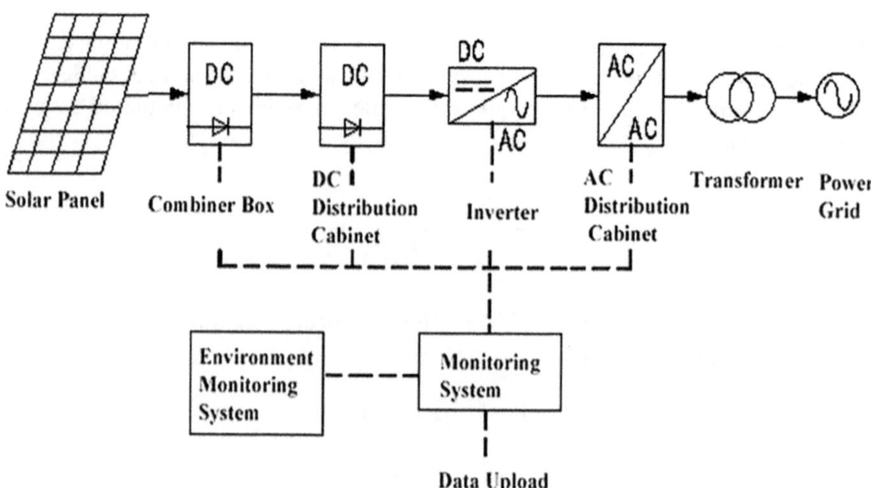

FIGURE 8.1 Typical components of a solar PV system.

TABLE 8.4
Theoretical Data Sheet of Modules in Solar Power Substation in Benban PV Site Project with Total Installed Capacity of 64,516.8 KWP [87, 88]

	STC								NOCT					
Module Types	V_{oc}(V)	I_{sh}(A)	I_m(A)	V_m(V)	P_{max} WP)	n_{STC}(%)	Calculated Area of module (m^2)	Calculated $P_{mmaxden}$ (Watt/m^2	V_{oc} (V)	I_{sh} (A)	I_m(A)	V_m(V)	P_m(WP)	n_{NOCT} (%)
Solar world 325XL (Monocrystalline NP solar cells)	46.1	9.4	8.8	37	325	16.29	1.99	163.3	40.2	7.88	7.28	34	247.5	12.4
Trina Tallmax 320 (Multicrystalline NP cells)	45.5	9.1	8.6	37.1	320	16.5	1.94	164.9	42.1	7.39	6.92	34.3	237.4	12.3
Trina Tallmax 325 (Multicrystalline NP cells)	45.6	9.1	8.7	37.2	325	16.7	1.95	166.6	42.2	7.42	7	34.4	241	12.35

TABLE 8.5
Module Allocation of the Solar Power Substation in Benban PV Site Project with Total Installed Capacity of 64,516.8 KWP [89]

Module	Total number of modules for each type	Total number of strings for each type
Solar world 325XL (Monocrystalline NP solar cells)	28,800	1440
Trina Tallmax 320 (Multicrystalline NP cells)	85,440	4272
Trina Tallmax 325 (Multicrystalline NP cells)	84,480	4224
Number of modules/string	20	
Total number of modules in substation	198,720	
Total number of strings in substation	9936	

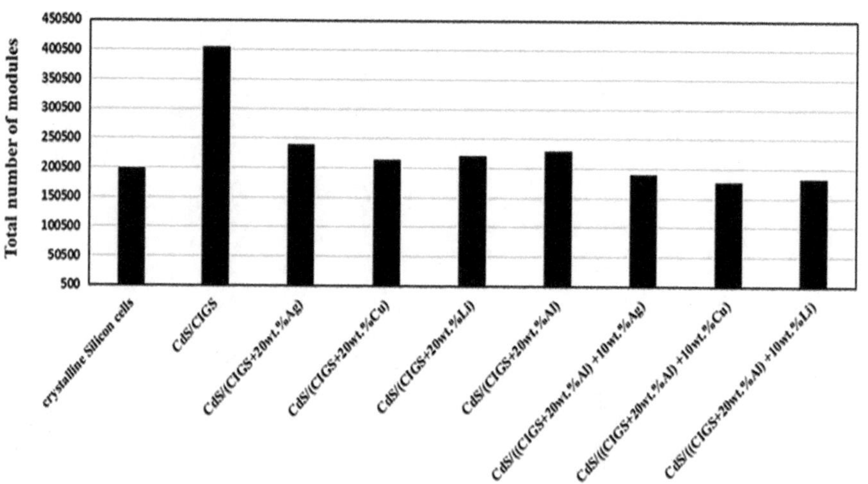

FIGURE 8.2 Total number of modules required for traditional and proposed multiple-nanocomposite solar power substation with total installed capacity of 64,516.8 KWP.

Figure 8.5 shows the comparison study between the maximum power of traditional NP and the proposed CdS/(metal/CIGS) multiple nanocomposites thin-film solar modules. The figure shows that the PV modules based on a (CIGS + 20wt.%Al) + 10wt.%Ag, (CIGS + 20wt.%Al) + 10wt.%Cu, and (CIGS + 20wt.%Al) + 10wt.%Li submicro absorbing layer achieved higher maximum power than the modules based on mono- and polycrystalline NP cells.

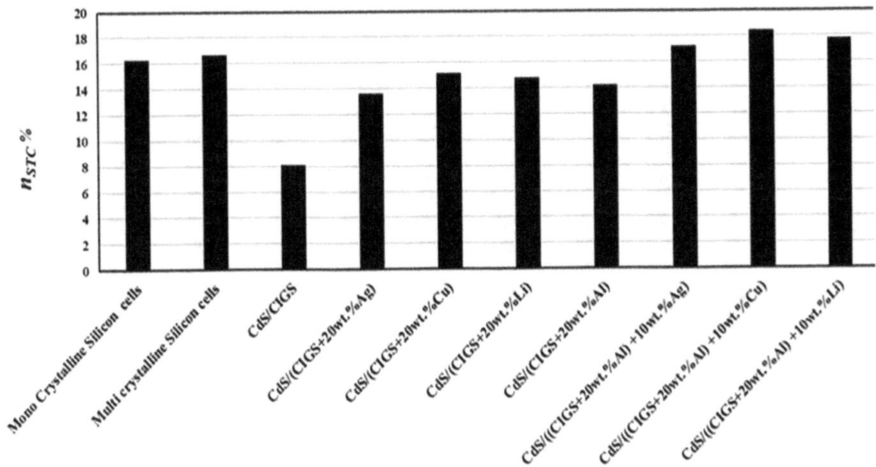

FIGURE 8.3 Efficiency at STC for traditional and proposed multiple-nanocomposite solar power substation.

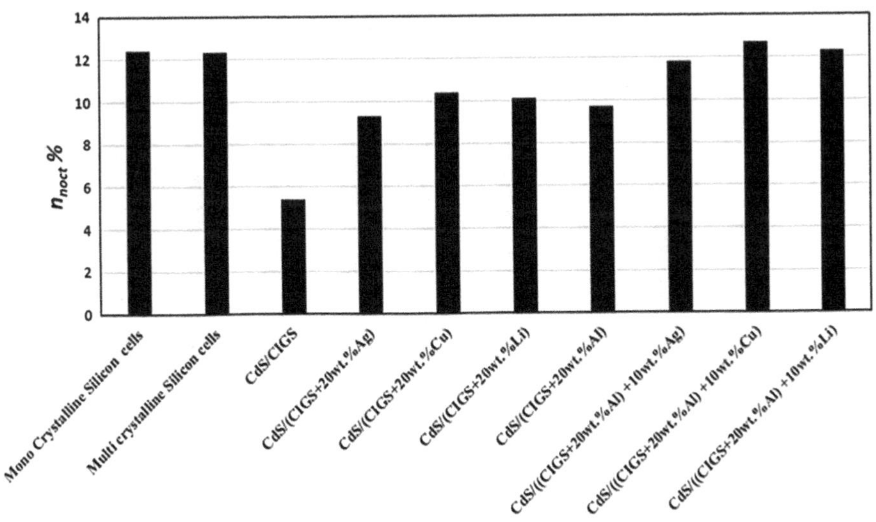

FIGURE 8.4 Efficiency at NOCT of traditional and proposed multiple-nanocomposite solar power substation.

On the other side, Figure 8.6 shows a comparison study between the high temperature at which the electrical efficiency drops to zero for the traditional and the proposed CdS/(metal/CIGS) cells/modules; that temperature depends on the cell's materials. As shown in the figure, the predictable high temperature at which the electrical efficiency drops to zero, calculated for the CdS/(metal/CIGS) nanocomposite thin-film solar cells is higher than the same temperature for polycrystalline and monocrystalline cells.

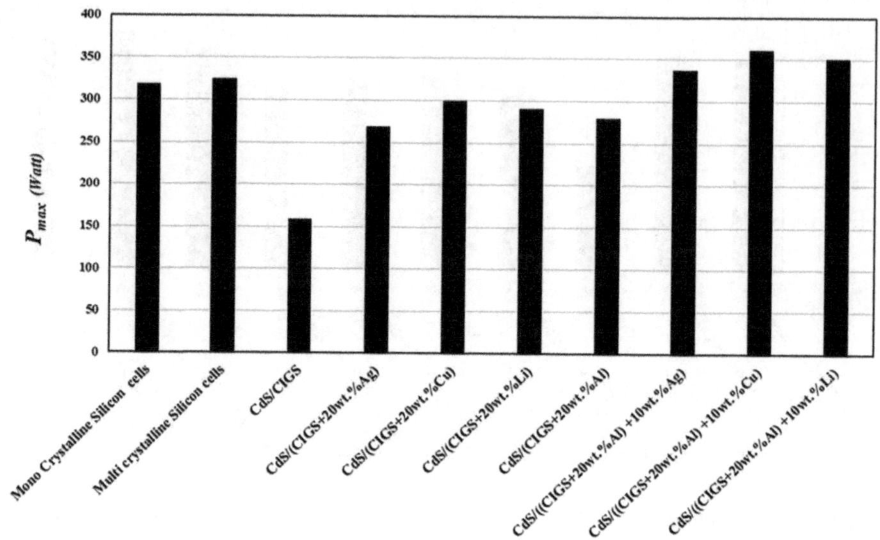

FIGURE 8.5 Maximum output power point of traditional NP and proposed CdS/(metal/CIGS) multiple nanocomposite solar modules using the same module area 1.95 m^2.

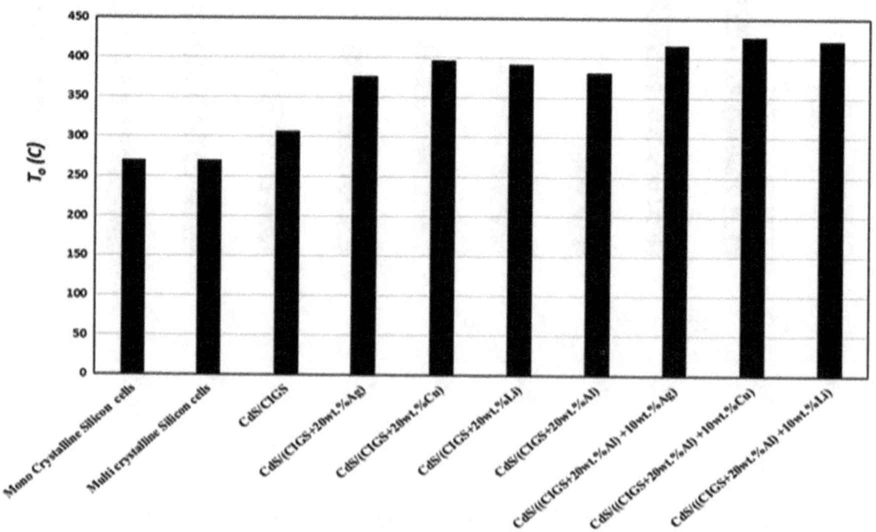

FIGURE 8.6 High temperature at which the electrical efficiencies of the traditional and proposed CdS/(metal/CIGS) multiple nanocomposite cells/modules drops to zero.

8.5 SYNTHESIS OF NANOTECH WIND FARM MODULES

Most WTBs are made from fiber-reinforced polymer; it is the most expensive part of the machine; carbon composite blades are one of the most important components of the energy production process, but just like other composite products, they blades are susceptible to damage. Leading edge erosion caused by rain, dust, salty vapor, hailstones, and insects significantly reduces the lifetime of WTBs; this in turn decreases the turbines' annual energy production. WTBs are able to resist erosion owing to modern elastomeric erosion-resistant coatings. Several types of graphene NPs have been modified into polyurethane (PU) to make these coatings including graphene NPs with hydroxyl-functionalized surfaces and silica-based sol-gels with hydrophobic surface properties. Graphene nanoplatelets and silica-based sol-gels provide improved mechanical properties over PU because they have a low elastic modulus and high tensile strength and can withstand large deformations. Functionalized graphene nanoplatelets and silica-based sol-gels were used to improve the mechanical properties of PU. The newly developed PU coatings were primarily intended to protect the leading edge of wind turbine blades against erosion. Rubber balls were used to conduct erosion tests on the new coatings; water-jet erosion tests conducted shortly. These coatings have a wide range of applications such as for use in line-marking paints and roof coatings with anticorrosive/corrosion properties. Furthermore, water droplets can roll on the surface of multilayer coatings and clean themselves, thanks to their hydrophobicity [90, 91].

A variety of reasons exist for coating a substrate, such as making the substrate more biocompatible; increasing the material's thermal, mechanical, or chemical stability; increasing the wear resistance; improving durability; decreasing friction; inhibiting corrosion; or changing the overall properties of the material in terms of physicochemical and biological aspects [92]. Protective layers such as tape or paintable and elastic coatings are used to protect the blades from leading edge erosion; the impact energy is absorbed by these layers without cracking. A coating absorbs and distributes energy as a result of collision frequency; the current blade coating system typically comprises a putty layer, a primer, a flexible topcoat, and a putty layer to fill pores in the composite substrate [93].

A covalent modification in graphene typically leads to a band gap opening and changes in conductivity due to a break in the extended conjugation of p-electrons. The addition of functional groups to the graphene sheet gives the graphene controlled chemical properties; along with improving the solubility and ability of the material, covalent modification can also make it easier to process the material [94]. Deicing and anti-icing technology for WTBs are divided into two important categories based on their operational modes [95, 96]. There are several active deicing methods available, including mechanical, thermal, and ultrasonic vibration. However, they are often more expensive, more energy-intensive, more inefficient, and more short-lived, as well as requiring special equipment. This makes these methods unsuitable for engineering applications [97–100]. Blade icing can have serious effects on wind turbine aerodynamics and safety performance in cold regions; applying coatings is an effective way to prevent WTB icing. Hydrothermal and liquid phase methods were used to provide superhydrophobic anti-icing nano-coatings; the adhesion strength to ice was 60.2% lower than an uncoated surface. A wind tunnel test involving the NACA0018 airfoil blade model with and without MoS2/ZnO coatings tested whether the blade model icing effect was affected by different ambient temperatures and wind speeds [101].

8.6 TRENDS AND RECOMMENDATIONS

8.6.1 Trends in Hydrogen Energy Modules

Pt-based catalysts have been improved with a variety of synthesis methods over the past few decades. However, it remains difficult to simultaneously enhance the durability of Pt-based catalysts and their activity because catalysts with high activity tend to be less stable than less-active catalysts. In the future, multiple strategies are likely to be combined to create catalysts that are more efficient for PEMFCs, which will result in improved performance.

There are also few practical applications of Pt-based catalysts that can readily be demonstrated at the MEA level since most Pt-based catalysts are evaluated with the rotating disc electrode, but successful ORR performance on this test does not necessarily translate into high performance on a MEA test; the latter is conducted in a more complex environment and can have a greater bearing on performance. Thus, there must be standardized evaluation parameters such as evaluating battery testing or designing electrodes that take advantage of gas diffusion. It is beneficial for the MEA to have an ordered structure that allows it to reduce the amount of catalyst, and providing the appropriate carrier can enhance the electrode's stability. Further reductions in PEMFC cost will require the design of Pt-based catalysts that are not only capable of delivering high stability and activity but also require a low Pt load so that they can continue to be used in PEMFCs at a reasonable price in the future.

A suitable reaction environment can be provided for electrochemical reactions by the rational design of catalytic layers. In a gradient catalytic layer structure, different reaction positions are distributed among electrodes resulting in different electrochemical reactions that can be studied in more depth; this can lead to a better understanding of electrochemical reaction processes. PEMFCs are rapidly becoming one of the most widely used types of battery in everyday life and are expected to become an important component of our future transportation system as well as a major source of electric power. They have the potential to be a major development direction for membrane electrodes in the future.

8.6.2 Trends in Solar Energy Modules

Due to crystalline films' minimum grain size, quantum confinement can be illustrated by enhancing band gap values in multiple-nanocomposite films with different volume fractions. The band gap improvement can increase the solar cell efficiency; the multiple-nanocomposite films with various volume fractions have large absorption coefficients and appropriate energy band gap range.

There is a recent trend on the market of using more promising absorbing layers for solar cells with a higher band gap to improve their efficiency. As the volume fraction of the nanocomposite increases, the band gap of the nanocomposite is considerably smaller, which means it is suitable for absorbing visible light in solar cells. Using multiple nanocomposites in solar cells can help improve their efficiency by increasing short-circuit current. In addition to their high absorption coefficient, metal NPs can shift their absorption edge to become visible in the visible region, causing the formation of films that facilitate charge separation and charge carrier collection in PbS semiconductors.

Furthermore, multiple-nanocomposites films with various ratios were found to have a visible and near-infrared transmittance. The thin PbS films are also secularly reflective and therefore can reflect a large amount of the radiant energy that falls on them. Optical systems can use these films as absorbing layers since their low transmittance and reflectance indicate they are effective radiation absorbers. Due to thin films' high absorption coefficient, it is possible to manufacture thin-film cells with high efficiency even with absorb layers that are up to 0.5 m thick, thus making thin film cells suitable for high efficiency applications.

Nanocomposite films prepared with different ratios of Cu as second fillers exhibit good optical conductivity, and they have low refractive indices. These films in absorbing layers in NP thin-film cells achieve high optical performance that increases cell efficiency. Efficient light management guarantees high absorption of the incoming light by thin-film solar cells; reducing the thickness of the absorbing layer results in reduced material usage, processing time, and cost. Using metal NPs increases the absorption of the absorbing layer, which results in a higher generation rate of electron holes. Moreover, the decreasing space charge region and reverse saturation current lead to a decrease in recombination as the energy band gap approaches a maximum value. In CdS/PbS thin film solar cells, a decrease in the thickness of the absorb layer and an increase in electron–hole generation give rise to improved performance and lower recombination rates in the cells. This enables the use of less material for the absorbed layer; the improvements in the performance of NP solar cells continue to be driven by advances in material science.

A comparison of experimental and theoretical models indicates that multiple nanocomposites exhibit high optical properties when compared with individual NPs; the experiments and theoretical results indicate high absorption characteristics near the optimum energy band gap of a good absorbing layer. The absorption edges of several nanocomposite materials and films agreed in both theoretical and experimental studies with the bulk absorption edge of PbS. Material as powder exhibits a high blue shift at the onset of the absorption edge, both theoretically and experimentally. There is a link between particle size and a blue shift in the onset of absorption edge in films; decreasing particle size shifts the absorption edge to lower wavelengths; the blue shift is mainly due to quantum dimensional effects of the NPs, which cause the wider band gap, as well as NPs' surface effect, which causes crystal lattice aberration and a low crystal constant as a result of the large surface force. A thin film of row powders on glass can be effectively used to raise optical band gaps to achieve a wide range of energy band gaps.

As an alternative, individual and multiple Al and Cu metal NPs are used to decrease the refractive index of bulk PbS. Individual and multimetal NPs decrease the reflectance of PbS base matrix, resulting in the decrease in refractive index. Theoretically, metal/PbS nanocomposites exhibit enhanced optical properties as their volume fraction increases; however, experimentally, these enhancements depend on the kind and concentration of the metal component. QD CdTe, CIGS, and PbS shells ertr designed to reduce the dielectric constant and refractive index of QD Al, Cu, Li, Cs, or Au cores. When the volume fraction of metal-semiconductor absorbing layers increased, their dielectric constants decreased, thereby increasing semiconductor energy band gaps, decreasing space charge regions, and reducing reverse saturation currents of HJQD thin-film solar cells with QD window layers and metal-semiconductor absorbing layers.

Metal core materials such as Al are improving the efficiency of traditional absorbing shell materials such as CdTe, CIGS, and PbS. Depending on the size of the QD core used, the performance of models can be enhanced by adding a QD shell layer made of Cu, Li, or Au core. By reducing the reverse saturation current and optimizing the energy band gap, absorption, and electron–hole generation rate in QD metal-semiconductors cores/shells of thin-film solar models, the cells are able to increase their output power density, short circuit current density, open circuit voltage, energy conversion efficiency, fill factor, and external quantum efficiency.

QD window layer core/metal shell and core/multimetal shell absorbing layer modules with individual and multiple NPs in modern HJQD CdS/CIGS and CdS/CdTe modules have performed well under standard test conditions and nominal operating cell conditions (NOCT) simultaneously. In modules based on these nanocomposites, solar irradiance increased open circuit voltages, short circuit currents, and energy conversion efficiency. It can be noted that increasing solar irradiance led to a slight increase in short-circuit current at constant module temperature. Modern HJQD CdS/CIGS, CdS/CdTe modules were installed instead of traditional polycrystalline and monocrystalline silicon solar cells to create a 64,156.8 KWP solar power substation using QD window layers, core/metal shells, and core/multimetal shell absorber layers. At NOCT, the silicon solar cells proved highly efficient, requiring fewer materials for modules and strings. In thin-film solar cells operated at 25°C to 127°C, open circuit voltages decrease linearly with module temperature. Therefore, power point and efficiency decrease linearly with module temperature.

8.6.3 Trends in Wind Energy Modules

There is a great need for tools for erosion lifetime prediction in the development of rain erosion coatings for WTBs as well as for identifying suitable combinations of coatings and composite substrates for a rain erosion coating. For wind turbine energy production to be cost competitive, the leading edges of the turbine blades should be protected from rain erosion. There is a limit to the thickness of a sol-gel-derived coating of 10 ml, with hardly any effect on the blade weight; Sol–gel coatings with higher hydrolysis water content, increased hardiness and flexibility, anda decreased molar ratio of organic to inorganic materials will make them more resistant to rain erosion. Alternatively, if the coating is not well adhered to the substrate, then, reduced water erosion resistance will be observed due to poor adhesion. Adding CNPs, especially CNTs and graphene, enhances the coating's erosion resistance. However, CNPs must be functionalized properly for effective dispersion, which alters the coating's properties.

Hydrothermal and liquid phase methods were used to prepare MoS2/ZnO nanomaterials for superhydrophobic surfaces; the surfaces formed micro-nano rough structures. The surfaces of MoS2/ZnO/PDMS superhydrophobic nano-coatings had pores and roughness that acted like an air cushion, thus reducing adhesion between liquids and coatings and between coatings and ice. A wind tunnel test that examined icing behavior showed that the blades coated with nanomaterials made up of MoS2/ZnO/PDMS had less icing than those that were uncoated under the same conditions.

REFERENCES

1. Hanif MM, Malik MZ, Usman M, Adnan M, Ashraf S, Mehmood R, ur Rehman MO and Zahoor S, "Role of Physics and New Insights in Development of Energy System through Nanotechnology", *Saudi Journal of Engineering and Technology, Scholars Middle East Publishers, Dubai, United Arab Emirates*, 2022, 94–98.
2. Bondavalli P, *The Graphenes Cousins from Dream to Reality. Graphene and Related Nanomaterials*. Amsterdam, Netherlands: Elsevier, 103–136, 2018.
3. Limjoco V, "Super Battery," *ScienCentral News*, 2006.
4. Hasan Balali M, Nouri N, Omrani E, Nasiri A and Otieno W, "An Overview of the Environmental, Economic, and Material Developments of the Solar and Wind Sources Coupled with the Energy Storage Systems", *International Journal of Energy Research,* 2008; 32:1454–1456, Wiley InterScience (www.interscience.wiley.com). DOI: 10.1002/er.1484.
5. Balali MH, et al., "Development of an Economical Model for a Hybrid System of Electricity Generation and Storage", *2015 International Conference on Renewable Energy Research and Applications (ICRERA)*, 2015, pp. 1108–1113.
6. Khare V, Nema S and Baredar P, "Solar–Wind Hybrid Renewable Energy System: A Review", *Renewable and Sustainable Energy Reviews,* 2016, 58, 23–33.
7. Bet PE and Thornton HE, "The Climatological Relationships between Wind and Solar Energy Supply in Britain", *Renewable Energy*, 2016, 87, 96–110.
8. Mamia I and Appelbaum J, "Shadow Analysis of Wind Turbines for Dual Use of Land for Combined Wind and Solar Photovoltaic Power Generation", *Renewable and Sustainable Energy Reviews*, 2016, 52, 713–718.
9. Nouri N, "Water Withdrawal and Consumption Reduction Analysis for Electrical Energy Generation System", Theses and Dissertations, 2015, Paper 1068.
10. Wu CY and Mathews JA, "Knowledge Flows in the Solar Photovoltaic Industry: Insights from Patenting by Taiwan, Korea, and China", *Research Policy*, 2012, 41(3), 524–540.
11. Serrano E, Rus G and Garcia-Martinez J, "Nanotechnology for Sustainable Energy", *Renewable and Sustainable Energy Reviews*, 2009, 13(9), 2373–2384.
12. Sethi V, Pandey M and Shukla MP, "Use of Nanotechnology in Solar PV Cell", *International Journal of Chemical Engineering and Applications*, 2011, 2(2), 77.
13. Bruni G, Cordiner S and Tosti S, "A Novel Procedure for the Preliminary Design of Dense Metal Membrane Modules for Hydrogen Separation", *International Journal of Hydrogen Energy*, 26 November 2016, 41(44), 20198–20209.
14. Myagmarjav O, Tanaka N, Nomura M and Kubo S, "Module Design of Silica Membrane Reactor for Hydrogen Production via Thermochemical IS Process", *International Journal of Hydrogen Energy*, 23 April 2019, 44(21), 10207–10217.
15. Hwang GJ, Onuki K, Shimizu S and Ohy H, "Hydrogen Separation in H2eH2OeHI Gaseous Mixture Using the Silica Membrane Prepared by Chemical Vapor Deposition", *Journal of Membrane Science*, 1999, 162, 83e90.
16. Myagmarjav O, Ikeda A, Tanaka N, Kubo S and Nomura M, "Preparation of an H2-Permselective Silica Membrane for the Separation of H2 from the Hydrogen Iodide Decomposition Reaction in the Iodineesulfur Process", *International Journal of Hydrogen Energy*, 42, 6012–6023, 2017.
17. Pasquini L, "Design of Nanomaterials for Hydrogen Storage", *MDPI, Energies*, 13, 3503, 2020.

18. Narayan TC, Baldi A, Koh AL, Sinclair R and Dionne JA, "Reconstructing Solute-Induced Phase Transformations within Individual Nanocrystals", *Nature Materials*, 2016, 15, 768–774.
19. Surrey A, Nielsch K and Rellinghaus B, "Comments on 'Evidence of the Hydrogen Release Mechanism in Bulk MgH2'", *Scientific Reports*, 2017, 7, 44216.
20. Hamm M, Bongers MD, Roddatis V, Dietrich S, Lang KH and Pundt A, "In situ Observation of Hydride Nucleation and Selective Growth in Magnesium Thin-Films with Environmental Transmission Electron Microscopy", *International Journal of Hydrogen Energy*, 44, 32112–32123, 2019.
21. Lundberg O, Bodegard M and Stolt L, "Influence of the Cu(In,Ga)Se2 Thickness and Ga Grading on Solar Cell Performance", *Progress in Photopholatics: Research and APblication*, 2003, 12.
22. Peng B, Zhang Q, Liu X, Ji Y, Demir HV, Huan CHA, Sum TC and Xiong Q, "Fluorophore-Doped Core-Multishell Spherical Plasmonic Nanocavities: Resonant Energy Transfer Toward a Loss Compensation", *ACS Nano*, 2012, 6250–6259.
23. Wang Y, Yang T, Tuominen MT and Achermann M, "Radiative Rate Enhancements in Ensembles of Hybrid Metal-Semiconductor Nanostructures", *Physical Review Letters*, 102, 163001, 2009.
24. Zhang N, Liu S and Xu YJ, "Recent Progress on Metal Core Semiconductor Shell Nanocomposites as Apromising Type of Photocatalyst", *Nanoscale*, 2012, 2227–2238.
25. Holmström P, Thylen L and Bratkovsky A, "Dielectric Function of Quantum Dots in the Strong Confinement Regime", *Journal of Applied Physics*, 2010, 8.
26. Hamaguchi C, *Basic Semiconductor Physics* (2nd edition), Springer, 2009, p. 201.
27. Jimenez J and Tomm W, *Spectroscopic Analysis of Optoelectronic Semiconductors*, Springer Series in Optical Science, 2016, p. 27.
28. Harrison P, *Quantum Wells, Wires and Dots: Theoretical and Computational Physics of Semiconductor Nanostructures* (3rd edition, Chapter 13), Wiley Interscience, 2011.
29. Ridley BK, *Quantum Processes in Semiconductors* (Chapter 16), Oxford University Press, 2013, p. 418.
30. El-Nahass M, Youssef GM and Sohaila Z, *Structural and Optical Characterization of CdTe Quantum Dots Thin Films*, Elsevier, 2014, p. 7.
31. Thabet A, Abdelhady S and Mobarak Y, "Design Modern Structure for Heterojunction Quantum Dot Solar Cells", *International Journal of Electrical and Computer Engineering*, June 2020, 10(3), 2918–2925.
32. Ebrahim K, "Quantum Dots Solar Cells", *InTech Solar Cells New Approaches and Reviews*, 2015, 31.
33. Singh P and Ravindra N, *Temperature Dependence of Solar Cell Performance -An Analysis* (Vol. 10), Elsevier, 2012.
34. Mannan MA, Anjan MS and Kabir MZ, *Modeling of Current–Voltage Characteristics of Thin Film Solar Cells* (Vol. 6), Elsevier, 2011.
35. Thabet A, Abdelhady S, Ebnalwaled AA and Ibrahim AA, "Improvement Optical and Electrical Characteristics of Thin Film Solar Cells Using Nanotechnology Techniques", *International Journal of Electronics and Telecommunications*, 2019, 65(4), 625–634.
36. Thabet A, Abdelhady S, Ebnalwaled AA and Ibrahim AA, "Innovative Industrial Cu(In,Ga)Se2 Thin Film Solar Cell with High Characterization Using Nanoparticles Structure", *Indonesian Journal of Electrical Engineering and Informatics (IJEEI)*, June 2019, 7(2), 382–392.
37. Tripathy SK, "Refractive Indices of Semiconductors from Energy Gaps", *Optical Materials*, 2015, 46, 240–246.
38. Qu D, Liu F, Yu J, Xie W, Xu Q, Li X and Huang Y, "Plasmonic Core-Shell Gold Nanoparticle Enhanced Optical Absorption in Photovoltaic Devices", *Applied Physics Letters*, 2011, 98, 113119.

39. Lu W, Wang B, Zeng J, Wang X, Zhang S and Hou J, "Synthesis of Core/Shell Nanoparticles of Au/CdSe via Au-Cd Bialloy Precursor", *Langmuir*, 2005, 21(8).
40. Bhandari K, Roland P, Mahabaduge H, Haugen N and Grice C, *Thin Film Solar Cells Based on the Heterojunction of Colloidal PbS Quantum Dots with CdS* (Vol. 7), Elsevier, 2013.
41. Bai Z, Yang J, DePba S and Wang D, "Thin Film CdTe Solar Cells with an Absorber Layer Thickness in Micro- and Sub-Micrometer Scale", *Applied Physics Letters*, 2011, 4.
42. Liang F, Gou J, Kapat J, Gu H and Song G, "Multifunctional Nanocomposite Coating for Wind Turbine Blades", *International Journal of Smart and Nano Materials*, 2011, 2(3), 120–133.
43. Boncel S, Kolanowska A, Kuziel AW and Krzy I, "Carbon Nanotube Wind Turbine Blades—How Far Are We Today from Laboratory Tests to Industrial Implementation?", *ACS Applied Nano Materials*, 2018, doi: 10.1021/acsanm.8b01824.
44. Yao X, Hawkins SC and Falzon BG, "An Advanced Anti-Icing/De-Icing System Utilizing Highly Aligned Carbon Nanotube Webs", *Carbon*, 2018, 136, 130–138.
45. Fischer T, Ruhling J, Wetzold N, Zillger T, Weissbach T, Goschel T, Wurfel M, Hubler A and Kroll L, "Roll-to-Roll Printed Carbon Nanotubes on Textile Substrates as a Heating Layer in Fiber-Reinforced Epoxy Composites", *Journal of Applied Polymer Science*, 2018, 135, 45950.
46. Jiang G, Chen L, Zhang S, Huang H, "Superhydrophobic SiC/CNTs Coatings with Photothermal Deicing and Passive Anti-Icing Properties", *ACS Applied Materials & Interfaces*, 2018, 10, 36505–36511.
47. Thomasa L and Ramachandra M, "Advanced Materials for Wind Turbine Blade–A Review", *Materials Today: Proceedings*, 2018, 5, 2635–2640.
48. Durai Prabhakaran RT, *Future Material for Wind Turbine Blades – A Critical Review*, Section of Composites and Materials Mechanics, Department of Wind Energy, Technical University of Denmark, 2012.
49. Loos M, Yang J and Feke D, Manas-zloczower I, "Enhanced fatigue life of carbon nanotube-reinforced epoxy composites", *Polymer Engineering and Science*, 2012, 52, 1882–1887.
50. Boncel S, Kolanowska A, Kuziel AW and Krzy#ewska I, "Carbon nanotube wind turbine blades: How far are we today from laboratory tests to industrial implementation?", *ACS Applied Nano Materials,* 2018, 1, 6542–6555.
51. Bhanushali H and Bradford DP, "Woven Glass Fiber Composites with ligned Carbon Nanotube Sheet Interlayers", *Journal of Nanomaterials*, 2016, 9705257, 2016.
52. Li H, Zhao H, Tao B, Xu G, Gu S, Wang G and Chang H, "Pt-Based Oxygen Reduction Reaction Catalysts in Proton Exchange Membrane Fuel Cells: Controllable Preparation and Structural Design of Catalytic Layer", *Nanomaterials*, 2022, 12, 4173.
53. Fan X, Luo S, Zhao X, Wu X, Luo Z, Tang M, Chen W, Song X and Quan Z, "One-Nanometer-Thick Platinum-Based Nanowires with Controllable Surface Structures", *Nano Research*, 2019, 12, 1721–1726.
54. Yi S, Jiang H, Bao X, Zou S, Liao J and Zhang Z, "Recent Progress of Pt-Based Catalysts for Oxygen Reduction Reaction in Preparation Strategies and Catalytic Mechanism", *Journal of Electroanalytical Chemistry*, 2019, 848, 113279.
55. Li W, Wang D, Zhang Y, Tao L, Wang T, Zou Y, Wang Y, Chen R and Wang S, "Defect Engineering for Fuel-Cell Electrocatalysts", *Advanced Materials*, 2020, 32, 1907879.
56. Wang X, Li Z, Qu Y, Yuan T, Wang W, Wu Y and Li Y, "Review of Metal Catalysts for Oxygen Reduction Reaction: From Nanoscale Engineering to Atomic Design", *Chemistry*, 2019, 5, 1486–1511.
57. Li C, Tan H, Lin J, Luo X, Wang S, You J, Kang YM, Bando Y, Yamauchi Y and Kim J, "Emerging Pt-Based Electrocatalysts with Highly Open Nanoarchitectures for Boosting Oxygen Reduction Reaction", *Nano Today*, 2018, 21, 91–105.

58. Bates FS, "Polymer-Polymer Phase Behavior", *Science*, 251, 898, 1991.
59. Li LG, Che YK, Gross DE, Huang HL, Moore JS and Zang L, "Temperature controlled, Reversible, Nanofiber Assembly from an Amphiphilic Macrocycle", *ACS Macro Letters*, 2012, 1, 1335.
60. Liu GH, Shi GH, Sheng HY, Jiang YY, Liang HJ and Liu SY, "Doubly Caged Linker for AND-Type Fluorogenic Construction of Protein/ Antibody Bioconjugates and in situ Quantification", *Angewandte Chemie International Edition,* 2017, 56, 8686.
61. Kawazoe T, Hashimoto K, Kitazawa Y, Kokubo H and Watanabe M, "A Polymer Electrolyte Containing Solvate Ionic Liquid with Increased Mechanical Strength Formed by Self-Assembly of ABA-Type Ionomer Triblock Copolymer", *Electrochimica Acta*, 2017, 235, 287–294.
62. Joahim PS, Stefan M, Christoph H and Martin M, "Ordered Deposition of Inorganic Clusters from Micellar Block Copolymer Films", *Langmuir*, 2000, 16, 407–415.
63. Kim SJ, Kim DJ, Heo SY, Ahn H, Ryu DY and Kim JH, "Micron-Thick, Worm-Like, Organized TiO2 Films Prepared Using Polystyrene-b-poly(2-Vinyl Pyridine) Block Copolymer and Preformed TiO2 for Solid-State Dye-Sensitized Solar Cells", *Electrochimica Acta*, 2013, 105, 15.
64. Roland S, Gamys CG, Grosrenaud J, Boiss S, Pellerin C, Robert EP and Bazuin CG, "Solvent Influence on Thickness, Composition, and Morphology Variation with Dip-Coating Rate in Supramolecular PS-b-P4VP Thin Films", *Macromolecules*, 2015, 48, 4823.
65. Gan Y, da Z Wang, Shi Y, Guo CQ, Tan HY, Lu ZX and Yan CF, "Synthesis of Density-Multiplied Pt-NP Arrays and Their Application in Fuel Cell by Self-Assembly of Di-Block Copolymer", *Electrochimica Acta*, 2018, 283, 1–10.
66. Klingenhof M, Hauke P, Brückner S, Dresp S, Wolf E, Nong HN, C. Spöri, Merzdorf T, Bernsmeier D, Teschner D, R. Schlögl and Strasser P, "Modular Design of Highly Active Unitized Reversible Fuel Cell Electrocatalysts", *ACS Energy Letters*, 2021, 6, 177–183.
67. Ren X, Lv Q, Liu L, Liu B, Wang Y, Liu A and Wu G, "Current Progress of Pt and Pt-Based Electrocatalysts Used for Fuel Cells", *Sustain. Energy Fuels*, 2019, 4, 15–30.
68. Jadallah AA, Mahmood DY and Abdulqader ZA, "Modeling and Simulation of a Photovoltaic Module in Different Operating Regimes", *Acta Physica Polonica A*, 2015, 128, B-461–B-464.
69. Yağan YE, Vardar K and Ebeoğlu MA, "Modeling and Simulation of PV Systems", *IOSR Journal of Electrical and Electronics Engineering (IOSR-JEEE)*, 2018, 13, 1–11.
70. Kamarzaman NA and Tan CW, "A Comprehensive Review of Maximum Power Point Tracking Algorithms for Photovoltaic Systems", *Renewable and Sustainable Energy Reviews*, 2014, 37, 585–598.
71. Belkassmi Y, Gueraoui K, Tata O, Rafiki A, Elmaimouni L and Hassanain N, "Modeling and Simulation of Photovoltaic Module Based on One Diode Model Using Matlab/ Simulink", *Science Direct*, 2015, 14, 864–877.
72. Ma G, He J, Rajiv K, Tang SH, Yang Y and Nogami M, "Observation of Resonant Energy Transfer in Au: CdS Nanocomposite", *Applied Physics Letters*, 2004, 84, 4684–4686.
73. Sharma A and Gupta B, "Metal–Semiconductor Nanocomposite Layer Based Optical Fibre Surface Plasmon Resonance Sensor", *Journal of Optics*, 2007, 9, 180–185.
74. Liu X, Liang S, Nan F, Pan Y, Shi J, Zhou L, Feng S, Wang J, Feng X and Quan Q, "Stepwise Synthesis of Cubic Au-AgCdS Core-Shell Nanostructures with Tunable Plasmon Resonances and Fluorescence", *Optics Express*, 2013, 21, 24793–24798.
75. Zhou L, Fu XF, Yu L, Zhang X, Yu XF and Hao ZH, "Crystal Structure and Optical Properties of Silver Nanorings", *Applied Physics Letters*, 2009, 94, 153102.
76. Yang K, Liu M, Persson K, Han Y and Zheng H, "Strain Mediated Interfacial Dynamics during Au–PbS Core–Shell Nanostructure Formation", *ACS Nano*, 10, 6235–6240, 2015.

77. Lu W, Wang B, Zeng J, Wang X, Zhang S and Hou J, "Synthesis of Core/Shell Nanoparticles of Au/CdSe via Au-Cd Bialloy Precursor", *Langmuir*, 2005, 21, 3684–3687.
78. Ciulla G, Brano VL, Franzitta V and Trapanese M, "Assessment of the Operating Temperature of Crystalline PV Modules Based on Real Use Conditions", *International Journal of Photoenergy*, 2014, 11, 18315.
79. Villalva MG, Gazoli JR and Filho ER, "Comprehensive Approach to Modeling and Simulation of Photovoltaic Arrays", *IEEE Transactions on Power Electronics*, 2009, 24, 1198–1208.
80. Krismadinata, Rahim AN, Ping HW and Selvaraja J, "Photovoltaic Module Modeling Using Simulink/Matlab", *Procedia Environmental Sciences*, 2013, 17, 537–546.
81. Schwingshackla C, Petittaa M, Wagnera JE, Belluardoc G, Moserc D, Castellia M, Zebischa M and Tetzlaff A, "Wind Effect on PV Module Temperature: Analysis of Different Techniques for an Accurate Estimation", *Energy Procedia*, 2013, 40, 77–86.
82. Koehl M, Heck M, Wiesmeier S and Wirth J, "Modeling of the Nominal Operating Cell Temperature Based on Outdoor Weathering", *Solar Energy Materials & Solar Cells*, 2011, 95, 1638–1646.
83. Ayoub M and Quaiyum A, *Handbook of Research on Industrial Informatics and Manufacturing Intelligence: Innovations and Solutions*, ISBN:978-1-4666-0294-6, p. 464, 2012.
84. Dubey S, Sarvaiya JN and Seshadri B, "Temperature Dependent Photovoltaic (PV) Efficiency and Its Effect on PV Production in the World A Review", *Energy Procedia*, 2013, 11.
85. Thabet A, Abdelhady S and Mobarak Y, "Design Modern Structure for Heterojunction Quantum Dot Solar Cells", *International Journal of Electrical and Computer Engineering*, 2019, 10(3), 2918–2925.
86. New and Renewable Energy Authority, "Benban 1.8gw Pv Solar Park, Egypt Strategic Environmental & Social Assessment Final Report", February 2016.
87. Solar World AG Reserves, "Sun Module SW325XL MONO", Solar World Real Value Data Sheet, 2016.
88. Traina Solar Reserves, "The Tall Max Framed 72 Cell Module", Traina Solar Data Sheet, 2018.
89. Infinity Vogt Services S.A.E, "Benban Single Line Diagram", ib vogt GmbH Develops Renewable Energy Projects, Project Number. 2243, Serial Number. EF.101.1.1, 2017.
90. Dashtkar A, Frost-Jensen Johansen N, Mishnaevsky L Jr., Williams NA, Hasan SW, Wadi VS, Silvello A and Hadavinia H, "Graphene/sol-gel Modified Polyurethane Coating for Wind Turbine Blade Leading Edge Protection: Properties and Performance", *Polymers and Polymer Composites*, 2022, 30, 1–18.
91. Liu B, Liu Z, Li Y and Feng F, "Rain Erosion-Resistant Coatings for Wind Turbine Blades: A Review", *Polymers and Polymer Composites*, 2019, 27(8), 443–475.
92. Caruso RA and Antonietti M, "Sol_gel Nanocoating: An Approach to the Preparation of Structured Materials", *Chemistry of Materials*, 2001, 13(10), 3272–3282.
93. Zhang S, et al., "Erosion of Wind Turbine Blade Coatings—Design and Analysis of Jet-Based Laboratory Equipment for Performance Evaluation", *Progress in Organic Coatings*, 2015, 78, 103–115.
94. Park J and Yan M, "Covalent Functionalization of Graphene with Reactive Intermediates", *Accounts of Chemical Research*, 2013, 46, 181–189.
95. Chen J, Luo Z, An R, Marklund P, Björling M and Shi Y, "Novel Intrinsic Self-Healing Poly-Silicone-Urea with Super-Low Ice Adhesion Strength", *Small*, 2022, 18, 2200532.
96. Liu Z, Feng F, Li Y, Sun Y and Tagawa K, "A Corncob Biochar-Based Superhydrophobic Photothermal Coating with Micro-Nanoporous Rough-Structure for Ice-Phobic Properties", *Surface and Coating Technology*, 2023, 457, 129299.

97. Liao C, Yi Y, Chen T, Cai C, Deng Z, Song X and Lv C, "Detecting Broken Strands in Transmission Lines Based on Pulsed Eddy Current", *Metals*, 2022, 12, 1014.
98. Mu Z, Li Y, Guo W, Shen H and Tagawa K, "An Experimental Study on Adhesion Strength of Offshore Atmospheric Icing on a Wind Turbine Blade Airfoil", *Coatings*, 2023, 13, 164.
99. Li L, Liu Y, Tian L, Hu H, Hu H, Liu X, Hogate I and Kohli A, "An Experimental Study on a Hot-Air-Based Anti-/De-Icing System for Aero-Engine Inlet Guide Vanes", *Applied Thermal Engineering*, 2020, 167, 114778.
100. Li Y, Shen H and Guo W, "Effect of Ultrasonic Vibration on the Surface Adhesive Characteristic of Iced Aluminum Alloy Plate", *Applied Science*, 2022, 12, 2357.
101. Liu B, Liu Z, Li Y and Feng F, "A Wind Tunnel Test of the Anti-Icing Properties of MoS2/ZnO Hydrophobic Nano-Coatings for Wind Turbine Blades", *Coatings*, 13, 686, 2023.

9 Nanotech Green Energy Generation Systems

Nanotechnology has seen rapid progress and growth in its use in green energy generation systems based on its potential benefits in a variety of fields and industries across a wide range of applications. The rapid shift from conventional to nanoscale technology in the energy field has resulted in nanotechnology gaining popularity because it deals with different kinds of materials including graphite and graphene nanomaterials, nanocrystalline and nanocomposites. Nanotechnology in the energy sector has proven essential for converting, storing, and conserving renewable energy efficiently and economically.

9.1 INTRODUCTION

Among all the sources of environmental pollution, fossil fuels have been identified as the most significant one; renewable sources of energy have the obvious advantage of not using fuel, so they eliminate carbon dioxide production. A significant part of the current global energy problem is due to the insufficient supply of fossil fuels and the increased emission of greenhouse gases that result from the use of fossil fuels. Nanotechnology has played an important role in the development of selected green energy technologies via the conversion of sunlight directly into electricity using solar cells, the conversion of solar energy into hydrogen fuel through the separation of water into its constituents, and the storage of hydrogen in solid state, as well as the use of fuel cells in order to produce electricity. Future generations of renewable energy could replace fossil fuels using nanotechnology [1].

A major advantage of nanotechnology in sustainable energy is that it increases the efficiency of PV solar cells while reducing electricity production cost to unprecedented levels. In fuel cells, nanostructured materials increased hydrogen adsorption ability; this led to improved hydrogen production, storage, and electricity generation [2]. The use of nanomaterials allows generating clean and cheap energy, which is an ever-increasing field in the global economy. A nanomaterial can play a crucial role in increasing the efficiency of solar cells, fuel cells, and wind turbines.

Nanotechnology makes it possible to significantly reduce the costs of costly components such as solar cells and hydrogen production and storage elements if the developed countries want to reduce the environmental impact of burning fossil fuels to produce green energy generation systems. The wind energy sector, however, has also emerged as both a viable alternative to traditional energy sources and a key source of sustainable energy. Nonetheless, wind energy still faces numerous challenges with regard to its sustainability as a key source of energy. Wind energy competes with other conventional sources for market share based on factors such as manufacturing cost, environmental factors, and site-specific characteristics [3].

DOI: 10.1201/9781003512486-11

A carbon nanotube-containing epoxy is used for making windmill blades to increase electricity generation. The blades of these devices can be made stronger and lighter using epoxy that contains nanotubes; these blades are longer, thereby increasing the amount of electricity production and reduction power loss in transmission lines. The use of nanotechnology can improve wind power utilization via high strength, lightweight rotor blade materials, tribological coatings, and wear protection layers on bearings and gear boxes. Along with the development of nanomaterials for light protection, there are devices for energy storage that are nano-optimized for better charging efficiency and storing wind energy [4]. Besides, nanotechnology has the potential to play a significant role in increasing the number of hydrogen energy generators, solar power stations and wind farms that are currently available.

9.2 VISION OF GREEN ENERGY GENERATION

9.2.1 Vision of Hydrogen Energy

Hydrogen energy has become a popular source of energy; it is cleaner than fossil fuels, and it can be used instead of these fuels as an alternative source of energy [5]. Hydrogen also has higher energy density and is less harmful to the environment than fossil fuels. Hydrogen atoms consist of a proton and an electron, and their power is dispersed in a way that is invisible to observers. Hydrogen atoms are colorless and odorless, and hydrogen energy's gravimetric density is typically about seven times that of fossil fuels [6]. Hydrogen could become the main source of energy; however, a number of challenges will have to be addressed after it has been produced, including transport and storage [7]. These types of problems are solved through technological advancements and studies within the next few years; it is proven that hydrogen energy does not pose a threat to the environment or the climate in any way [8].

As transportation vehicles such as cars, planes, railways, and ships move towards hydrogen energy in the future, petrol will be replaced by hydrogen energy [9]. As technology advances, hydrogen production, storage, and transportation can become easier and cheaper. It is generally used in fertilizer production and refineries. Hydrogen is an excellent and cheap solution for all aspects of our lives since it can be converted into electrical energy; it is expected that hydrogen will be transported in liquid form by ships or pipelines. Our energy needs can be met easily through hydrogen energy as well as other renewable sources of energy; these sources of energy can be stored and used whenever and wherever our needs. A major advantage of hydrogen is its long-term storage capacity at low cost, whereas electrical energy can be stored for a long period from other sources of energy [10].

Advancements and development in hydrogen technology are already satisfied to be incorporated into national strategies. As the energy transition progresses, hydrogen may play a very important role with regard to cost reduction and fuel efficiency, but the main challenges of the energy transition are more related to setting up society and users for hydrogen's presence in normal day-to-day use and applications. In a relatively short time, hydrogen-based technologies and applications could move from cutting edge to everyday occurrence, based on government strategies and predictions [11].

Hydrogen is silent to operate and generates power at a low voltage using fuel cells. Using hydrogen or hydrogen-rich fuels and an oxidizer such as oxygen, fuel cells generate electricity. A compressor or blower is used to cool the stack and transfer oxygen to it and to eliminate water produced in the reaction, and the apparatuses are used to move the fuel through the system and eliminate water. Some fuel cells use an electrolyte such as hydrogen–oxygen with an acid, but hydrogen enters the anode section in others. A positive hydrogen ion is generated at the anode electrode when hydrogen molecules lose electrons at the anode. In a subsequent step, electrons move through the electrolyte layer and onto the cathode, where they combine with oxygen and produce water [12].

Anode and cathode are connected electrically by means of electrically conductive materials when electricity is required. In addition to hydrogen, other materials that can serve as ionic energy carriers include carbon and oxides of different elements: Based on the theory of chemical oxidation, any material that has the capacity to undergo chemical oxidation at the anode of a fuel cell can be used as a fuel source at the anode of the fuel cell in the presence of continuous supply (as a fluid) [13]. The electrochemical reaction releases some electricity, and the rest of the energy released comes from heat; therefore, this process is more efficient for electrical devices and automobiles than for personal computers. A fuel cell emits no nitrogen oxide since it reacts only with hydrogen and oxygen, which are the main components of air pollution. Heat is produced by the regular combustion process; the combustion of gasoline in cars and other engines produces oxides of nitrogen, which are toxic substances that are released into the atmosphere when open flames are ignited. Due to recent advances in nanotechnology, many interesting nanomaterials have been developed that are making fuel cells cheaper, lighter, and more efficient [14].

9.2.2 Vision of Solar Energy

Since solar energy has a clean and nearly unlimited supply of power, it is considered one of the most promising future renewable energy options. The different methods of utilizing solar energy have so far included solar thermal, solar NP, and solar thermoelectric approaches; so far there are a variety of ways to utilize solar energy. PV energy conversion from solar energy is one of the most widely used approaches among all the solar energy applications, and solar energy can be converted directly into electricity using this technology [15, 16].

The rapid development of solar cell technology and its remarkable potential to meet the world's energy needs attract the attention of many scientists and engineers. In spite of this, solar cells remain confined to a limited range of applications due to their high cost and low conversion efficiency with increasing ambient temperatures. A significant role will be played by thin-film solar cells in the NP market in the future to overcome such limitations; this technology is designed to minimize material consumption and in turn reduce cost. Increasing optical absorption and different light-trapping technologies have been extensively used to enhance thin-film solar cells' light absorption [17–20].

There has been considerable effort devoted to developing structures that trap energy within thin-film solar cells; many new structures, including graded refractive index photonic crystals, have been proposed in order to increase light harvesting in solar cells [21–23]. Recently, the third-generation technology of nanostructured light trapping has been shown to be quite promising for improving the efficiency of thin-film solar panels; however, it is still under evaluation because thin-film solar cells currently have a very low efficiency.

The limitation of solar cells led us to develop a technology that traps light, which has been one of the hotspots in solar cells and is now applicable to both organic and inorganic cells. However, the current light-trapping methods only consider optical effects; a solar cell is a photoelectric coupling device, and therefore, the electrical effects of light trapping are just as significant as the optical effects [15, 16, 24–26]. A wide variety of natural resources are available to generate electricity, including wind energy, water kinetic energy, fuel chemistry, and nuclear energy from radioactive materials. An important factor in a nation's monetary and financial growth is its use of natural energy sources. There is a growing demand for energy in different developing countries, which has led to a huge investment in conventional and non-conventional energy sectors [27–29].

This type of plant is usually controlled centrally and the capacity and dynamic characteristics of the plant can also be taken into account at the national level in order to be matched with the overall demand, which fluctuates continuously but does follow a fairly predictable pattern. As part of the generation dispatch system, a number of generating units are selected to meet a particular demand with enough spare capacity to meet the demand, besides some extra. Whenever extra power is needed, there must be a unit that has the lowest incremental fuel cost to meet the demand until the incremental fuel cost of that unit exceeds that of another unit for optimal economical operation. The generating unit must be able to deliver the minimum power it can handle after it has been committed in order to minimize generation costs [27, 28].

9.2.3 Vision of Wind Energy

Wind energy has been used as a renewable energy source for a long period of time because it is quite cheap and can be used to power industrial machines and systems. Wind energy is a clean and renewable source of energy that is becoming an increasingly important part of the ecosystem of self-powered energy systems. There is a great deal of potential for wind energy to be utilized and converted into self-powered power [30].

Sunlight and wind are two forms of renewable energy that can be utilized in significant numbers, and there is a growing trend toward sustainable renewable energy sources in response to the rapidly shrinking supply of viable fossil energy sources [31]. Materials with a high strength-to-mass ratio are being developed continuously to create wind turbine blades (WTBs) that are durable and cost-effective. As reinforcement for polymer-based nanocomposites, CNTs exhibit exceptional mechanical and multifunctional properties because of their unique structural makeups. CNT/polymer nanocomposites have been developed in significant amounts, and they are currently in the process of being developed for applications in wind energy. Nanocomposites made of CNTs and polymers, which can be used as matrixes for FRP structures or as coatings/sizings for fibers, have great potential for improving the stiffness, strength, and fatigue resistance of blade materials. The multifunction properties of nanocomposites based on CNTs and polymers offer a new concept for the development of blade materials that were designed to counteract lighting strikes, thermal shocks, and humidity in the course of their service life. A promising technique for evaluating blade reliability and lifetime is to use nanocomposites containing CNTs and polymer, but there are many fundamental issues and challenges to be addressed before CNT/polymer nanocomposites can be used in wind blade materials [32].

There are structural cracks between the root and the airfoil of composite WTBs. Structural compromise can be caused by abrupt changes in thickness or stress concentration at crack locations or lack of resin at crack locations. As a general rule, this type of problem results from fatigue, which is a key component of blade design. However, it can also be related to ineffective materials and inefficient production methods [33]. WTBs have been extensively analyzed among other wind turbine components, and today, as the need for renewable energy grows, companies now focus on rotor blades with a length of 80 meters. There are many environmental effects that can adversely affect the blade material in addition to the aerodynamic, inertial, and fatigue loads; ultraviolet degradation of the surface, dust accumulation in sandy locations, ice accumulation in colder climates, insect collisions, and moisture infiltration are just some examples of environmental effects that can affect the blade material. To ensure that the blades reach their intended lifespan, all of these considerations must be taken into account. Furthermore, an exponential increase in composite blade manufacturing produces an unrecyclable amount of material; all of these issues pose challenges for choosing the appropriate material for wind blades, ensuring that they are able to address the numerous issues while maintaining their structural integrity [34].

9.3 GREEN ENERGY POWER GENERATION

9.3.1 Hydrogen Energy Power Plants

Fuel cells may be divided into a variety of types depending on the electrolyte and operating temperature; for example, alkaline fuel cells operate between 60°C and 90°C, while chemical fuel cells range in efficiency from 50% to 60% [35–37].

Fuel cells based on polymer electrolytes or ion exchange membranes are one type of proton exchange membrane fuel cells (PEMFCs). These fuel cells can deliver more power while consuming less energy. The electrolyte of PEMFCs is composed of a solid polymer, while the electrodes of the cells are made of porous carbon containing Pt catalysts; the elements of the cell contain oxygen, hydrogen, and water; therefore, PEMFCs do not require corrosive fluid to operate. A PEMFC relies on an expensive catalyst (typically Pt) in order to separate hydrogen electrons and protons during hydrogen oxidation. An additional issue is that Pt catalysts are incredibly sensitive to carbon monoxide poisoning. This needs to be addressed by the addition of an additional reactor or providing CO-resistant catalysts if hydrogen is extracted from alcohol or hydrocarbon fuels [38].

A phosphoric acid fuel cell (PAFC) is a type of fuel cell that operates with liquid phosphoric acid as its electrolyte; it has an efficiency of 55% and operating temperatures of 160°C to 220°C. With the same weight and volume, PAFCs are larger, more massive, and less powerful than other fuel cells. In addition to using expensive Pt as a catalyst, the main reasons for the high cost of PAFCs is their size, mass, and lack of power [39].

Fuel cells powered by molten carbonate operating at temperatures between 620°C and 660°C are generally made of a porous, inert ceramic called lithium aluminum oxide ($LiAlO_2$); the expected efficiency is 60% to 65%. Furthermore, it is possible to significantly reduce the cost of fuel cells by using nonprecious metal catalysts. PAFCs, PEMFCs, and molten carbonate FCs differ from alkaline fuel cells in that they do not require the use of an external reformer to convert a more energy-dense fuel into hydrogen, which is an advantage to these types of fuel cells. This type of reformer is cheaper because it has a high operating temperature; however, since molten carbonate FCs operate at high temperatures and contain corrosive electrolytes that accelerate component breakdown, their durability is a major issue. To meet the growing need for longer-lasting batteries, it is necessary to look for materials that are resistant to corrosion [40–44]. Typically, solid oxide fuel cells operate at high temperatures between 880°C and 1000°C. These fuel cells can utilize coal-derived gas directly since it is inert to carbon monoxide, allowing it to be utilized directly in them. However, high-temperature working conditions can lead to slower startup times, as well as a need for appropriate thermal shielding to keep personnel safe and maintain heat while meeting the high durability requirements of the materials. Among the most important technical challenges, finding affordable materials that are durable when working under cell conditions is by far the greatest challenge [45].

9.3.2 Solar Energy Power Plants

Solar energy is one of the earliest forms of electricity generation from primitive times because it was free, readily available, and easily gathered through the transfer of heat from the sun to the earth. Solar electricity systems use solar radiation to generate electricity via NP thermal systems. NP solar energy systems produce hot water and air, cook food, dry materials, etc., through taking advantage of the sun's heat; solar electric energy systems utilize solar radiation to create electricity. NP effects are generated by beam and diffused solar radiation in the solar electric system.

PVs work by absorbing solar radiation, which is a combination of photons, and converting it into electrical power by producing electrons. When the conduction band has been occupied by electrons in the conduction band, holes in the valence band will be left once the electrons have left the conduction band. The fundamentals of a p-n junction state that some electrons are attracted to the n-side to merge with holes on the p-side. A similar phenomenon occurs when holes on the near p-side merge with electrons on the nearby n-side at the same location [46–50]. PV power generation systems are used in the following applications:

Off-grid domestic NP systems

An off-grid system is a system that can run independently of the main power grid and can supply electricity to remote locations and villages that do not have access to the main grid of electricity. This kind of system has been installed in a number of locations around the world, and it is often the best method for meeting the electricity needs of off-grid communities. A typical off-grid domestic system can produce between two and three kilowatts of power, and it typically works on distances beyond two and three kilometers from existing distribution and transmission lines [46, 47, 50], which extends the distribution grid cost-effectively.

Off-grid nondomestic NP systems

Commercial and industrial buildings have access to such systems, since these systems are the best arrangements when a small amount of electricity is of great value and can have great consequences. Hence, they are likely to compete with other small sources of electricity generation in terms of their cost in the future. A wide range of applications can be served by off-grid nondomestic solar systems, such as communications, water pumping and refrigeration in agriculture sectors, and navigational aids. These nondomestic solar systems provide power at a low operation and maintenance cost [46–48].

Grid-connected distributed NP systems

Commercial and industrial buildings or other loads connected to the utility grid can be powered by these solar power systems. Residential houses and commercial and industrial buildings often receive electrical power from these systems because they are usually integrated into the structure of the building. It is unnecessary to install a battery storage unit because these systems are directly connected to the approved electrical network. In general, these systems are more cost-effective than off-grid installations; they typically range from kW to MW in size, and if on-site generation is greater than load demand, there is feedback into the grid of the surplus. A grid-connected solar system is ideal for residential use during the summer season since it can be utilized depending on the load pattern at home [46, 48].

Grid-connected centralized NP systems

In general, these systems are installed for one of two reasons: They are either intended as a replacement for conventional centralized power generation systems, or they are designed to strengthen the utility distribution systems. The main components of a PV system are suitable for operation either standalone or in a grid-connected configuration; there are two kinds of NP power generation systems [46–50]. A solar panel converts solar irradiance into electrical power using the solar irradiance; any material can be used as a PV module to supply individual loads independently from the grid.

Since PV systems generate DC voltage, they need a DC/AC inverter to be able to supply AC loads as well as being connected to the grid. A DC-to-AC inverter converts DC current from the PV modules to the grid so it can be used for AC loads [46, 50, 51]; an inverter is the device that communicates between the PV modules and the grid.

Power electronics switching devices used in line-based inverters are typically thyristors that are not self-commutated and therefore rely on grid voltage to operate (commute). Unlike other types of inverters, line-commutated inverters draw reactive power from the grid (for the purpose of switching the thyristors); thus, they suffer from higher losses and lower efficiency overall if used with standalone PV systems.

This type of inverter uses self-commutating power electronic switching devices (usually MOSFETs when the application is applied to low power applications and IGBTs with medium to high power applications). These switching devices are switched on and off using gate signals, so grid voltage is not needed to operate these inverters. These inverters operate independently without reactive power, resulting in fewer power losses and a higher overall efficiency. Therefore, these inverters can be used on both grid-connected and standalone PV systems. Voltage source inverters (VSIs) are usually used in PV systems to control the output voltage and current. These regulators use a control signal to follow the desired reference voltage, while the current depends on the load to which the inverter is connected. In contrast, it is also possible to use a current-controlled VSI, which provides a current control of the output current in response to a reference signal [46, 49–51].

Power factor and fault current can be controlled with the current-controlled VSI, which is widely used for grid-connected systems [46, 49–51]. The PV modules and the loads in PV systems are integrated with DC-DC converters as an interface between them; DC-DC converters play an important role in the efficiency of solar PV [51–53]. To convert voltage levels of one DC source to another, DC-to-DC converters make use of electronic circuitry that is designed to convert voltage levels from low to high DC transmissions [54, 55].

In addition, the I-V characteristics of NP cells are nonlinear, and they also depend on factors like climate and temperature; for this reason, it is essential to work out where the maximum power can be extracted from the solar panel so that it can be used effectively. Among the P-V characteristics of the solar panel, there arises a point at which the entire panel operates at the maximum efficiency known as the maximum power point (MPP). Consequently, maximum power point tracking (MPPT) algorithms and techniques were developed to ensure maximum power from the PV module under all conditions [51, 52].

Proper MPPT is also employed along with DC-DC converters in order to maximize the power. There are now six accepted generations of DC-DC converters: first, classical/traditional; second, multiquadrant; third, switched-component SI/SC; fourth, soft-switching ZCS/ZVS/ZT; fifth, synchronous rectifier SR; and sixth, multiple energy storage elements resonant [56–58]. MPPT refers to the process of determining the peak power of solar panels to ensure maximum yield. MPPT is always able to maximize the amount of power being extracted from the solar panels, regardless of the atmospheric conditions [51, 52]. Even though solar energy is one of the most promising renewable energy resources for the future, its efficiency depends on the amount of solar irradiation falling on the PV panels and the temperature and shading that affect the efficiency of power transfer [50]. By locating the MPP under rapid atmospheric changes and partial shading conditions, soft computing techniques have an advantage over conventional MPPT; there is a continuous fluctuation around MPP in conventional techniques [51].

Several factors influence MPPT performance, including tracking time, convergence time, and oscillations near MPP, which are considered when analyzing the performance of advanced MPPT techniques [50–52]. Solar panel voltage, current, temperature, and solar irradiation are the data used in the conventional methods to track the MPP of the system. The conventional MPPT techniques are effective under uniform environmental conditions; for example, when solar insolation and temperature are constant, there is only one maximum power point, whereas when these conditions change, the point changes as well. MPPT is one of the most common methods used to adapt to current and future inputs, but there are various methods associated with it, including perturb and observe, incremental conductance, short-circuit current, and open-circuit voltage [50–52].

DC-to-DC converters can be controlled in several different ways to track the MPP; the solutions range from a simple current-control system that assumes the PV or load voltage to remain constant, and the MPPT try to maximize the current to a more sophisticated system that optimize the output power [50–52]. Controlling the DC-to-DC converter consists of changing the duty ratio (mark-to-space ratio) of the converter to produce maximum output power by controlling the PV module voltage accordingly. As the battery voltage is a major factor in determining the operating point of standalone PV systems, the operating point is usually set as close to the MPP as possible. A battery is not permitted to fully discharge, so the voltage is kept fairly constant and close to the MPP in practice as shown in Figure 9.1. Hence, MPPT is not essential for standalone systems as the PV modules are typically operated around the MPP the majority of the time, making it desirable (to maximize the PV energy output).

According to the data in Tables 9.1–9.3 and Figures 9.2 and 9.3, the J-V and P-V characteristics of the CdS/(CdTe + 20wt.%Al) nanocomposite thin-film solar cell made with multiple NPs enhanced the short circuit current density and open circuit voltage of CdS/CdTe compared with cells made with individual NPs. Adding multiple NPs (20wt.%Al + 10wt.%Cs) in CdTe absorber layer was the best solution for increasing the short circuit current density, open circuit voltage, and output power of the cell.

Figures 9.4 and 9.5 show the J-V and P-V characteristics of CdS/(PbS + 20wt.%Al) nanocomposite thin-film solar cell using multiple NPs. It is noticed that using individual NPs or NP composites improved J-V and P-V characteristics of the CdS/(PbS

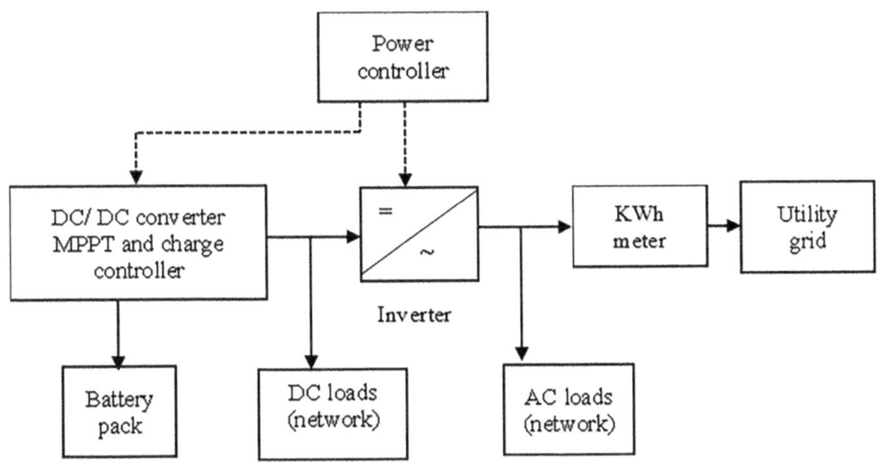

FIGURE 9.1 Typical main components of a PV system.

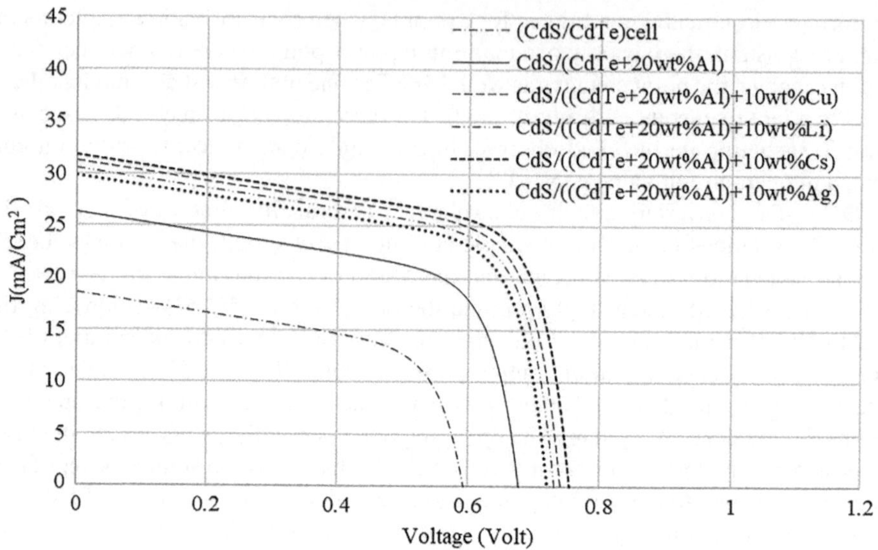

FIGURE 9.2 J-V characteristics of CdS/(CdTe + 20wt.%Al) nanocomposite thin-film solar cells.

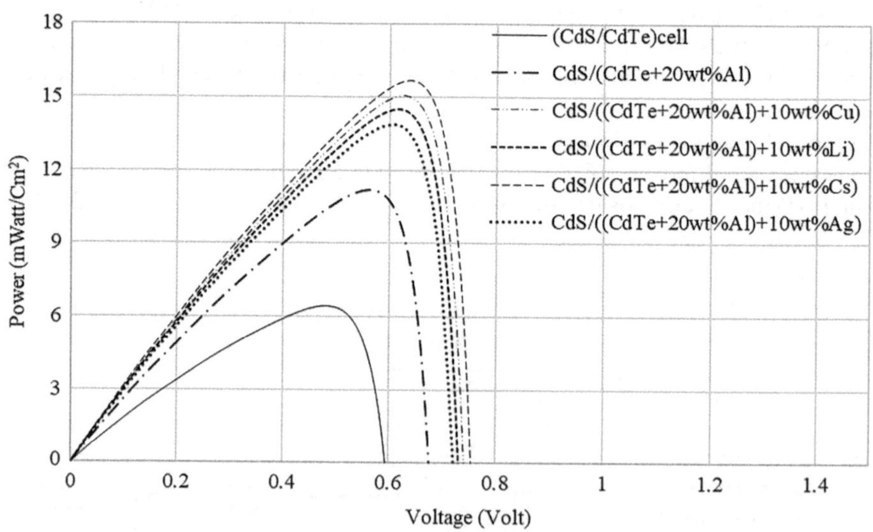

FIGURE 9.3 P-V characteristics of CdS/(CdTe + 20wt.%Al) nanocomposite thin-film solar cell.

+ 20wt.%Al) cell. Moreover, adding 10wt.% of Cs, Cu, Li, or Au as a second NP filler enhanced the output power, short circuit current density, and open circuit voltage of the CdS/(PbS + 20wt.%Al) nanocomposite thin-film solar cell. In a PbS absorbing layer, 20wt.%Al and 10 wt. %Cs added into the base matrix represented the best

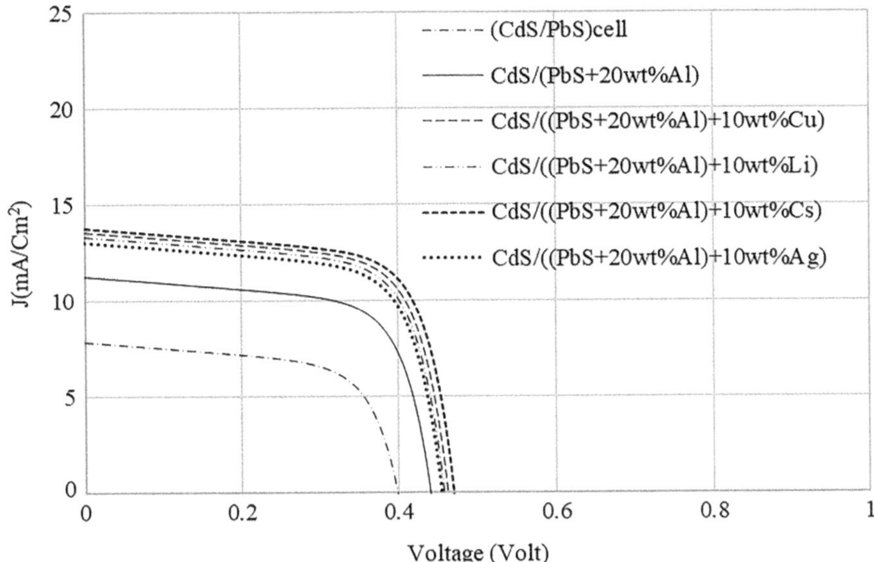

FIGURE 9.4 J-V characteristics of CdS/(PbS + 20wt.%Al) nanocomposite thin-film solar cells.

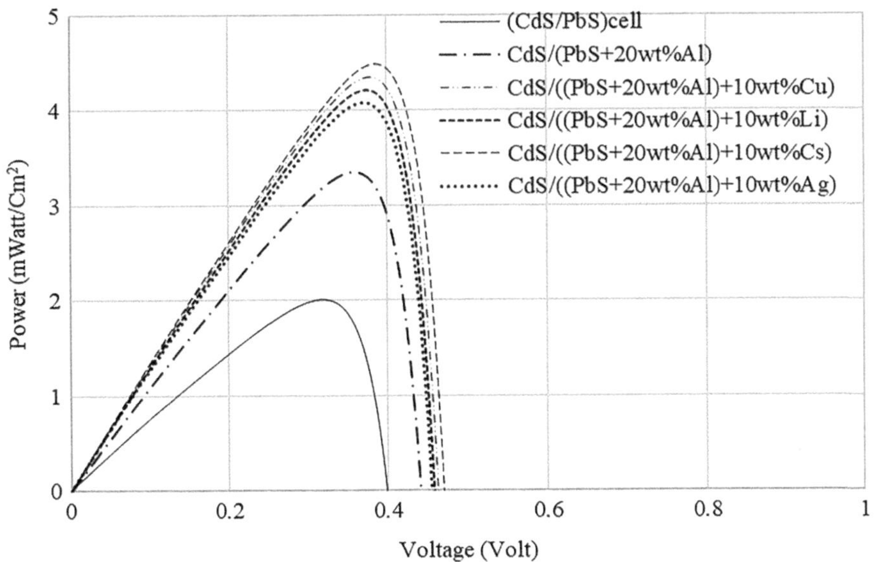

FIGURE 9.5 P-V characteristics of CdS/(PbS + 20wt.%Al) nanocomposite thin-film solar cells.

TABLE 9.1
Temperature Effect on Efficiencies of ITO/CdS/PbS/Al and SnO_2/CdS/CdTe/Cu Thin-Film Solar Cells Using Metal Nanocomposites

	Efficiency % SnO2/CdS/CdTe/Cu cells				
Solar Cells	**300k**	**330k**	**350k**	**370k**	**400k**
CdS/CdTe	6.15	5.26	4.69	4.13	3.32
CdS/(CdTe+20wt.%Al)	11.86	9.79	8.89	8.02	6.76
CdS/(CdTe+20wt.%Al) +10wt.%Ag)	13.86	12.24	11.19	10.17	8.71
CdS/(CdTe+20wt.%Al) +10wt.%Cu	15.07	13.31	12.19	11.09	9.53
CdS/(CdTe+20wt.%Al) +10wt.%Li	14.49	12.80	11.71	10.66	9.14
	Efficiency % ITO/CdS/PbS/Al cells				
CdS/PbS	2.01	1.74	1.58	1.42	1.21
CdS/(PbS+20wt.%Al)	3.37	2.98	2.74	2.50	2.16
CdS/(PbS+20wt.% Al)+10wt.%Ag	4.16	3.73	3.45	3.17	2.78
CdS/(PbS+20wt.% Al)+10wt.%Cu	4.39	3.92	3.61	3.31	2.89
CdS/(PbS+20wt.% Al)+10wt.%Li	4.27	3.82	3.53	3.25	2.84

filler for enhancing the performance of the CdS/(PbS + 20wt.%Al) nanocomposite thin-film solar cell.

Thermal condition is an effective parameter in selecting the design, production, and operation of nanocomposites for green energy generation. Table 9.1 shows the temperature effect on the efficiency of nanocomposite thin-film solar cells, ITO/CdS/PbS/Al and SnO_2/CdS/CdTe/Cu, using individual and multiple-metal NPs. Increasing the temperature decreased the efficiency of CdS/CdTe and cells, and all results depict that adding multiple NPs improved the efficiency performance of CdS/(PbS + 20wt.%Al) and CdS/(CdTe + 20wt.%Al) nanocomposite thin-film solar cells at high temperature. The Cs NPs were ineffective at high temperature due to their low melting point.

9.3.3 Wind Energy Farms

The wind turbine is a mechanical device that converts kinetic energy into mechanical energy that is then utilized by a generator to generate electricity by rotating a shaft mounted on the generator. The use of wind energy does not emit any harmful pollutants, making it one of the cleanest forms of energy. There are several components to a wind turbine unit [55]:

1. **Anemometer:** Data is transmitted to the controller based on wind speed measurements.
2. **Blades**: Most wind turbines are bi- or tribladed; when wind blows over the blades, the blades move upward and rotate.
3. **Brake**: Disc brakes are applied mechanically, electrically, or hydraulically to stop the rotor in an emergency.

4. **Controller:** When the wind speed reaches a certain level, the controller shuts off the machine and starts it up again.
5. **Gear box**: For efficient generator operation (electricity production), wind turbines usually require a step-up gearbox. To produce electricity, generators need to rotate at a rotational speed. In order to reduce gearbox cost and weigh, engineers are exploring "direct-drive" generators that do not require gearboxes and operate at lower rotation speeds.
6. **Generator**: Electricity generated by an induction generator with a 60-cycle run.
7. **High-speed shaft**: Drives the generator.
8. **Low-speed shaft**: The rotor turns the low-speed shaft at about 30 to 60 rotations per minute.
9. **Nacelle**: Consists of a gearbox, shafts with low and high speeds, a generator, controller, and brake and is usually mounted on top of a tower. Some nacelles can house helicopters as well.
10. **Pitch**: A rotor's speed is controlled by turning the blades out of the wind, or pitching them, to prevent the rotor from exceeding its maximum or minimum torque.
11. **Rotor**: The blades and the hub together are called the rotor.
12. **Tower**: A taller tower enables turbines to capture and generate more energy. Towers can be made of tubular steel, concrete, or steel lattice.
13. **Wind direction**: A "downwind" turbine runs away from the wind, whereas an "upwind" turbine works with the wind behind it; other turbines are designed to work toward the wind, facing away from it.
14. **Wind vane**: Orients the turbine properly to the wind by measuring wind direction and communicating with the yaw drive.
15. **Yaw drive**: Downwind turbines are not required to have a yaw drive because the wind blows the rotor downwind. During the change in wind direction, upwind turbines maintain the rotor facing into the wind by using a yaw drive.

A new form of hydrophobic coating will be developed based on recent advances in nanotechnology and material engineering. Anaerobic digestion uses lignocellulosic biomass, a raw material currently undervalued [56, 57]. A natural gas residue that is improperly disposed of can lead to environmental pollution, while hydrophobic nanomaterials prepared from it can potentially improve wind turbine blade anti-icing properties. WTBs coated with nanocarbon coatings were less prone to icing than uncoated blades, and they were easier to deice. Corn straw biomass left over from processing a corn crop made a coating that could effectively reduce icing on WTBs [58].

9.4 SUSTAINABILITY OF GREEN ENERGY GENERATION SYSTEMS

9.4.1 Hydrogen Energy Systems

Hydrogen energy generation poses significant challenges for material science, such as production, purification, storage, and conversion. Solving these challenges will help advance this area. Hydrogen can be produced in a variety of ways, including water electrolysis, allowing pure hydrogen to be produced, and catalytic conversion of natural gases and alcohols. The use of composite materials in the design or manufacturing of chemical processes, such as oxide-supported metal particles or alloys in conversion processes, is commonly accepted as a method of catalysis; however, the processes also produce carbon oxides. The most commonly used low-temperature fuel cells are based on fuel cell power systems, and carbon dioxide emissions are regulated while carbon monoxide is poisonous to the catalysts [59].

Low-temperature proton-conducting membrane FCs are almost always equipped with Pt or Pt alloys as catalysts due to their high efficiency in electrocatalytic processes. Researchers have attempted to replace Pt with low-cost transition metals, but their degradation upon contact with the proton-conducting membrane leads to rapid poisoning and rapid termination of the process. Reducing catalyst particle size increases the rate of electrochemical processes, which decreases Pt consumption, but small catalyst particles degrade rapidly [60]. Scientists have used cathodic and anodic processes to obtain hydrogen using nanomaterials, electrochemical, photocatalytic, and electrophotocatalytic [61–63]. A catalyst is recommended for water splitting to produce hydrogen using semiconductors with high stability during interaction with water [64].

Hydrogen can be generated using electrical, thermal, biological, and photonic methods, and hydrogen generation systems have been developed for a wide range of applications such as heaters and coolants for refineries, glass manufacturers, iron factories, transport systems, and industrial setups for extended periods of time [65]. Nanomaterials that have been utilized to create hydrogen include carbon nanomaterials, solid inorganic–organic hybrid materials, graphite oxide, MoS_2 nanomaterials, Ni, QDs, decomposed methane, metal sulfides, TiO_2-based materials, and metal NPs. Future research should focus on the development of composites based on these advanced nanomaterials in order to improve the efficiency and quality of energy production in the future [66].

Metal oxide nanomaterials and nanotechnology have also been extensively used in numerous fields, including energy, the environment, health care, and electronics, due to their high surface area; tunable band gap; enhanced catalytic activity; and enhanced electrical, optical, and mechanical performance. Metal oxide nanomaterials have proven to be highly effective in improving the performance of batteries, supercapacitors, and solar cells in a sustainable and efficient manner, enabling the mass production and storage of energy. These nanomaterials have been used in environmental applications such as water purification, air filtration, and pollutant degradation, thereby contributing to the preservation of clean air and water resources. Metal oxide nanomaterials and nanotechnology can play an important role in addressing some of the most pressing global challenges and contributing to the promotion of sustainable development in many sectors.

In the health care sector, metal oxide nanomaterials play a role in drug delivery, diagnostics, and regenerative medicine. Various sustainable energy applications and environmental applications can be derived from metal oxide nanomaterials synthesized using green chemistry techniques. Enhancing green synthesis methods, exploring new materials for sustainable energy, and improving metal oxide nanomaterial applications, this research can pave the way for a cleaner, more sustainable future by offering viable solutions to a significant number of global energy and environmental challenges; through studying these materials, studies can pave the way for a cleaner and more sustainable future [67].

Metal oxide nanomaterials are crucial to unlocking the full potential of metal oxide nanomaterials and contributing to the implementation of innovative, ecofriendly nanotechnologies in emerging areas including photocatalytic water splitting, solar cells, and thermoelectric materials. Using ecofriendly synthesis techniques and examining innovative materials can promote a cleaner, more sustainable future, too. Metal oxide nanomaterials exhibit immense potential for addressing urgent energy and environmental issues [67].

A nanotechnology-based insulating material can save 33% in energy over a conventional material. These insulating materials can be used on solid panels or even as a thin film on any surface to reduce heat loss. A new technology based on nanotechnology is bringing fuel cells to the next level of efficiency through hydrogen sensors and nanotechnological membranes. Likewise, wind panels and solar cells are more efficient owing to nanomaterials. The high insulation ability of nanomaterials can be used to prevent power losses in transmission lines and transmission cables, and energy savings can be achieved through the use of increased hydrogen storage rates and reduced loss rates in power transmission lines and transmission cables [68–70].

Nanofluids can be used as coolants to reduce radiator sizes and facilitate easier positioning of radiators; as a result of their high thermal performance, it is possible to reduce the amount of working fluid that must be used, shrink the cooling system pump, and operate truck engines at high temperatures to achieve higher horsepower. As a result of the use of nanofluids as a working fluid in radiators, there is a possibility that the volume of the radiator system may be reduced by 10% [71, 72].

9.4.2 Solar Energy Systems

Solar cells based on alternative materials are increasingly being developed since NPs has a low absorption coefficient and an indirect band gap. Polycrystalline NP solar cell structures require higher thicknesses because of their low absorption; they have been employed in applications using non-NP intermediate band gap materials [73–76]. Thin-film solar cells have a number of unavoidable shortcomings; among them is the poor optical absorption because of the thinner active layer, and it limits the power conversion efficiency of these solar cells with temperature increase. There have been extensive advancements in light-trapping technologies capable of enhancing the photon absorption abilities of thin-film solar cells to boost their optical absorption capabilities [77–81].

Several studies have identified QDs as a promising third-generation solar cell. Since several properties can be changed due to a decrease in NP size, QDs are a great alternative to using these in all solar cells to decrease cost, increase absorption and increase efficiency. As a result of a decrease in dot size, several properties can be altered as well [82, 83]. QDs have been applied in a variety of types of solar cells [84–88]. The Schottky solar cell is a combination of organic and inorganic elements. Optoelectronic devices can be enhanced by metal core/shell nanostructures since they enhance charge separation and light absorption, increasing semiconductor light harvesting efficiency. Significant performance improvement is observed when considering the functional differences between the metal and semiconductor components, as well as tuning the shell and core nanostructures during synthesis. Furthermore, the structure of the interfacial region influences the electronic properties of the core shell NPs [89, 90].

In metal-semiconductor nanocomposites, the charge transfer mechanism increases the photoluminescence intensity to 40 times that of CdS NPs. In addition, it is possible to use the surface plasmon waves on the gold surface to excite Au electrons that can transfer to the CdS conduction band and recombining with holes in the CdS valence band. The atomic structure of a nanomaterial can determine its electronic and optical properties; this leads to improved band gaps and defect emissions in CdS nanostructures [91]. There are many ways to control NP properties such as size, shape, composition, and atomic structure. The combination of properties of various compositions and shell structures in these core/shell NPs has led to their increasing popularity; metal-semiconductor core/shell structures are particularly interesting due to the bifunctional properties of metals and semiconductors [92]. Today, they have been used to improve the performance of traditional CdS/QDPbS HJ-QDSc models, thin-film PV CdS/CdTe and CDS/CIGS solar cells using QD metal-semiconductor cores and shells. Au/shell absorber layers with QD window layers achieved high efficiency using less material [93–95].

9.4.3 Wind Energy Systems

A new generation of nanocomposites has been developed with advanced nanomaterials used in wind turbine blades to make them lighter and stronger. Nanocomposites with the light weight and strength that wind turbines need can be developed using graphene, a layer of carbon atoms that resembles a chain-link fence on a nanoscale. Graphene platelets are nanometer-thick particles of graphene that have become increasingly popular as additives for advanced composites due to their superior performance over epoxy composites that are infused with CNTs or other NPs and nanofillers. Graphene and CNTs have similar chemical compositions, exceptional strength, and mechanical properties. However, graphene has the ability to lend itself more readily to a host material in comparison with CNTs. Adding graphene to a composite of 0.1% of its weight can increase its strength and stiffness as much as adding CNTs to 1% of its weight to increase its strength and stiffness to a similar degree. A composite containing graphene filler can resist fatigue crack propagation by two orders of magnitude in comparison with an epoxy material without graphene fillers [96].

CNTs infused into polymers determine the overall mechanical properties of the composite by determining the weak link between the CNTs and the polymer. It is more cost-effective to mass produce graphene than CNT because graphite is readily available and low-cost [97]. In response to the numerous developments associated with nanotechnology, there has been a growing concern regarding nanomaterials' potential adverse environmental impacts. Using graphene-based nanocomposites in airplanes could save 1000 tons of gasoline in weight reduction. With its excellent properties, nanocomposite graphene can reduce WTB weight and improve energy conversion efficiency. However, graphene nanocomposite is harmful to the environment due to its toxic properties, thermal conductivity, and fire-retardance, which are extremely difficult to remove from waste [98].

9.5 HYBRID GREEN ENERGY SYSTEMS

9.5.1 Hybrid Green Hydrogen Energy

Hydrogen is regarded as one of the most efficient forms of green, safe, and sustainable energy because of its affordability and lack of environmental impact. Additionally, the conversion of sunlight and water resources into hydrogen fuel is an important aspect that we need to keep in mind. Since the development of nanotechnology, there has been intense focus on the separation of water; however, the various chemical processes used to synthesize NPs for water splitting are toxic to humans and the environment. There have been extensive efforts in recent years to produce NPs from plants and other microbes as a green method of synthesizing them. Prior to designing NP catalysts, it is imperative to acknowledge the mechanism of photo water splitting. Additionally, NPs should be synthesized and tuned using green methods; in one-line statements, NPs utilize the sun's energy to split water into hydrogen and oxygen [99].

There are two types of bands on NPs: conduction bands and valence bands. The conduction bands are formed by the absorption of photons from the sun, and the electrons are able to jump between the bands. Water molecules are reduced and oxidized in the presence of these electrons and holes at the surfaces of NPs as described:

$$H_2O_{(l)} \rightarrow H_{2(g)} + \frac{1}{2}O_{2(g)} \tag{9.1}$$

The half-cell reactions are as follows:

$$H_2O_{(l)} + 2h^+ \rightarrow \frac{1}{2}O_{2(g)} + 2H^+ \tag{9.2}$$

$$2H^+ + 2e^- \rightarrow H_{2(g)} \tag{9.3}$$

A CN/BG hybrid electrocatalytic system based on boron-doped graphene with porous, active graphite carbon nitride layer was designed and synthesized; this hybrid electrocatalytic system has a great potential for HER. On graphene, a porous CN layer can be deposited to provide defects and edges that can serve as active sites for proton adsorption and reduction. This hybrid CN/BG system was tuned to maximize HER activity and stability by optimizing its composition, structure, surface electronics, and chemical properties. A novel electronic and chemical feature of the hybrid surface was tuned so that it could produce outstanding HER performance comparable with that of some traditional metal catalysts as well as carbon-based metal-free catalysts that have comparable HER performance. An electronic interaction between CN and BG components facilitates the HER process, as shown by surface chemical XPS and surface structural TEM analysis. Moreover, it is possible to influence a catalyst's electrocatalytic performance significantly by adjusting its structure, composition, electronic properties, and surface chemical properties at the nanoscale; it is able to provide new insights into carbon-based hybrid catalysts for HER [100].

A major challenge in the fields of chemistry and energy is to develop electrocatalysts that generate hydrogen that are active, stable, and low-cost; nonprecious metal catalysts are regarded as promising alternatives to noble metal catalysts used in HER, but they are still practically unfeasible largely due to unsatisfactory activity and durability. Through calcination of W-polydopamine compounds precursors, an easy two-step method has been developed to prepare WOx nanowires with high oxygen vacancies; the hybrid material with an ultralong and uniform structure of 1D nanowires, as well as rough and raised surfaces, can be used to maximize the surface area of a material.

The nanowires also exhibit metal properties and are capable of being a new kind of robust catalyst for electrochemical reactions in acidic media. This simple route allows the nano-structuring, doping, and hybridization of nanocarbon with nanocatalysts to be performed, which is an effective strategy to improve the catalytic activity of an electrocatalyst [101]. A range of UV and visible light active metal oxides have also been used for photocatalytic hydrogen production. Various metal oxide based photocatalysts are useful for developing visible light active photocatalysts including metal ion doping, nonmetal ion doping, metal–metal, metal–nonmetal cooped systems, and dye-sensitized systems. It still needed to develop a photocatalyst with a narrower band gap of 2 eV; it must have a high photocatalytic activity and must be stable over time. The fabrication of a narrow band gap semiconductor that effectively utilizes visible light is essential in order to develop a highly efficient visible light-driven photocatalyst; checking the recombination of charge carriers generated by the photogeneration process is very important for photocatalytic efficiency [102]. Wind, solar power, bioenergy, and hydrogen are all clean energy sources that complement each other within an energy system; when one energy source becomes insufficient, the other energy source must be utilized to supply the system. As a result of the transition to renewable energy, every clean energy source must complement each other.

Hydrogen is principally produced by water, but it can also be produced by the decomposition of waste plants and wastewater, which is important for the use of fuel technology in today's vehicles. Many of the largest difficulties restricting the use of fuel technology in today's vehicles are related to hydrogen production, storage, and transportation [103]. Recently, graphene@TiO_2 nanomaterial was used as a sacrificial reagent in a mass ratio of 1:20 to investigate the photocatalytic, thermocatalytic, and photothermocatalytic activities, and at 110°C, the hydrogen evolution activity of the nanomaterial was 24.8 times that at room temperature. Thermally assisted photocatalysis, which is the main driving force of synergistic photothermal catalysis, had less thermocatalytic activity at 90°C than photothermocatalysis at this temperature. Hydrogen production can be enhanced because of the gas–liquid flow in a micro-boiling state, the rapid migration and diffusion of the sacrificial reagent molecules at higher temperatures, and the excitation of these absorbed molecules at these higher temperatures. This technique can produce hydrogen without chemicals in a system powered by solar light for a novel and efficient hydrogen production process [104].

9.5.2 Hybrid Green Solar Energy

Several solar-pumped hydro storage systems have been developed using DC-coupling pumps and have seen a number of refinements; in spite of the fact that solar energy is only available during daytime, solar energy is considered less intermittent than wind. Through pumping from a lower reservoir, pumped hydro storage (PHS) is widely used to store energy from the grid at times of low demand and price and release it into hydroelectric turbines during peak demand and price periods to regain the stored energy. A PHS can also provide additional benefits by storing surplus solar produced during midday peak periods for later use during evening peak demands. In the first generation of solar-PHS systems, centrifugal pumps driven by DC/AC motors displayed an efficiency range of 25%–35%. In the second generation, positive displacement pumps based on low PV energy demonstrated an efficiency range of 70%–80%. Currently, the modern solar pumps can lift water to more than 200m with an output of more than $250\ m^3/day$ using the latest electronic technology [105, 106]. Solar power systems with battery, fuel cells, PHS, flywheel, or compressed air storage have proven to be more reliable when solar power is the only primary energy source. The following are applications for hybrid solar fuel–wind energy [107]:

a. **Solar NP-Wind-Battery Hybrid Systems (PV-W-B)**
   PV-W-B comprises solar NP panels and small wind turbine generators for producing power. For the most part, such systems are capable of operating at lower power levels. The regular energy generation limits for PV-W-B systems are in the range of 1 to 10 kW.
b. **Solar NP-Wind-Diesel-Battery Hybrid System (PV-W-D-B)**
   This system incorporates solar panel, diesel generator, wind generator, and energy storage device.
c. **Solar NP-Fuel Cell Hybrid System (PV-FC)**
   PV and fuel cell energy frameworks are synchronized with electrolyzer to produce hydrogen and manage phase charging-dispatch control.
d. **Solar NP-Wind-Fuel Cell System (PV-W-FC)**
   This framework comprises PV panels, wind power system, and fuel cell system. Electrolyzer is utilized to assimilate the quickly fluctuating output power with load and produce hydrogen.

9.5.3 Hybrid Green Wind Energy

A novel wind-solar hybrid approach that integrates a dust-cleaning function generated by oscillations for actual typical agricultural scenarios has been thoroughly proven as a useful component of smart agriculture. Wind-solar hybrid energy harvesting systems have been one of the major components of smart agriculture. With this approach, wind power generation can compensate for the sharp oscillations in solar power caused by day–night cycles and weather fluctuations. In addition to cleaning PV panels during rainy weather, the approach will also slow down attenuation trends of solar power generation as much as possible maintenance-free. The control circuit also means that there is no need to conduct regular maintenance on the panels [108].

Plants that need controlled climate conditions are grown in greenhouses that have walls and roofs composed mostly of transparent materials. Certain greenhouse processes require human involvement and care for them to remain operational on a continuous basis, such as irrigation and ventilation; ideally, the greenhouse should be fully automated in order to achieve this goal as effectively as possible while minimizing human intervention and support. To maintain the plant's desired temperature, its ventilation is equipped with an integrated temperature control system. Water is pumped through a submersible pump and flows downward, flowing first through the water piping system and then down to the vertical plant mount tubing. This modern irrigation system watered plants in a more efficient manner, helping them receive water more quickly, which increases their growth rate [109].

An ANFIS-based energy management system was more effective in controlling hybrid system resources and managing their energy than a conventional energy management system. A converter with multiple ports offers more efficiency than a converter with a single port in terms of switching losses, cost, power density, and thermal management than a converter with just one port. Under ANFIS control, battery and hybrid system efficiency can be enhanced significantly over traditional STATE controls by injecting additional energy into the grid. A teamwork has been presented a technical overview of hybrid PV/wind converter configurations, along with the system's ability to work as a grid-connected or standalone power [110].

Hybrid NCPV/HAWT systems are suitable for eco-houses located in deserts or rural areas where electricity and drinkable water are not readily available. Wind turbines and nanocrystal NP panels are clean and renewable forms of energy and are also pollution-free. Electricity from the public utility is much cheaper than the electricity that is generated from the renewable energy resources, but it is not available in desert zones. There is great benefit from using hybrid systems to generate electricity in deserts and rural areas when they are cost-competitive with conventional sources [111].

9.6 TRENDS AND RECOMMENDATIONS

Nanotechnology is being applied in a wide range of different aspects of renewable energy; in the case of PV/T systems, nanotechnology is being used to increase efficiency by enhancing the heat transfer ability of the nanofluids that are used in heat transfer. Nanotechnology is also improving the structures and materials of PV modules that are used in their production. In fuel cell technology, nanotechnology is producing lower-cost fuel cells with higher efficiency and higher distribution and storage of hydrogen fuel, and nanocatalysts improve biofuel efficiency. Nanotechnology can provide materials with higher durability and lower weight; moreover, there is a growing trend of applying nanotechnology to wind and ocean energy generators to improve their performance [12].

9.6.1 Trends in Hydrogen Energy Power Generation

Hydrogen supply directly affects the safety and stability of power plant operations; continuous and stable hydrogen supply is of great significance. At nuclear power plants, hydrogen storing tanks must be as far as possible away from any air vents. In addition, it is advisable to install the hydrogen leakage sensor away from the storage tank section in the corner rather than on the corner line [112]. Many car manufacturers have introduced hybrid or electric cars with liquid or gaseous hydrogen fuel cells or even internal combustion engines to operate on a hydrogen fuel. This is a well-established application for hydrogen, one of the most abundant substances in the universe. A large number of ship design firms have introduced designs for auxiliary power sources or hybrid propulsion systems that utilize liquid hydrogen to power combustion engines or fuel cells. It is possible to use hydrogen as a marine fuel in internal combustion engines; however, existing diesel engines will need to be modified to utilize hydrogen. Developing fuel cells powered by hydrogen gas turbines presents an interesting challenge both from the standpoint of adopting the new type of fuel for waterborne vehicles and from the standpoint of hydrogen's different properties from those of other types of fuels [113].

9.6.2 Trends in Solar Energy Power Generation

Solar cell systems with nanostructures exhibit a variety of properties and have been used for converting solar energy into electricity or fuels in new ways. Nanostructured solar cells have several advantages, but they still have two main limitations: one, the sun's radiated energy cannot be harvested during the night, and two, the amount of radiated energy available throughout the day is not the same. The operation of a NP device requires a large amount of energy because solar energy fluctuates throughout the day as a result of the intensity of the sun's rays. To reach the surface of the earth, solar radiation has to overcome many hurdles, including travel time from the sun and weather conditions, as well as location during summer and winter seasons. Winter offers minimal sun radiation; some researchers have been working on developing high-efficiency solar cells with energy storage capability that can be used at night to overcome the limitations of solar technology [1–3].

Despite advantages of nanostructure devices for commercial applications, nanostructured solar cells can suffer from poor surface passivation due to improper etching. As NP technologies operate for longer periods, the total cost of the system decreases. Stability issues can occur due to chemistry or device configuration; using encapsulation routes to resolve stability issues enables the commercial viability of nanostructured solar cells. Two key features of nanostructures for NP applications have been identified and are outlined in the following paragraphs: a) a noticeable decrease in material consumption and/or associated cost and b) the ability to provide devices with a greater limiting efficiency based on Shockley–Queisser analysis. If both approaches are combined, household energy consumption can be significantly reduced in terms of cost per kWh; rated cost per peak watt is the unit that measures household energy consumption.

Solar cells convert a fraction of energy to total energy irradiance to calculate power conservation efficiency. The total consumption cost per peak watt is two times higher than module cost per peak watt and inversely related to power conservation efficiency. Efficiency of solar cells is classified based on thermodynamics, reflectance, conductivity, and separation of charge carriers. Hence, power conversion efficiency refers to the efficiency with which incident light power is converted into electrical power. Several other parameters can also be used to check the efficiency limits of power conversion, such as temperature coefficients, shadow angles, and temperature-dependent curves for measuring power conversion efficiency.

Quite a few factors influence external quantum efficiency, such as the level of fill factor, the amount of open circuit voltage, and the reflectance ratio. The separation and absorption mechanisms between single and multijunction materials are different. Single junctions are formed when there is only one layer of light-absorbing material on the surface of the material. Until now, the solar cell family consists of three generations. Currently, a wafer-based solar cell and a NP technology are the main components of the first generation; thin-film solar cells are the component of the second generation. The third generation is considered to be an emerging NP technology based on the thin-film solar cell technologies. Various organic, inorganic, and organometallic compounds are under examination in the stages of third-generation solar cells' development [1, 24]; these cells offer high efficiencies but also stability issues. Perovskites and dye-synthesized solar cells have great futures due to their increasing efficiency. Nanostructured solar cells have great potential to achieve such mentioned objectives due to their high-power conversion efficiency (PCE), price cost, and long life.

9.6.3 Trends in Wind Energy Generation

Nanotechnology aims to improve and change material property, behavior, and cost characteristics; wind power nanotechnology is in early stages of research and development of industrial manufacture. A wind energy farm produces a maximum amount of electricity, and it can be dispatched to any location in the world at any time at a minimum cost with no pollution; in such a manner, the process of generating electricity from natural resources without disrupting the ecological balance is a novel approach to electricity generation. There have been several applications of these nanomaterials and nanostructures in the areas of distributed resources, renewable energy sources, thermal storage batteries, sensors, and power electronics [114]. Nanotechnology is an integral part of cloud computing for smart grids; computer architecture is another, with nanomaterials being used in complementary metal-oxide-semiconductor technology [115, 116]. A modern computer can be affected in the following ways by nanotechnology [117, 118]:

a. Nanofabrication of integrated circuits is the technology used to produce nanodevices, following one of two methods: top down, which changes the behavior of a certain material or device, or bottom-up, which changes the molecules' nanostructure to produce absolutely new materials or devices.
b. CNT material is used to create electronic components (of the computer), such as transistors, diodes, relays, and logic gates.
c. Nonvolatile RAM (NVRAM) is made up of tiny nanoengineered ferroelectric crystals. It is possible for NVRAM to retain data even when the power goes out due to the crystals' inability to dissolve naturally.
d. QD material consists of crystals (that can emit only one waveform length of light) used as quantum bits forming the basis of quantum computers.

Furthermore, wind energy plants are becoming more popular due to their commercial viability. Renewable energy sources that use wind energy are proved to be the fastest growing in the world.

The amount of electrical power produced by one wind-powered generator is limited, and, hence, a large number of wind turbines is usually deployed in order to create a wind farm. There is one disadvantage of wind energy, namely the uncertainty that surrounds its availability, as with many other renewable energy resources. There are also a lot of considerations to be taken into account when balancing the cost of creating a wind turbine (or a wind farm) as opposed to the revenue generated by the construction of the turbine over its lifetime. In wind energy conversion systems, a mechanical gearbox is typically employed in large wind turbines so as to increase the speed of the generator and, thus, increase its efficiency. There are several types of wind turbines available today, but the most common type is the variable-speed wind turbine, which has shown low energy consumption and high reliability. Due to their higher efficiency and reliability, variable-speed wind turbines have become more popular, particularly in small-scale applications [15, 53]. A large wind turbine's rotor is usually horizontal-axis, providing greater wind energy capture capabilities. In turbulent wind environments, however, the control of such an axis becomes increasingly difficult due to fluctuations in wind speed, which makes it difficult to maintain a constant position of the rotor and work at maximum power. Therefore, the installation of vertical-axis turbines in such regions is a good option in such cases.

REFERENCES

1. Mao S, Chen X, "Selected nanotechnologies for renewable energy applications", *Int. J. Energy Res.*, 31, 619–636, 2007.
2. Serrano E, Rus G, Garcıa-Martınez J, "Nanotechnology for sustainable energy", *Renew. Sustain. Energy Rev.*, 13, 2373–2384, 2009.
3. Tejeda J, Ferreira S, "Applying systems thinking to analyze wind energy sustainability", *Procedia Comput. Sci.*, 28, 213–220, 2014.
4. Johna SS, Malviyaa P, Sharmaa N, Manjhia VK, Sudhakar K, "Nanotechnology for solar and wind energy applications", 2015 International Conference on Green Computing and Internet of Things, 1223-12-27, 2015.
5. Singh R, Altaee A, Gautam S, "Nanomaterials in the advancement of hydrogen energy storage", *Heliyon*, 6(7), e04487, 2020.
6. Yang J, Sudik A, Wolverton C, Siegel DJ, "High-capacity hydrogen storage materials: Attributes for automotive applications and techniques for materials discovery", *Chem. Soc. Rev.*, 39(2), 656–675, 2010.
7. Murray LJ, Dincă M, Long JR, "Hydrogen storage in metal–organic frameworks", *Chem. Soc. Rev.*, 38(5), 1294–1314, 2009.
8. Baykara SZ, "Hydrogen: A brief overview on its sources, production and environmental impact", *Int. J. Hydrogen Energy*, 43(23), 10605–10614, 2018.
9. Farrell AE, Keith DW, Corbett JJ, "A strategy for introducing hydrogen into transportation", *Energy Policy*, 31(13), 1357–1367, 2003.
10. Tarhan C, Cil MA, "A study on hydrogen, the clean energy of the future: Hydrogen storage methods", *J. Energy Storage*, 40, 102676, 2021.
11. Kova A, Paranos M, Marciu D, "Hydrogen in energy transition: A review", *Int. J. Hydrogen Energy*, 46, 10016–10035, 2021.

12. Ahmadi MH, Ghazvini M, Nazari MA, Ahmadi MA, Pourfayaz F, Lorenzini G, Ming T, "Renewable energy harvesting with the application of nanotechnology: A review", *Int. J. Energy Res.*, 1–24, 2018.
13. Varcoe JR, Kizewski JP, Halepoto DM, Poynton SD, Slade RCT, Zhao F, "Fuel cells-alkaline fuel cells: Anion-exchange membranes", *Encycl. Electrochem. Power Sources*, 329–343, 2009.
14. Raduwan NF, Shaari N, Kamarudin SK, Masdar MS, Yunus RM, "An overview of nanomaterials in fuel cells: Synthesis method and application", *Int. J. Hydrogen Energy*, 47, 18468–18495, 2022.
15. Sharma S, Ali K, *Solar Cells from Materials to Device Technology* (Chapters 1 and 2), Springer, pp. 1–50, 2020.
16. Lee T, Ebong A, "A review of thin film solar cell technologies and challenges", *Renew. Sustain. Energy Rev.*, 12(3), 1286–1297, 2016.
17. Shruti S, Ashutosh S, "Solar cells: In research and applications—a review", *Mater. Sci. Appl.*, 6, 1145–1155, 2015.
18. Yan D and Xuan Y, "Effect of temperature on performance of nanostructured silicon thin-film solar cells", *Sci. Direct*, 11(4), 109–119, 2015.
19. Yan D, Xuan Y, "Role of surface recombination in affecting the efficiency of nanostructured thin-film solar cells", *Opt. Express*, 21(4), 1065–1077, 2013.
20. Braun A, Katz A, Cordon J, "Basic aspects of the temperature coefficient of concentrator solar cell performance parameters", *Prog. Photovolt.*, 21(3), 1087–1094, 2013.
21. Da Y, Xuan Y, "Role of surface recombination in affecting the efficiency of nanostructured thin-film solar cells", *Opt. Express*, 21(6), 1065–1077, 2013.
22. Bozzola A, liscidini M, Andreani L, "Photonic light-trapping versus Lambertian limits in thin film silicon solar cells with 1D and 2D periodic patterns", *Opt. Express*, 20(2), 224–244, 2012.
23. Buencuerpo J, Munioz E, Dotor M, Postigo A, "Optical absorption enhancement in a hybrid system photonic crystal thin substrate for photovoltaic applications", *Opt. Express*, 20(4), 452–464, 2012.
24. Deceglie M, Ferry V, Alivisatos A, Atwater H, "Design of nanostructured solar cells using coupled optical and electrical modeling", *Nano Lett.*, 2012, 12(6), 2894–2900.
25. Deinega A, Eyderman S, John S, "Coupled optical and electrical modeling of solar cell based on conical pore silicon photonic crystals," *J. Appl. Phys.*, 10(113), 1–10, 2013.
26. Reyes L, Rangel R, Jose S, *Nanocomposites for Photonic and Electronic Applications Book*, Elsevier, pp. 175–183, 2020.
27. Khare V, Khare C, Nema S, Baredar P, "Tidal energy systems: Design, optimization and control", *Sci. Direct*, 1–39, 2019.
28. Forman A, "Energy justice at the end of the wire: Enacting community energy and equity in wale", *Energy Policy*, 107, 649–657, 2017.
29. Venkateswari R, Sreejith S, "Factors influencing the efficiency of photovoltaic system", *Renew. Sustain. Energy Rev.*, 101, 376–394, 2019.
30. Elahi H, et al., "Characterization and implementation of a piezoelectric energy harvester configuration: Analytical, numerical and experimental approach", *Integr. Ferroelectr.*, 212(1), 39–60, 2020, https://doi.org/10.1080/10584587.2020.1819034.
31. Tock JY, Lai CL, Lee KT, Tan KT and Bhatia S, "Banana biomass as potential renewable energy resource: A Malaysian case study", *Renew. Sustain. Energy Rev.*, 14(2), 798–805, February 2010.
32. Ma PC, Zhang Y, "Perspectives of carbon nanotubes/polymer nanocomposites for wind blade materials", *Renew. Sustain. Energy Rev.*, 30, 651–660, 2014.
33. Marin JC, Barroso A, Paris F, Canas J, "Study of damage and repair of 300KW wind turbine", *J. Energy*, 1068–1083, 2008.
34. Ahmad S, Haq IU, "Wind blade material optimization", *Appl. Mech. Mater.*, 66–68, 1199–1206, 2011.

35. Varcoe JR, Kizewski JP, Halepoto DM, Poynton SD, Slade RCT, Zhao F, "Fuel cells-alkaline fuel cells: Anion-exchange membranes", *Encycl. Electrochem. Power Sources*, 329–343, January 2009.
36. Bidault F, Brett DJL, Middleton PH, Abson N, Brandon NP, "An improved cathode for alkaline fuel cells", *Int. J. Hydrogen Energy*, 35(4), 1783–1788, February 2010.
37. Mulder G, "Fuel cells-alkaline fuel cells: Overview", *Encycl. Electrochem. Power Sources*, 321–328, January 2009.
38. Peighambardoust SJ, Rowshanzamir S, Amjadi M, "Review of the proton exchange membranes for fuel cell applications", *Int. J. Hydrogen Energy*, 35(17), 9349–9384, September, 2010.
39. Aindow TT, Haug AT, Jayne D, "Platinum catalyst degradation in phosphoric acid fuel cells for stationary applications", *J. Power Sources*, 196(10), 4506–4514, May 2011.
40. Yuh CY, Farooque M, "Fuel cells-molten carbonate fuel cells: Materials and life considerations", *Encycl. Electrochem. Power Sources*, 497–507, 2009.
41. Zhang H, Lin G, Chen J, "Performance analysis and multiobjective optimization of a new molten carbonate fuel cell system", *Int. J. Hydrogen Energy*, 36(6), 4015–4021, March 2011.
42. Kim JE, et al., "Using aluminum and Li2CO3 particles to reinforce the α-LiAlO2 matrix for molten carbonate fuel cells", *Int. J. Hydrogen Energy*, 34(22), 9227–9232, November 2009.
43. Paoletti C, Carewska M, Lo Presti R, Phail SM, Simonetti E, Zaza F, "Performance analysis of new cathode materials for molten carbonate fuel cells", *J. Power Sources*, 193(1), 292–297, August 2009.
44. Randstrom S, Lagergren C, Capobianco P, "Corrosion of anode current collectors in molten carbonate fuel cells", *J. Power Sources*, 160(2), 782–788, October 2006.
45. Yokokawa H, Tu H, Iwanschitz B, Mai A, "Fundamental mechanisms limiting solid oxide fuel cell durability", *J. Power Sources*, 182(2), 400–412, August 2008.
46. Khare V, Khare C, Nema S, Baredar P, "Tidal energy systems: Design, optimization and control", *Sci. Direct*, 1–39, 2019.
47. Forman A, "Energy justice at the end of the wire: Enacting community energy and equity in wale", *Energy Policy*, 107, 649–657, 2017.
48. Venkateswari R, Sreejith S, "Factors influencing the efficiency of photovoltaic system", *Renew. Sustain. Energy Rev.*, 101, 376–394, 2019.
49. Richhariy G, Kumar A, Samsher, "Photovoltaic solar energy conversion: Technologies, applications and environmental impacts", *Sci. Direct*, 27–78, 2019.
50. Putrus G, Bentley E, *Electric Renew. Energy Syst.* (Chapter 20), Elsevier, pp. 487–518, 2016.
51. Premkumar M, Sowmya R, "An effective maximum power point tracker for partially shaded solar photovoltaic systems", *Energy Rep.*, 5, 1445–1462, 2019.
52. Verma D, "Maximum power point tracking (MPPT) techniques: Recapitulation in solar photovoltaic systems", *Renew Sustain Energy Rev*, 54, 1018–1034, 2016.
53. Ranabhat K, L.Patrikeev, Revina A, Andrianov K, Lapshinsky V, "An introduction to solar cell technology", *Eng. Environ. Sci.,* 405(4), 481–491, 2016.
54. Singh GK, "Solar power generation by PV (photovoltaic) technology: A review", *J. Energy*, 53, 1–13, 2013.
55. Ahmed M, Khan WA, Ul-Hasnain S, Ahmed Z, "Improving wind turbine performance using nano materials", *Accessed J.*, 1, 1–5, 2021.
56. Nguyen-Tri P, Tran HN, Plamondon CO, Tuduri L, Vo DVN, Nanda S, Mishra A, Chao HP, Bajpai AK, "Recent progress in the preparation, properties and applications of superhydrophobic nano-based coatings and surfaces: A review", *Prog. Org. Coat.*, 132, 235–256, 2019.
57. Xu X, Sun Y, Sun Y, Li Y, "Bioaugmentation improves batch psychrophilic anaerobic co-digestion of cattle manure and corn straw", *Bioresour. Technol.*, 343, 126118, 2022.

58. Feng F, Wang R, Yuan W, Li Y, "Study on anti-icing performance of biogas-residue nano-carbon coating for wind-turbine blade", *Coatings*, 13, 814, 2023, https://doi.org/10.3390/coatings13050814.
59. Filippov SP, Yaroslavtsev AB, "Hydrogen energy: Development prospects and materials", *Russ. Chem. Rev.*, 90(6), 627–643, 2021.
60. Pavlov VI, Gerasimova EV, Zolotukhina EV, Don GM, Dobrovolsky YA, Yaroslavtsev AB, "Degradation of Pt/C electrocatalysts having different morphology in low-temperature PEM fuel cells", *Nanotechnol. Russ.*, 11, 743–750, 2016.
61. Perez J, Paganin VA, Antolini E, "Particle size effect for ethanol electro-oxidation on Pt/C catalysts in half-cell and in a single direct ethanol fuel cell", *J. Electroanal. Chem.*, 654, 108–115, 2011.
62. Peres-Alonso FJ, McCarthy DN, Nierhoff A, Hernandez-Fernandez P, Strebel C, Stephens IEL, Nielsen JH, Chorkendorff I, "The effect of size on the oxygen electroreduction activity of mass-selected platinum nanoparticles", *Angew Chem.*, 51, 4641–4643, 2012.
63. Meier JC, Galeano C, Katsounaros I, Witte J, Bongard HJ, Topalov AA, Baldizzone C, Mezzavilla S, Schueth F, Mayrhofer KJJ, "Design criteria for stable Pt/C fuel cell catalysts", *Beilstein J. Nanotechnol.*, 5, 44–67, 2014.
64. Yeh TF, Syu JM, Cheng C, Chang TH, Teng H, "Graphite oxide as a photocatalyst for hydrogen production from water", *Adv. Funct. Mater.*, 20(14), 2255–2262, 2010.
65. Maleki A, Khajeh MG, Rosen MA, "Two heuristic approaches for the optimization of gridconnected hybrid solar–hydrogen systems to supply residential thermal and electrical loads", *Sustain. Cities Soc.*, 34, 278–292, 2017.
66. Nkele AC, Offiah SU, Chime CP, Ezema FI, "Review on advanced nanomaterials for hydrogen production", *IOP Conf. Ser. Earth Environ. Sci.*, 1178, 012001, 2023.
67. Danish MS, "Exploring metal oxides for the hydrogen evolution reaction (HER) in the field of nanotechnology", *RSC Sustain.*, 1, 2180–2196, 2023.
68. Serrano E, Rus G, Garcia-Martinez J, "Nanotechnology for sustainable energy", *Renew. Sustain. Energy Rev.*, 13, 2373–2384, 2009.
69. Abdalla AM, Elnaghi BE, Hossain S, Dawood M, Abdelrehim O, Azad AK, "Nanotechnology utilization in energy conversion, storage and efficiency: A perspective review", *Adv. Energy Convers. Mater.*, 1, 30–54, 2020.
70. Narsimha P, Kumar PR, Pandiyan KRR, Suryawanshi PL, Vooradi R, Kishore KA, Sonawane SH, "Synthesis of nanomaterials for energy generation and storage applications", in *Nanotechnology for Energy and Environmental Engineering*, Green Energy and Technology, Springer, Cham, pp. 215–229, 2020.
71. Saidur R, Leong KY, Mohammed HA, "A review on applications and challenges of nanofluids", *Renew. Sustain. Energy Rev.*, 15, 1646–1668, 2011.
72. Khanlari A, Sozen A, Tuncer AD, Şirin C, Kumaş K, Gungor A, "Nanoakışkanların Isıl Sıstemlerde Kullanımı", *Mühendislik Alanında Akademik Çalışmalar*, 175–188, 2020.
73. Rasukkannu M, Velauthapillai D, Ponniah V, "A promising high-efficiency photovoltaic alternative non-silicon material: A first-principle investigation", *Scripta Mater.*, 156, 134–137, 2018.
74. Blachowicz T, Ehrmann A, "Recent developments of solar cells from PbS colloidal quantum dots", *Appl. Sci.*, 16, 1743, 2020.
75. Rasukkannu M, Velauthapillai D, Bianchini F, Vajeeston P, "Properties of novel non-silicon materials for photovoltaic applications: A first-principle insight", *Mater. J.*, 11, 2006, 2018.
76. Rex Rosario S, Kulandaisamy I, Deva K, Arulanantham AMS, Valanarasu S, Youssef MA, Awwad S, "Deposition of p-type Al doped PbS thin films for heterostructure solar cell device using feasible nebulizer spray pyrolysis technique", *Physica B: Phys. Condensed Matter*, 575, 411704, 2019.

77. Da Y, Xuan Y, "Effect of temperature on performance of nanostructured silicon thinfilm solar cells", *Sience Direct*, 11, 109–119, 2015.
78. Da Y, Xuan Y, "Role of surface recombination in affecting the efficiency of nanostructured thin-film solar cells", *Opt. Express*, 21(106), A1065–A1077, 2013.
79. Braun A, Katz A, Cordon JM, "Basic aspects of the temperature coefficient of concentrator solar cell performance parameters", *Prog. Photovolt.*, 21, 1087–1094, 2013.
80. Da Y, Xuan Y, "Role of surface recombination in affecting the efficiency of nanostructured thin-film solar cells", *Opt. Express*, 21(106), A1065–A1077, 2013.
81. Bozzola A, Liscidini MA, "Photonic light-trapping versus Lambertian limits in thin film silicon solar cells with 1D and 2D periodic patterns", *Opt. Express*, 20, A224–A244, 2012.
82. Rex Rosario S, Kulandaisamy I, Deva K, Arulanantham AMS, Valanarasu S, Youssef MA, Awwad S, "Deposition of p-type Al doped PbS thin films for heterostructure solar cell device using feasible nebulizer spray pyrolysis technique", *Physica B: Phys. Condensed Matter*, 575, 411704, 2019.
83. Bang JH, Kamat PV, "Quantum dot sensitized solar cells. A tale of two semiconductor nanocrystals: CdSe and CdTe", *ACS Nano*, 3, 1467–1476, 2009.
84. Beard MC, Knutsen K, Yu P, Luther J, Song O, Metzger W, Ellingson R, Nozik A, "Multiple exciton generation in colloidal siliconnanocrystals", *Nano Lett.*, 7, 2506–2512, 2007.
85. Olson J, Rodriguez Y, Yang L, Alers GB, Carter SA, "CdTe schottky diodes from colloidal nanocrystals", *Appl. Phys. Lett.*, 96, 242103, 2010.
86. Debnath R, Greiner M, Kramer I, Fischer A, Tang J, Barkhouse D, Wang X, Levina L, Lu Z, Sargent EH, "Depleted-heterojunction colloidal quantum dot photovoltaics employing low-cost electrical contacts". *Appl. Phys. Lett.*, 97, 023109, 2010.
87. Huang X, Huang S, Zhang Q, Guo X, Li D, Luo Y, Shen Q, Toyodo T, Meng Q, "A flexible photoelectrode for CdS/CdSe quantum dot sensitized solar cells (QDSSCs)", *Chem. Commun.*, 47, 2664–2666, 2011.
88. Kamat PV, "Boosting the efficiency of quantum dot sensitized solar cells through modulation of interfacial charge transfer", *Acc. Chem. Res.*, 45, 1906–1915, 2012.
89. Liu CP, Wang HE, Ng TW, Chen ZH, Zhang WF, Yan C, Tang YB, "Hybrid photovoltaic cells based on ZnO/Sb2S3/P3HT heterojunctions", *Phys. Status Solidi*, 249, 627–633, 2012.
90. Yang K, Liu M, Persson K, Han Y, Zheng H, "Strain mediated interfacial dynamics during Au–PbS core–shell nanostructure formation", *ACS Nano*, 10, 6235–6240, 2016.
91. Qin W, Hou JB, "Effect of interface atomic structure on the electronic properties of nano-sized metal–oxide interfaces", *Nano Lett.*, 15, 211–217, 2015.
92. Lin H, Chen Y, Wu J, Wang D, Chen C, "Carrier transfer induced photoluminescence change in metal-semiconductor core-shell nanostructures", *Appl. Phys. Lett.*, 88, 161911, 2006.
93. Lu W, Wang B, Zeng J, Wang X, Zhang S, Hou J, "Synthesis of core/shell nanoparticles of Au/CdSe via Au-Cd bialloy precursor", *Langmuir*, 21(8), 3684–3687, 2005.
94. Bhandari KP, Roland PJ, Mahabaduge H, et al., "Thin film solar cells based on the heterojunction of colloidal PbS quantum dots with CdS", *Sol. Energy Mater. Sol. Cells* 117, 476–482, 2013.
95. Bai Z, Yang J, DePba S, Wang D, "Thin film CdTe solar cells with an absorber layer thickness in micro- and sub-micrometer scale", *Appl. Phys. Lett.*, 99, 143502, 2011.
96. Eldada L, "Nanotechnologies for efficient solar and wind energy harvesting and storage", *Proc. SPIE*, 7764, 776408, 2010.
97. Rafiee MA, Rafiee J, Wang Z, Song H, Yu ZZ, Koratkar N, "Enhanced mechanical properties of nanocomposites at low graphene content", *ACS Nano*, 3, 3884 –3890, 2009.

98. Babatunde DE, et al., “Environmental and societal impact of nanotechnology”, *IEEE Access*, 8, 4640–4667, 2020.
99. Basheer AA, Ali I, “Water photo splitting for green hydrogen energy by green nanoparticles”, *Int. J. Hydrogen Energy*, 44, 11564–11573, 2019.
100. Yang L, Wang X, Wang J, Cui G, Liu D, “Graphite carbon nitride/boron doped graphene hybrid for efficient hydrogen generation reaction”, *Nanotechnology*, 29(34), 45705, 2018.
101. Liu C, Qiu Y, Xia Y, Wang F, Liu X, Sun X, Liang Q, Chen Z, “Noble-metal-free tungsten oxide/carbon (WOx/C) hybrid nanowires for highly efficient hydrogen evolution”, *Nanotechnology*, 28, 445403, 2017.
102. Martha S, Sahoo PC, Parida KM, “An overview on visible light responsive metal oxide based photocatalysts for hydrogen energy production”, *RSC Adv.*, 5, 61535–61553, 2015.
103. Sıdkı Uyar T, Besikci DG, “Integration of hydrogen energy systems into renewable energy systems for better design of 100% renewable energy communities”, *Int. J. Hydrogen Energy*, 42, 2453–2456, 2017.
104. Ma L, Luo B, Geng J, Huang Z, Guo L, “Efficient photo-thermo-catalytic hydrogen production performance over a graphene-titanium dioxide hybrid nanomaterial”, *Int. J. Hydrogen Energy*, 46, 2871–2877, 2021.
105. Liu W, Liu C, Gogoi P, Deng Y, “Overview of biomass conversion to electricity and hydrogen and recent developments low-temperature electrochemical approaches”, *J. Pre-Proof*, 6, 1351–1363, 2020.
106. Nachtane M, Tarfaoui M, Goda I, Rouway M, “A review on the technologies, design considerations and numerical models of tidal current turbines”, *Renew. Energy*, 157, 1274–1288, 2020.
107. Prasad A, Jain K, Gairola A, “Pumped storage hydropower plants environmental impacts using geomatics techniques: An overview”, *Int. J. Comput. Appl.*, 81, 41–48, 2020.
108. Xia L, Ma S, Tao P, Pei W, Liu Y, Tao L, Wu Y, “A wind-solar hybrid energy harvesting approach based on wind-induced vibration structure applied in smart agriculture”, *Micromachines*, 14, 58, 2023.
109. Merin JV, Reyes RC, Sevilla RV, Caballero JMF, Cuesta TRPD, Paredes RJM, “Automated greenhouse vertical farming with wind and solar hybrid power system”, 2019 IEEE 11th International Conference on Humanoid, Nanotechnology, Information Technology, Communication and Control, Environment, and Management (HNICEM), Laoag, Philippines, pp. 1–6, 2019.
110. Indragandhi V, Subramaniyaswamy V, Logesh R, “Resources, configurations, and soft computing techniques for power management and control of PV/wind hybrid system”, *Renew. Sustain. Energy Rev.*, 69, 129–143, 2017.
111. Farouk AA, Mohamed AA, “Hybrid nanocrystal photovoltaic/wind turbine power generation system in building”, *J. Adv. Res. Mater. Sci.*, 40, 8–19, 2018.
112. Wang K, Zhang X, Miao Y, He B, Wang C, “Dispersion and behavior of hydrogen for the safety design of hydrogen production plant attached with nuclear power plant”, *Int. J. Hydrogen Energy*, 45(39), 20250–20255, 2020.
113. Ammar NR, Alshammari NFSH, “Overview of the green hydrogen applications in marine power plants onboard ships”, *Int. J. Multidiscip. Curr. Res.*, 6, 84–86, 2018.
114. Abdelsalam HA, Abdelaziz AY, “A brief overview of nanotechnology applications in smart power grid”, *Electr. Power Compon. Sys.*, 42, 306–314, 2014.
115. Gamrat C, “Challenges and perspectives of computer architecture at the nano scale”, IEEE Computer Society Annual Symposium on VLSI (ISVLSI), Lixouri, Kefalonia, UK, pp. 8–10, 5–7 July 2010.

116. Geer D, "Nanotechnology: The growing impact of shrinking computers", *IEEE Pervas. Comput.*, 5, 7–11, 2006.
117. Ullah Z, "Nanotechnology and its impact on modern computer", *Global J. Res. Eng.*, 12, 2012.
118. Mieno T, Matsumoto N, Tomie T, Inoue H, "Production of carbon nanotubes and nano-clusters by the J × B arc-jet discharge method", 12th IEEE Conference on Nanotechnology (IEEE-NANO), Birmingham, UK, pp. 1–6, 20–23 August 2012.

10 Economic Studies for Nanotech Green Energy

The scientific and technological advancement of nanotechnology today is a major factor in international competitiveness. Using nanotechnology knowledge throughout the green energy sector can produce innovative products and features that are more cost-effective than existing products in the global market and that bring environmental benefits. Nanotechnology is a new industry and science that will play a significant role in global market competition. Nanotechnology green energy opportunities are directly related to regulatory and technical support, and nanomaterials are integral to advancing green energy from an economic standpoint as well. Nanomaterials have a transformative impact on shaping a sustainable and economically robust energy future, as well as market growth, creating jobs, improving efficiency, enhancing energy security, and contributing to sustainability.

10.1 INTRODUCTION

Economic and environmental factors are driving the gradual transition to a cleaner and more sustainable energy economy. However, in spite of the advances in technology for energy storage, even the state-of-the-art technologies do not completely meet the economic and technical requirements for conventional applications. Economic development in the energy sector is relying heavily on nanomaterials, which are used in a wide range of fields, including generation, distribution, and consumption of energy. The use of nanomaterials is expected to increase even further in the future.

Energy carriers like hydrogen and electricity can solve environmental problems and ensure adequate energy security in a sustainable energy economy. They are preferably renewable and have a large domestic availability. A consensus has emerged regarding a vision for a new sustainable energy economic system that will become reality in a few years' time and that will be based on a diverse spectrum of low-carbon or carbon-free primary energy sources [1, 2].

It is crucial that scientists and engineers develop sustainable energy solutions as society transitions from a fossil fuel-driven economy to one based on more sustainable energy sources. However, during this time, it is also critical that scientists and engineers find efficient ways of producing, refining, and using fossil fuels. A significant part of the solution should be to increase the efficiency of traditional power sources and to increase the recycling rate. In addition, renewable energy sources such as solar, wind, and ocean power will be exploited with nanotechnologies so that we can reduce our dependence on depleting petroleum resources in the future. In order to find new sources of power and improve efficiencies, breakthrough technologies are required, regardless of the specific nanotechnology or specific energy application [3].

DOI: 10.1201/9781003512486-12

The use of nanotechnology techniques offers new approaches to fundamental questions concerning green energy generation (i.e. hydrogen, solar, wind and so on), and the use of nanomaterials can enable the efficient and economical storage and transport of hydrogen atoms. Solar energy conversion into electricity has become a standard with dye-sensitized nanocrystalline electrochemical NP systems; the dye molecules that absorb across the visible spectrum have recently been developed for these devices as sensitizers. In addition to cost reduction and simplification of dye solar cell manufacturing, the development of solid-state heterojunction dye solar cells continues to hold potential; nanotechnology has the potential to increase the efficiency of solar cells, but it is likely to have the greatest potential for reducing manufacturing cost [4, 5].

Wind energy is considered one of the most promising renewable energy sources and has been extensively used; the advantages of wind energy include low environmental impacts, low cost, and a sustainable energy source. CNTs combined with epoxy are used in the manufacturing of windmill blades to lend lightness and strength, which increases the generation of electricity, and they are used in the wind energy sector, too. A further example of a nanomaterial application in promoting renewable energy is nano-optimized power storage devices that utilize nanomaterials to enhance the power of wind energy; conductive nanomaterials are another application, for increasing the production of lighting [6]. In green energy, nanomaterials have gained much attention because of their unique properties and performance. Sustainable nanomaterial design and development must consider environmental sustainability, economic viability, and material functionality [7, 8].

10.2 THEORETICAL MODELS FOR GREEN ENERGY SYSTEMS

10.2.1 Hydrogen Energy Systems

A compact, lightweight, safe, and cost-efficient way to store hydrogen is one of the key elements to a successful hydrogen economy. One of the most feasible solutions to store hydrogen in hydrogen-powered systems has actually been identified as a solid-state storage system that uses metal hydrides. Metal hydrides are slow kinetics and cannot release hydrogen when temperatures are low, so most of them cannot store large amounts of hydrogen, but there is sufficient evidence to indicate that metal hydrides are capable of storing hydrogen effectively and that there would be substantial benefits to doing so. Transition metals are the basic components of all reversible hydrides that operate at room temperature and at atmospheric pressure. There are many electropositive elements that are reactive, including those in the lanthanide and actinide families along with members of the Ti and vanadium families. Different metal hydrides react at different temperatures depending on the reaction thermodynamics; typically, if the reaction is reversible and a metal or alloy reacts with hydrogen, the interaction can be expressed as follows [9, 10]:

$$M + \frac{x}{2} H_2 \leftrightarrow MH_x + Q \tag{10.1}$$

where Mis is a metal, an intermetal alloy, or a solid solution, x is a positive whole number; MHx is the formed hydride, and Q is the heat of reaction.

In the design and manufacturing world, nanotechnology innovations are transforming the way of creation products and materials for commercial use through innovative processes. A new nanotechnology commercialization model was developed based on the business perspective and was analyzed from the perspective of the organization and customers. Nanotechnology commercialization makes sense for any company planning to use nanotechnology-based products or services [11, 12].

10.2.2 Solar Energy Systems

The cost of solar power production and the sale price of electricity are relatively high because it is a newly implemented renewable energy industry. To reduce production cost and improve economic competitiveness, it is necessary to improve power generation technology [13]. A system's economic competitiveness is determined by its production cost of electricity; the life cycle cost of electricity is calculated by dividing the total life cycle cost of PV by the amount of energy that the system produces over its lifetime [14]:

$$LCOE = \frac{Total\,Life\,Cycle\,Cost}{Total\,Life - time\,Energy\,Production} \tag{10.2}$$

Therefore, Equation 10.2 can be disaggregated for solar generation as follows:

$$LCOE = \frac{Initial\,Ivestment - \sum_{n=1}^{N} \frac{Depreciation}{(1 + Discount\,Rate)^{n}}.(Tax\,Rate) + \sum_{n=1}^{N} \frac{Annual\,Costs}{(1 + Discount\,Rate)^{n}}.(1 - Tax\,Rate) - \frac{Residual\,Value}{(1 + Discount\,Rate)^{N}}}{\sum_{n=1}^{N} \frac{Initial \frac{KWh}{kwp(1 - System\,Degradation\,Rate)^{n}}}{(1 + Discount\,Rate)^{n}}} \tag{10.3}$$

Increasing technological advances have resulted in a reduction in the cost of NP modules [15, 16], making it possible to continually improve energy storage devices with the aim of reducing the impacts of nature, season, and climate on PV power generation, as well as reducing the overall operation and maintenance costs for PV power generation [17]. A payback period, internal rate of return (IRR), net present value, and cost of capital (LCOE) are used to evaluate economic performance of investment in solar energy generation. PV industry analysis is conducted [18].

In order to calculate the feasibility of project investment, IRR and payback period are used to assess the investment income capacity of the power generation industry. As an investor, it is imperative to first select a site with adequate solar radiation and second incorporate technological advancements that can improve the power generation capability of the plant and reduce cost. Afterwards, it is necessary to transform its financing mode so that it can increase the percentage of its loans in a manner that will improve economic efficiency. Furthermore, it is also necessary to safeguard the environment implementing a flexible mode of operation, adjusting the self-use ratio, and preventing the waste of electric energy [19, 20].

Performing routine maintenance influences the module's performance and profitability; PV plant maintenance operations typically include cleaning PV modules and performing routine preventive maintenance of PV components such as modules, inverters, wiring, and solar tracking systems. Moreover, it can be necessary to carry out extraordinary maintenance operations when failures are detected by the monitoring system of the plant; in particular, dirt accumulation (soiling) can significantly impact the performance of PV modules if they are left unattended for a long period of time: A study was conducted in southern Italy to monitor two PV plants, and they showed variations in their power depending on the soiling characteristics and cleaning methods used [21]. A report out of Egypt showed that even when solar modules are cleaned every day without pressurized water, after 45 days, they can lose 50% of their power [22]; soiling decreases power depending on the frequency of cleaning, rainfall, and the angle of the PV module [23].

PV plant investment profitability needs to be taken into account not only in terms of investment cost, but also in terms of operating and maintenance cost. As long as a feed-in tariff is present, maintenance and operating cost is usually ignored or a rough estimate is provided, since it ensures a satisfactory return on investment and good payback time. A significant aspect of the economic evaluation of an investment is the cost of operating and maintaining it, and there should be specific models developed to evaluate the impact of maintenance activities on the performance in addition to the cost of the investment when incentives are marginal. It is important to take into account a number of indexes when evaluating the performance of solar panels [24].

The main index, the efficiency at standard test conditions (STC), is the most useful without direct measurements (Eq.), and can be calculated as the product of the nominal DC power (or peak power) Pnom at STC (measured in watts) by the STC irradiance GSTC (measured in watts). (1) is the ratio of the nominal DC power (or peak power) Pnom at STC (measured in watts) by the surface area of modules A (in square meters).

$$\eta_{STC} = \frac{P_{nom}}{A.G_{STC}} \tag{10.4}$$

The STC differs depending on whether the technology is concentrating or nonconcentrating. In real-life scenarios, PV modules do not always deliver the nominal DC power PDC as it can be affected by variations in module temperature Tm and/or irradiance G (expressed in W/m2), which can all affect measured DC power. An efficient PV module can be defined as one with a DC efficiency hDC determined by the product of PDC and A's surface area times G's irradiance.

$$\eta_{DC} = \frac{P_{DC}}{A.G} \tag{10.5}$$

IEC standard 61724 specifies performance indexes that refer to a period of time, as opposed to a continuous index that evaluates module performance immediately;

these indices evaluate module performance instantly; instead of evaluating the electrical power and irradiance in this case, the corresponding energy values must be evaluated over the defined period of time (a day, a month, or a year).

10.2.3 Wind Energy Systems

Economic uncertainty has not yet been considered as an important factor when calculating the optimal size of the units for a co-generation system that combines electricity and heat production on an off-grid basis. In addition to providing precise results when simulated, analyzed, and optimized, Hybrid Optimization Model for Electric Renewables (HOMER) software has been used to explore hybrid renewable energy systems from economic, technological, and environmental perspectives. Scholars determined that a combination of PV/DG/battery along with a levelized electricity cost (LCOE) and a renewable fraction was the most efficient architecture, especially when taking into account the total net present cost (TNPC) [25, 26]. In the USA, the National Renewable Energy Laboratory developed HOMER Pro, which analyzes various types of hybrid energy systems, including off-grid and on-grid systems, to determine which are most appropriate in any given situation [27]. This software has been extensively applied in research because of its high speed and accuracy [28] and similarity to meta-heuristic optimization algorithms [29].

Nanostructured materials provide potential advantages in a variety of renewable energy applications that rely primarily on connectors to separate loads, such as NP, thermoelectric, and electrochemical energy storage. Engineers and scientists can make electrochemical systems more efficient and power dense by applying nanostructures and increasing chemical activity levels. For instance, lighter nanomaterials with greater resistance improve the efficiency of wind turbine rotor blades; coating the blades provided better corrosion protection for tidal energy equipment, and in drilling machines, nanocomposites can be used in geothermal energy systems to provide higher resistance [30].

In recent years, the ecofriendly, sustainable, affordable, free, and easily accessible nature of wind energy has made it one of the most efficient and reliable sources of energy. Analysis of hourly wind speed fluctuations showed that higher wind speeds appeared at lower heights during the daytime, achieving the peak in the afternoon, whereas the pattern alters at upper heights. Mainly, the hourly wind speeds stay constant at all four heights. Grey wolf optimizer algorithm (GWOA) was developed as an intelligent swarm optimization technique based on leadership hierarchy during group hunting [31].

The advantages of GWOA over earlier swarm optimization techniques is that its parameters can easily be evaluated. Most swarm strategies used to solve optimization problems cannot be controlled by the leader over the entire period. There are four general groups of grey wolves, with each group forming a strict social hierarchy. Fitting the WD from GWOA compares favorably with other methods based on the fitness test values for the all-data set. Among the methods that have been used, global wind energy analysis is the most appropriate approach for the estimation of parameter values for the Weibull distribution, and the area that was studied was suitable for the installation of wind turbines in order to produce electricity [32].

10.3 TRADITIONAL ECONOMIC STUDIES

10.3.1 Hydrogen Energy Systems

In the steam reforming industry, the cost of producing hydrogen is mainly determined by gas prices [33]. However, as fossil fuels run out, and renewable electricity becomes increasingly affordable, electrolytic hydrogen is becoming competitive and is likely to continue to expand in the near future. As for now, the cost of hydrogen production is approximately 3.2–5.2 EUR/kg when using alkaline electrolyzers, but it is approximately 4.1–6 EUR/KG if using proton exchange membrane electrolyzers [34]. A pipeline scenario costs 1.8–2.6 EUR/kg for compressing, storing, and dispensing hydrogen, whereas a distributed scenario costs 2.1–3 EUR/kg [35]. In spite of the potential benefits of producing hydrogen through electrolysis, electrical consumption should be taken into account which exceeds the total amount of electricity generated by the European Union [36].

Renewable energy sources such as hydrogen, solar, and wind power are potential solutions for lowering the prices of electricity. As renewable energy sources have been explored with cheaper cost than ever before, renewable electricity may become less expensive with time. Despite high transmission and distribution cost due to remote locations, hydrogen is a profitable form of renewable energy. Using wind and hydrogen to generate electricity on Corvo Island, 80% of the island's electricity needs can be met by local renewable energy [37]. Techno-economic analyses demonstrated that hydrogen power systems are economically feasible taking into account the direct and indirect costs of operating hydrogen-fueled facilities. Hydrogen-based fuels have a double impact on the industry, as they have the potential to be integrated into existing fuel infrastructures and eliminate greenhouse gas emissions since they are renewable energy sources [38].

Investigators conducted a national-level analysis of hydrogen market trends using data about hydrogen production and consumption; the analysis was divided into two phases. First they calculated Spearman's correlations; then they performed linear regression models with both one and multiple parameters (ordinary least squares method). Multiple regression models revealed that different characteristics in each country contributed to its vision for a hydrogen economy. Hydrogen production markets can be enhanced in a given country by increasing the share of renewable energy sources while limiting CO_2 emissions [39–42].

In addition to contributing to overall final energy consumption, hydrogen production will play a crucial role due to an increasing population and economic growth; hydrogen can be converted directly into electricity with greater efficiency than fossil fuels [43]. The total cost of energy is divided into three categories: electricity cost, annualized capital cost, and operating and maintenance cost. In addition to electricity, capital, and operating and maintenance cost, hydrogen's final price is determined by hydrogen demand from refueling stations. The cost of hydrogen delivered to the consumer is calculated using the following equation [44, 45]:

$$U = \frac{C_c + C_e}{Q} \tag{10.6}$$

$$C_c = (e+s+c).(CRF+op+m+i+t) \quad (10.7)$$

where U is the unit cost of hydrogen, Q is the annual hydrogen production from the dispensing station, C_e is the annual cost of electricity consumed by the electrolyzer and the balance-of-plant system, and C_c is the annualized capital cost including the annual operation and maintenance. Capital recovery factor, which calculates the annual payment stream for a specified period of time, is expressed as follows:

$$CRF = \frac{d}{1-(1+d)^{-\gamma}} \quad (10.8)$$

$$d = r + inf.(1+r) \quad (10.9)$$

where r is the after-tax real rate of return on investment and inf is the annual rate of inflation.

The following case studies are complemented with specific international examples for a global perspective on hydrogen penetration in the society; the first case study uses a global energy system model to determine how much hydrogen could be absorbed by society [46–48]. To estimate the impact of hydrogen diffusion, the second study uses a dynamic integrated climate ecosystem economics model (CEEM) that was developed by the researchers [49–52], and the third study examines the changes necessary to make on-road transportation in the United States more hydrogen-powered by increasing renewable power and water usage [53]. The fourth case study shows thatin the United Kingdom, hydrogen is effectively used for space heating [54–56].

Case study 1: Global energy model: A global model has been developed for 82 regions around the world and includes CO_2 restrictions, resource cost, energy conversion and energy use cost, capacity factors, renewable energy generation intermittency, electric transmission cost, and energy imports and exports between regions. This model optimizes the energy system under the outlined constraints based on a total system cost perspective (including both fixed and variable cost) for the period 2000–2050 in 10-year steps. A critical issue regarding the analysis of energy systems resulting from CO_2 emissions is that renewable energy must be increased rapidly, as well as carbon capture and sequestration infrastructure. A cost minimization constraint leads to a rapid and sustained increase in global energy system cost between 2020 and 2050 by a factor of 2.4.

Case study 2: Social welfare economics evaluation: In contrast to case study 1, where the social cost of carbon (SCC) emitted into the atmosphere was explored in relation to the financial cost of reducing carbon emissions at the global scale, this case study focuses on the financial cost of reducing carbon emissions at the local scale. There are several methods that can be used to estimate SCC, including integrated assessment [51]. In this study, the teamwork developed a dynamic integrated climate-ecosystem-economics model to calculate the impact caused by hydrogen energy diffusion. There is an exogenous variable in the model representing the cost of backstop technologies (US$/t$CO_2$), alternative technologies that prevent the depletion of resources. It is important to keep in mind that hydrogen is not a primary energy source, but when used as a fuel, it does not generate CO_2. There is a possibility that low carbon and low-cost energy could be realized in the future if hydrogen is produced efficiently from nuclear energy, renewable energy, or fossil fuels combined with carbon capture and storage systems. Therefore, hydrogen energy diffusion can be represented as a change in backstop cost since this combination reduces backstop cost [52].

Case study 3: hydrogen as an energy carrier for road vehicles: For this study, investigators took the United States as an example to determine whether hydrogen can be used as an energy carrier to support the current energy requirements for road vehicles. The teamwork has been compared relative energy efficiencies through the energy supply chain to estimate the amount of electricity required to produce hydrogen via electrolysis from renewable energies. They assessed potential water demands associated with hydrogen production based on the requirement that electrolysis requires water as a feedstock; for example, the process would require converting 0.7–2.1 TW of liquid fuel capacity to renewable electricity capacity. Additionally, producing hydrogen from renewable energy will improve overall efficiency by resulting in a 40% reduction in the transportation sector's energy requirements. Approximately 10% of the energy rejected in the overall economy will be saved, and more than 1500 million tons of CO_2 will be eliminated (an approximate 30% reduction in total U.S. CO_2 emissions) [53].

Case study 4: Decarbonizing the gas grid using hydrogen: In the UK, heating buildings consumes 40% of the irradiated energy and generates 20% of greenhouse gas emissions [54, 55]. Consumers have to pay additional cost of around US$22,008 per household if they wish to convert to hydrogen gas networks; this is compared with the estimated cost of air source heat pumps ranging between $3000 and $8200 and ground source heat pumps ranging between $9000 and $14,800 to install in UK homes. Ground source heat pumps are extremely efficient and can help minimize running cost to a significant extent (up to 250%) because they are able to produce electricity at the retail price of US$0.13/kWh, compared with US$0.04 and $0.15/kWh for heat from hydrogen boilers with 90% efficiency; however, without significant incentives, it is difficult for the market to accept the large capital cost associated with the heat pumps. A hydrogen conversion unit may be more economical than replacing a gas boiler, especially if the cost is similar [56].

10.3.2 Solar Energy Systems

LCOE is used to evaluate solar energy systems including considering time-varying meteorological and electric market conditions. In contrast with LCOE, electricity markets at different time scales do not take into account the dynamic flexibility provided by thermal energy storage. Concentrated solar power (CSP) technologies are assessed economically using two different methodologies. Depending on the design and the condition of the market, low cost of ownership and revenue may be contradictory metrics.

LCOE is the most common and easiest economic metric. It is defined as the amount of revenue (dollars per unit of energy) that must be earned with a particular generator design in order to recover the operating and investment cost over a particular time period. This analysis is commonly used to guide research and technology investment and to design appropriate policies and programs. LCOE uses ten days as its duration, and the calculation is based on a ratio between the sums of all costs over the lifetime of the equipment and the sum of the electrical energy produced over the expected lifetime [57]. An LCOE extension for hybrid fossil fuel-solar thermal systems is presented; future cash flows and energy flows are discounted at rate r [58]:

$$LCOE = \frac{\sum_{t=1}^{n} \frac{(I_t + M_t + F_t - H_t)}{(1+r)^t}}{\sum_{t=1}^{n} \frac{E_t}{(1+r)^t}} \tag{10.10}$$

where

I_t: Investment expenditures in the year t
M_t: Operations and maintenance expenditures in the year t
F_t: Fuel expenditures in the year t
H_t: Avoided heat production costs in the year t
E_t: Electricity generation in the year t
r: Discount rate
t: Year
n: Assumed lifetime of system (integer, in years)

The best way to understand revenue opportunities from multiscale electricity market signals is with a general economic assessment framework that replaces LCOE-based technology comparisons. LCOE analysis disregards the time-varying value of electricity, which consequently undervalues energy storage fundamentally in contrast with market-based approaches; the economic assessment framework demonstrated significant revenue opportunities for CSP systems participating in electricity markets. For CSP technologies to be competitive against other sources of renewable energy, there needs to be more comprehensive technoeconomic assessments [59].

Case study 1: PV systems in Northern Cyprus: In a new solar energy case study in Northern Cyprus, the best places to install PV systems were not based upon an evaluation of global solar radiation; little research has been conducted in terms of exploring the potential of large-scale grid-connected PV systems for generating electricity in Northern Cyprus. Solar energy in Northern Cyprus has much more potential than wind energy, and solar panels provide a useful tool for decreasing carbon dioxide emissions from electricity consumption. Using an integrated and comprehensive approach, researchers explored the feasibility of a 100 MW grid-connected wind farm or solar farm that is both grid-connected and environmentally friendly. Both technologies reduce electricity demand and greenhouse gas emissions in the selected region. In order to reduce peak demand in Northern Cyprus, small-size solar and wind systems were assessed on the basis of their economic viability. In contrast with wind energy systems, solar systems have a lower internal rate of return and are widely available in all parts of Northern Cyprus where solar potential is boundless, thereby making them an economical alternative. The feasibility of a grid-connected renewable energy project of 12 MW was demonstrated in Northern Cyprus with an annual cost of electricity at 0.0933$/kWh for the selected region and a reduction of over 2,906,917 tCO_2 avoided annually from the emissions of the power plant. The installation of PV systems in Northern Cyprus would have a positive impact on the environment and the economy [60–62].

Case study 2: PV systems in North Africa: Solar energy is abundant in North Africa, and it is close to the European continent; this meets the requirements for global energy interconnections to transport clean energy and electricity. A solar energy economic analysis is essential for promoting solar energy planning in North Africa and realizing clean energy transmission across continents. There is enormous potential for solar energy to be developed as a source for power generation in North Africa countries (Morocco, Tunisia, Algeria and Egypt) [63].

Technology for large-scale solar power generation includes solar PV and CSP; the technology is relatively mature, and through material and technical innovation, its development objectives are to improve the efficiency and reduce cost of the PV power generation system. In CSP technology, solar energy is transformed into thermal energy, which is then used to generate electricity through the dynamo driving the generator through the turbine. The use of a CSP system with thermal energy storage unit can produce power more smoothly and efficiently than PV. This can be accomplished by regulating the output of power and generating power after sunset to meet night load demands or to meet peak power regulation demands of power systems. This can result in increased energy utilization hours and improved system regulation efficiency. The economics of CSP are still very much in the demonstration stage, which translates into a high cost of consultation, engineering, procurement, and construction of CSP projects and equipment due to the limited experience and expertise for large scale applications. Solar thermal power generation equipment cost, construction cost, and LCOE reductions are motivated by the scale effect in terms of LCOE development trends of PV and CSP [64].

10.3.3 Wind Energy Systems

Electricity can be generated by conventional wind turbines when the wind speed reaches 3–5 m/s at the cut-in speed [65] and 12–15 m/s at the rated speed [66]. The majority of the African continent lacks the potential for wind speed, particularly in the sub-Saharan region. In addition, conventional wind turbines are very heavy and costly to purchase, install, and maintain; the Ferris wheel wind turbine (FWT) was developed in response to the need for a new type of wind turbine.

Case study 1: African continent: This model is applied to analyze the techno-economics of siting the FWT in low wind speed areas [67]; Probability density functions are used to estimate the wind potential. The Weibull, Rayleigh, Lognormal, and Gamma distributions are the most frequently used distributions for modelling wind speed [68–71]. The case study compares the technological and economic viability of the FWT with that of other existing commercial wind turbines and other energy sources. A variety of inputs were used in building this model, which include wind data from several African cities, wind turbine geometry and technical details, and economic data from both cities and wind turbines. Among the outputs of the model are the LCOE and levelized production of electricity in each location as well as the annual net energy generated, the turbines' capacity, the return on investment, and the simple payback period. This model can be used by investors and by governments to assess the effectiveness of wind energy investments, evaluate the potential at wind energy sites around the world, and attract government investment in wind energy. This generic model can also be used by anyone to determine the economic viability of wind farms planned in specific cities by using their own data; it provides decision-makers with more detailed understanding of wind farm economics, profit opportunities and risks associated with wind investments [72].

Case study 2: Offshore wind (OW) in European countries: There has been significant advancement in the offshore wind industry over the past decade, and the European wind market presently consists of an array of investors with varying levels of experience. A reasonable level of maturity has been reached by offshore wind investments over the past few years. A high return on investment is often achieved by investors through buying and selling stakes at various stages of the stake's service life. There are 92 wind farms in operation in European countries, and there are distinct clusters of investors that are expected to focus on the second half of the wind farm operational life, with a new cluster expected to concentrate on wind farms that are nearing their end. For instance, an investor might purchase an asset that is approaching the end of its life cycle at a low price and extend its life at an increased cost through higher maintenance and operation cost [73].

Case study 3: OW in Scotland: In the LCOE model, cost is considered throughout the life span of the asset. However, more investors are seeking to understand the profitability profiles from the purchasing instance through the exit of the investment at various phases of the OW farm. A time-profile of revenues yielded by an OW farm is required to estimate the profitability of an investment over its life cycle [74]. Researchers conducted an extensive study and assessment of the economic impacts of planned offshore wind developments in Scotland; for these planned developments, they estimated total expenditures in Scotland based on information about the planned projects and the local content and used that data as the basis for calculating the economic benefits of Scottish development. As a case study, this one was valuable for a number of reasons [75–77].

First, the Scottish government has set ambitious targets for renewable energy and carbon reduction (higher than those for the UK as a whole), claiming that this will lead to creating new jobs and securing billions of dollars of investment through 21st-century technology, which is a crucial aspect of the reindustrialization of Scotland according to the plan. In addition, Scotland has one of the highest offshore wind resources in the world, making up 25% of all European offshore wind resources, and offshore wind developments will be of vital importance if Scotland is to achieve its targets for renewable energy and carbon reduction. Alongside the environmental benefits of offshore wind, economic benefits are expected to be derived by increasing the share of offshore wind. To calculate them, the researchers employed two multisector modeling methodologies, input-output (IO) and computable general equilibrium (CGE) [75–76].

To date, Scottish offshore wind development, despite the abundance of resources available, has lagged other regions in the UK in terms of offshore wind development for technical and environmental reasons. Economic impact analysis is widely utilized to estimate the economic impact of a proposed project; however, there are some limitations to these models that can be overcome through the use of CGE models, such as the passive supply component. With the CGE model used, the researchers are given the choice of two alternative agent behaviors that produce different scales of economic impacts. Offshore wind developments in Scotland are beneficial because it results in an increase in output, GDP, and employment [75–77].

All of these are related to an economic impact that is positive. Using both IO and CGE modeling approaches to examine the impact of offshore wind capacity addition, the investigators demonstrated that increasing offshore wind capacity raises both employment levels and gross value added (GVA), where the amount and timing of this change depend critically on the type and maturity of the model. Based on the myopic simulation, the most significant overall results were obtained with a total GVA impact of £3.88 billion and a reduction of 82,393 person-years of employment in the myopic simulation [75–77].

Case study 4: Small-scale wind energy systems in South Africa: In one case study, the primary reasons for the inadequate growth of the technology were inefficiency and the low energy yield of small-scale wind energy systems. A research group studied various locations and regions of South Africa for their wind energy productivity and economic viability. According to economists and investment managers, the cost of electricity produced is by far one of the most significant factors when analyzing the investment potential of electricity generation facilities and the economic benefits they will reap. Small wind systems can be economically viable if they can generate electricity at low cost, but many factors can influence the unit cost of energy (expressed in terms of money per kW), which also varies in different countries for different sites [78–80].

It is important to analyze the economic parameters related to wind power and its economic viability before determining the cost associated with wind-generated electricity, such as investment cost, operation and maintenance cost, energy production, turbine lifetime, and interest rates. According to common wisdom, energy unit cost is equal to a fraction of total expenditures of energy for a given time interval based on the amount of energy produced, but the initial cost for the systems studied included both the specific cost of the turbine and the installation cost of the turbine system in its initial cost or investment cost. LCOE evaluation of the two turbine models showed that the electricity cost calculated at all sites was high and uncompetitive compared with the average tariff of R 1.53/kWh offered by Eskom, the national electricity company [78–80].

10.4 NANOTECH STUDIES

10.4.1 Hydrogen Energy Systems

Hydrogen energy and nanotech are emerging technologies that have attracted the attention of many countries and are expected to lead to the growth of their economies in the near future. As emerging sectors, the hydrogen energy and nanotech sectors share a major similarity: A relatively long period of time is required for technological development to improve the quality of diverse components and assemble them into complete systems. A wide range of technological domains uses nanotech's generic technology to develop products that are designed and integrated into a variety of applications and can be used across multiple domains. To build system reliability, repeated testing and demonstrations are urgent. The benefits of using small organizations in bridging ties between different technological domains in networks are more numerous than those of using hydrogen energy technology itself since boundary-spanning activities create more business opportunities than hydrogen energy technology itself does [81–83].

Alternatively, NPs can be synthesized so as to be used at a semi-industrial, full-scale, or industrial scale in order to produce biohydrogen. However, in order to use NPs at a semi-industrial, full-scale, or industrial scale, NPs of suitable types, sizes, and shapes have to be synthesized in special patterns. Furthermore, the development of a suitable fermenter design for the production of hydrogen must also be done to make the process more efficient; despite the fact that powdered NPs have been tested to produce biohydrogen by translating the NP shape into a nanotube, sheets or mixed arrangements will be a more effective means of facilitating their bonding and enhancing enzyme activity. An organic or nonmetal NP may also enhance biohydrogen production by immobilizing enzymes in nanotubes or carbon embedded nanotubes. As organic materials are nontoxic and degradable, they can be used in hydrogen production to decrease bacterial growth and death. For biohydrogen production involving NPs in dark fermentation to improve quality and yield, these are possible opportunities that can be practiced in real industrial settings [84].

Modern case studies use hydrogen fuel cells and hydrogen storage, both of which play an essential role in creating clean energy using hydrogen. A hydrogen fuel cell is a device that generates electric energy from a gas. A hydrogen storage device allows the storage and production of hydrogen for later usage. To make hydrogen a widespread energy carrier, it is necessary to make both technologies as technologically and economically efficient as possible. There are two main groups of hydrogen evolution reactions (HER), the first of which uses electricity to drive the process through water electrolysis when the reaction requires hydrogen to be split; the second uses photoelectrochemical water splitting when the reaction requires hydrogen to be split. As the latter HERs involves the use of catalysts, these include hydrogen fuel cells as well as a variety of HER catalysts made of metal oxides, metal alloys, metal phosphides, sulfurides, sulfurenides, carbides, and nitrides, as well as single atoms and functional carbon materials derived from the elements. The electron transfer reaction occurs at the surface of metal catalysts, and it is a two-step reaction that is initiated by the Volmer–Heyrovsky or Volmer–Tafel mechanisms. The HER in an acidic medium includes the following steps [85, 86]:

$$\textit{Volmer step:} \quad H^{+} + e^{-} \rightarrow H^{*} \tag{10.11}$$

$$\textit{Heyrovsky step:} \quad H^{*} + H^{+} + e^{-} \rightarrow H_{2} \tag{10.12}$$

$$\textit{Tafel step:} \quad H^{*} + H \rightarrow H_{2} \tag{10.13}$$

HER is a clean source of energy that is achieved with the use of metal oxide catalysts. In spite of the challenges associated with the HER, which include a high cost for catalysts and slow kinetics, this process can offer several advantages, such as the ability to generate only water vapor as a byproduct and use abundant resources in its production. Many strategies have been used to improve the activity, stability, and selectivity of metal oxides, including transition metal oxides and rare earth metal oxides. For a variety of fuel cells, environmental and energy-related applications have recently been demonstrated for metal oxide nanomaterials synthesized using green chemistry approaches; these materials offer viable solutions to the global energy and environmental challenges that are currently confronting the world today. A cleaner and more sustainable future can be achieved by exploring new materials that can be used in future sustainable energy applications, optimizing green synthesis methods, overcoming limitations in metal oxide nanomaterial applications, and exploring new methods for synthesizing green materials for next-generation energy systems [87].

Green economy has become an important concept in the debate because it addresses issues and challenges associated with environmental degradation, ever increasing energy demand, and health concerns [88–90]. The chemistry field plays an important role in the sustainability of the environment by developing lower-hazard products for industries, generating renewable energy sources, promoting renewable energy sources, and protecting the environment. Energy is essential to all developmental processes, which is why it is crucial to boost the green economy by producing and consuming sustainable energy. With nanotechnology in the energy sector, the teamwork can reduce the use of raw materials and increase the efficiency, life cycle, and stability of the cells in order to achieve a green economy. Green chemistry and green nanotechnology are intended to create applications of nanomaterials that are energy-efficient, environmentally friendly, and economically viable. The innovation seems to be making a real difference in reducing pressure on nonrenewable energy resources, and it is expected that they will continue to do so. In general, green nanotechnology should not only offer environmentally friendly options and solutions, but also pay attention to occupational health security, which is necessary to make them become widely accepted, adopted, and marketed so that the green economy can flourish [91].

10.4.2 Solar Energy Systems

Using solar energy as a renewable energy source is thought to be one of the most cost-effective ways to utilize renewable energy. This has led a number of industries to invest in this area given the fact that solar energy is a free and clean source of electricity that does not lead to any harm to the environment. All emerging green energy technologies, except PV and photovoltaic-thermal (PVT), are dominated by solar direct electricity generation systems. Since solar energy is abundant yet limitless, and solar cells are becoming more efficient and cost-effective due to recent advances in this field, they have become more popular. In modern nanotech solar energy, there are many ways in which nanotechnology is being used for NP systems, including incorporating NPs into PV cells [92], using nanofluids for PV thermal panels [93], and building a PV or PVT system using nano-improved phase change materials (PCMs) [94].

Nanotechnology can enhance the thermophysical properties of PCMs in a variety of ways. Recently, research has been conducted on the use of PCMs in PVT systems as well as cooling with nanofluids; generally, PCMs are used under solar cells, among cooling collector tubes, and above the insulation of the solar panels. Using PCM as a spectral filter can enhance the electrical and thermal efficiency of PVT under the same operating conditions [95, 96]. A concentrated PVT/PCM apparatus using PCM and nanofluid as spectral filters is clarified in [97].

NPs can improve PCE performance in this way, but the right particle types must be stable, nontoxic, and low cost. CNTs have the potential to significantly improve the efficiency of PCE and PCM in solar cells, but their cost makes them impractical for this application. In another modern application, desalination could be transformed into a greener industry if solar energy is applied and if solar thermal collectors and NP panels are developed further. There are many advantages of nanofluids, and one of them is their superior efficacy over normal base fluids. Thus, nanofluids are being used in solar stills; the most significant advantages of nanofluids are as follows:

1. NPs are easy to change in shape, material, and size, making them very efficient at absorbing solar energy and minimizing it.
2. Solar stills become more efficient and productive by using nanofluids because nanofluids have a much higher thermal conductivity than base fluids.
3. Using nanofluid increases the system's productivity and output temperature, whereas conventional stills need larger hot areas to transfer heat, thereby increasing their cost and size.
4. Nanofluids greatly improve solar radiation scattering and absorption.
5. The use of nanofluid in still basins can significantly reduce convection and radiation heat losses [98–101].

Incorporating NPs into a basin's water will tend to increase its thermal conductivity; it will increase its temperature, and it will also increase the coefficient of convective heat transfer which will directly result in an increase in the evaporation rate. Because the NPs store heat, they provide sufficient energy to the water so as to increase productivity at night. It is highly recommended that a wiper be mounted at the still basin so that the NPs will not sit at the bottom of the vessel and settle. Nanofluids are very effective in directly absorbing solar radiation due to the excellent match between their optical absorption spectrum and the solar radiation spectrum. It is possible to increase efficiency by about 60% if a nanofluid is used with a single-slope solar system; additionally, distillate output can be improved by using a nanofluid [102].

10.4.3 Wind Energy Systems

As nanotechnology is increasingly being used in energy conservation, it can play a vital role in reducing the specific energy consumption of desalination plants. Nanofluids can increase the heat transfer coefficient of desalination plants, enabling the production of more water from a given amount of water. In addition, concentration of brine discharge has adverse impacts on the marine ecosystem, and as such, it is vital to solve this problem in order to achieve the goal of sustainable desalination.

During the last several years, studies have focused on nanotechnology applications for desalination and brine treatment, as well as the role of renewable energy in desalination, among other things. There are a wide variety of renewable sources on this planet that are derived from the sun, such as wind power, solar energy, waves, and tidal energy. There is a generation of tidal energy due to the gravitational forces associated with the sun, moon, and the rotation of the earth [103].

In modern Nanotech applications, reverse osmosis (RO) and thermal desalination processes are best powered by solar and wind energy, while RO can also be achieved using geothermal and nuclear power. The right technique depends on the location, salinity, and availability of renewable energy resources. It is necessary to analyze the quality of the saltwater that is being pumped into the system as well as discharged and determined whether the seawater can be utilized for the chemical and fertilizer industries in the form of nanofluids. If both the technology and policy components can be aligned to reach sustainable desalination goals, high-efficiency, cost-effective, and environmentally friendly renewable desalination technologies can be developed with reduced environmental impacts using nanofluids as a support mechanism. The application of nanotechnology to dewatering has great potential to support the development of sustainable desalination systems powered by solar thermal energy [104].

The modern effort has been focused on improving efficiency and reliability of wind turbines to reduce the overall cost of wind power; maximizing wind energy capture and mitigating fatigue loads have been among the main goals of control design. A number of recent developments in remote wind measurement systems (e.g. LIDAR) have made it possible for the wind energy industry to implement advanced control algorithms. However, ambitious plans for installing more wind turbines are at considerable risk due to the cost associated with their generation; this challenge motivates great efforts to improve both the efficiency and reliability of wind turbines to reduce the cost of producing wind energy. In recent years, model predictive control (MPC) is an effective and efficient tool for the optimization of wind energy capture, fatigue load mitigation, and smoothing of wind power [105–109]. This model is a high-order nonlinear model of a horizontal axis wind turbine [110], which is not recommended in the economic MPC framework due to the high computational cost. The development of a model-plant mismatch rejection framework with an adaptive approach was achieved using LIDAR-assisted economic model predictive control mismatch [111].

10.5 SUSTAINABILITY AND INVESTMENTS

10.5.1 Hydrogen Energy Systems

The design of materials for hydrogen energy applications that are sustainable is given particular attention to the design of materials for the production, purification, storage, and conversion of hydrogen into energy. Oxide-supported metal or alloy NPs have become a crucial component in the production of hydrogen via natural gas or alcohol conversion. Several methods can be used to produce high-purity hydrogen, including electrolysis and membrane catalysis; fuel cells are used to produce hydrogen energy through the use of catalysts and proton-conducting membranes, carbon or oxide-supported core and shell structures. Membranes containing perfluorinated sulfonic acid provide high conductivity and selectivity as well as high performance. Thin-film materials can be designed to reduce operating temperatures and ohmic losses in high-temperature fuel cells, and ceramic membranes can be replaced with proton-conducting membranes [112].

A majority of the proton-conducting membrane FCs in low-temperature proton-conducting membranes use Pt or its alloys as catalysts, despite the high cost of Pt, because the metal illustrates the highest efficiency in electrocatalytic processes in FCs. Pt has been replaced by transition metals that are cheaper, but their degradation upon contact with the proton-conducting membrane leads to rapid poisoning of the electrons that flow through the cell. As the particle size of the catalyst decreases, the rate of electrochemical processes increases, thereby decreasing the Pt consumption [113]. Pt particles have a particle size of 2.5 nm with a diameter of 4 nm for both anodic and cathodic processes [114–116].

Hydrogen purification is achieved most effectively through Pa alloys, but the disadvantages of these membranes are their high cost and low efficiency in terms of hydrogen purification. This issue appears to be a major concern for many industries, and the most promising avenues of dealing with it seem to be designing porous composite materials with thin selective layers and producing high-purity hydrogen directly by using membrane catalysis to produce the gas directly. Modifying membranes by adding NPs can result in a significant improvement in proton conductivity and selectivity of the transport process. A lot of applications of nanotechnology are being developed in the areas of lighting, heating, renewable energy, energy storage, fuel cells, and hydrogen power generation and storage. Using better nanostructured materials for fuel cells, more efficient catalysts for splitting water, and better hydrogen adsorption have all made the production, storage, and transformation of hydrogen into electricity more effective for fuel cells that are simpler and cheaper. In addition, there has been an increase in the amount of hydrogen adsorption in fuel cells [117].

Nanotechnology has the potential to overcome the disadvantages of traditional capacitors, particularly the issue of low efficiency caused by the low surface area of electrodes; nanostructured materials can facilitate this process [118]. It is possible to increase the conductivity of nonaqueous liquid electrolytes by up to six times by adding NPs of alumina or zirconium. In this regard, nanocomposite polymer electrolytes have the potential to assist in constructing highly efficient, safe and environmentally friendly batteries. As an example, ceramic nanomaterials used as separators in polymer electrolytes are capable of increasing their electrical conductivity by 10 to 100 times more than those of solid polymer electrolytes without dispersed materials at room temperature. In rechargeable Li batteries, graphite NPs and CNTs have been proven to effectively protect against Li deposition and other safety issues by forming nanostructures of the anode [119, 120].

Electrochemical double-layer capacitors are being developed to improve the performance of their electrochemical electrodes built from activated carbon electrode materials to carbon-based nanostructures. A nanostructure like CNT in a capacitor can be used to achieve higher specific power, higher specific capacitance, and higher conductivity, and carbon aerogels improve capacitance, cyclability, and electrical conductivity since they have low charges and ions and are low-electronic resistance [121].

It is very significant to use nanotechnology to overcome some technological limitations associated with the use of various substitutes for nonrenewable energies; a lot of technological advances are needed to solve vital scientific and engineering challenges in order to move from a carbon-based energy economy to the one that is more sustainable. The efficiency problem cannot be solved by traditional materials at a feasible price in most cases. In the field of nanotechnology, novel materials with exceptional control over structure, size, and organization of matter are already overcoming some of these challenges [122]. Nanotechnology has a broad range of successful novel applications in many fields and has the capability of causing a great technological and economic revolution, one that has never been witnessed before [123]. This is an excellent example of how nanomaterial science can contribute to the development and well-being of future generations and current ones.

Developing new techniques and new materials with exceptional capabilities as better alternatives to the modifications to existing ones is vital. In nanotechnology, several products and discoveries have already reached the market, while others remain limited by technical limitations, cost-effectiveness, and potential risks. The short- and long-term impacts of nanotechnology are a major concern for scientists and environmentalists regardless of its innovative applications and promising potential. As nanotechnology marches forwards, proactive measures must be taken in order to mitigate the unwanted impacts of the technology even though it is unraveling; this is the key to the continuation of nanotech developments and the ability to maintain an economic position in which to compete at the global level [124].

10.5.2 Solar Energy Systems

Solar energy is a source of free, clean, and renewable energy that does not harm the environment; simple applications of new technologies such as nanotechnology are an affordable way of harnessing this form of energy. These facts have led many industries to invest in the field as a way to harness renewable energy. The first generation of solar cells was made based on semiconductor wafers that contained a single crystal of NP. However, a thin inorganic coating was used in the second generation of solar cells for the assembly of the cells. This is because amorphous thin-film solar cells are cheaper to manufacture, but the efficiency can be as low as 14%, which is lower than that of the conventional single-junction crystalline solar cells and can be as high as 27% for the first generation due to their simpler structure. Ideally, single-junction cells should have efficiencies of at least 33% [125], a limit which is defined by Shockley-Queisser thermodynamics; therefore, new solar cell technologies will be required in order to attain higher efficiency levels at lower costs than 33%, producing lower cost than currently available.

Solar cells are in the process of moving to the third generation [126], in which highly efficient and economical products are expected to be manufactured in the near future [127]. A polymer solar cell is a type of NP cell that can be used to replace a variety of different kinds of conventional solar cells. This type of cell has numerous noteworthy characteristics, such as low weight, flexibility, affordability, and the possibility of large-scale production. Using NPs and QDs, the solar industry has developed nanotechnology to print solar cells rapidly and easily. This type of cell is easily printed on flexible metal foil and therefore can be used to install the necessary circuits to supply a part of the energy requirements of buildings on their roofs by printing these cells to the desired size [128].

Solar energy is being considered a production source for electricity as a way of improving financial performance, protecting the environment, and developing a sustainable future. A sound scientific and technological basis is essential for the development of a sustainable economy; economics science has evolved to understand how new technologies can help transform societies and promote long-term prosperity. Technological developments have long been recognized as having a significant impact on economic systems. Sustainable and green energy production is now being emphasized globally through the use of renewable energy sources. Among the many potential uses of nanotechnology are for increasing energy efficiency, producing renewable energy, converting energy, storing energy, and optimizing energy usage. There are several applications of nanomaterials for creating energy sources such as solar cells and fuel cells, using them as an alternative to conventional clean energies sources such as organic batteries that are good for the environment [129].

Nanotechnology can optimize the use of solar energy by utilizing appropriate nanotechnology applications, such as polymer solar cells, NP coatings, low emission glass coated with nano-coating, smart windows (polyaniline nanowire arrays), thermal and anticorrosion polymer nano-colors, and nano-solar water heaters. A significant factor supporting the choice of eco-efficient architectural solutions is nanotechnology innovations in building materials and products. Certain types of nanostructured materials can already be identified to provide a significant contribution to the reduction of the environmental impact of industrial processes used in the construction sector in the form of a significant environmental benefit [130].

Modern investments should be analyzed using multidimensional linear regression (MDLR), a methodology that examines the relationship between a wide range of technological input variables, such as energy, finance, economics, sustainability, and regulation. MDLR is an effective and efficient method of processing a system of multiple stochastic variables as predictors. These variables are providing significant information about solar potentials and possible production of electricity [131, 132].

It is important to consider the financial aspects of electricity production when choosing technology; LCOE was applied to calculate the production price for solar energy [133]. Electricity, in particular, is a major driver of investment and economic activity because it drives human activity in the production of goods and services. The expected profit measurement is considered one of the key factors in the investment decision as it is a measure of cost-effectiveness. Solar capacities can also be explained in more subtle ways by taking into account the following [134]:

1. The feed-in tariffs could be high but only available to a relatively small number of producers, thereby limiting their impact on the market.
2. Investors who are willing may have to deal with alternative projects that use renewable sources other than solar.
3. Doing business list includes availability of electricity, but this availability is not controlling countries' ranking, on the list, and its significance could vary in time.
4. Electricity consumption may be insufficient to drive the installation of new solar capacities.

10.5.3 Wind Energy Systems

Through nanotechnology, wind turbines can be lighter, more durable, more fuel-efficient, and better thermally insulated. Additionally, nanomaterials offer significant benefits in fuel efficiency since they improve the process of catalysis from raw petroleum materials. They reduce fuel generated by higher-efficacy combustion and friction in engines. Nanotechnology is also used to uncover pipelines, gas elevator valves, and fractures in down wells. Through nano-biological engineering, it is possible to convert cellulose into ethanol that can be used as fuel in many ways.

Developing modern energy, and primarily renewable energy development, will be dependent on the transfer of technologies, nanotechnologies, and nanomaterials since these technologies play a significant role as catalysts for innovation in the economy. Economic development, along with national economic growth, requires the creation of powerful internal and external impulses. Technology transfer and nanotechnology are one of the most important drivers of innovative technologies and economic growth across virtually all industries, including the renewable energy sector [135]. Nanotechnology brings large-scale economic effects to the industry, and there are many ways to use it, including the increase of material efficiency and reduction of production cost in the energy sector [136].

Innovative systems of electric energy accumulation and storage, as well as new energy saving technologies, increase the economic efficiency of renewable energy. A significant decrease in wholesale energy prices can be attributed to a rise in renewable energy cost, which is in line with the interests of consumers. Innovative renewable energy technologies can reduce the negative effects of conventional energy on the environment and neutralize the negative costs associated with them [137]. A solar thermal collector becomes more thermally conductive when it contains metal or nonmetal NPs. The ability of NPs to strongly absorb solar radiation is something unique among nature's products compared to additives, such as water, oils, and even molten salt (which are transparent). In both situations, it is possible to maximize the efficiency of the solar collectors by adding a small amount of NPs. The preparation of nanofluids for renewable energy systems requires stable techniques to ensure that these enhanced properties remain stable throughout their operation life since these systems have demanding operational requirements [138].

Wind turbine efficiency is one of the most important factors to be taken into account when considering the environmental benefits of wind power. Wind turbines are a large source of heat that needs to be exhausted during operation; this heat can cause a significant rise in the temperature of electrical and mechanical components, thereby reducing their efficiency and their longevity. Moreover, high temperatures have been linked with the possibility of generator sets collapsing unexpectedly, resulting in extremely high maintenance cost, especially for offshore power plants. Quite recently, increasing research into wind turbine cooling systems is aimed at ensuring that wind turbines operate safely and efficiently; an advanced heat transfer system that utilizes nanofluids was developed to improve heat transfer efficiency as well as to dissipate waste heat from the surrounding environment [139]. This cooling system consists of a heat exchanger in the shape of a wind tower that dissipates the thermal load from the generator and the surrounding environment. It is important to note that most nanofluids are created using NPs that although readily available in the commercial market increase the overall production cost in addition to affecting the impacts and cost associated with the end-of-life disposal of those working fluids. Low-viscosity nanofluids are going to be the solution to our stability and pumping power problems, and these can also be used in high-temperature applications because they are going to be stable and pumping powerful. It is important to take into account that the NPs are highly flammable and that sonicating the nanofluid requires energy. Large-scale systems pose a significant environmental risk if spills or leaks occur [140].

10.6 TRENDS AND RECOMMENDATIONS

10.6.1 Economic Trends in Hydrogen Energy

Hydrogen is a promising candidate in the future of energy as several hydrogen production technologies are in various stages of development, and they can pave the way for future energy transitions if the researchers gain deeper understanding of their current status and future prospects. There are numerous advantages to the use of hydrogen for various sectors, such as transportation, heating, and alternative energy generation, even though storing the gas for its later use is a challenge. Additionally, hydrogen is highly efficient and low polluting, and it can be used in a wide range of commercial sectors.

Making the transition from conventional energy resources to renewable energy to meet current energy demands and reserve conventional resources for future generations requires a series of promising strategies; framing the future energy plan to deal with the energy crisis requires understanding of how technologies and markets will be developped in the future. For this transition to take place, understanding hydrogen production technologies' flexibility and compatibility is imperative. Hydrogen has an encouraging future among renewable energy resources; therefore, understanding these technologies' flexibility and compatibility is imperative. Without proper regulation and legislation for safety, nanotechnology may cause serious grievances resulting in irreparable damage to the environment and humanity as a whole. Therefore, a balance must be maintained between NP production, marketing, use, risk assessment, and management of nanotechnology.

10.6.2 Economic Trends in Solar Energy

The economic viability of designing PV solar power substations can be demonstrated by an economic analysis of nanocomposites and QDs used in these substations. Such analysis should consider the effects of the volumetric concentrations of different metal NPs within a multilevel core and multiple shell CdS/PbS module. As the volume fraction increases, the maximum power and efficiency of modules increase; the module cost per watt decreases when the volume fraction of metals in the absorbing layer increases. The best way to get high efficiency and save money in PV systems is to increase the concentration of Cu as an individual or multiple NP. In PV systems, the use of multiple NPs in the PbS absorb layer is more effective at reducing cost and enhancing the power point of modules because it results in a higher voltage point.

Economic modules have proved the effectiveness of a solar power substation with a total installed capacity and the high efficiency using QD radii of window and absorb layer. The efficiency, power, and cost of modules that utilize QD CdS and QD PbS windows and absorption depend on the radius of the layer material. CdS and PbS at smaller radii led to improved efficiency and maximum power at certain modules of 1.95 m^2, thereby reducing the cost of the modules. PV module economic analysis reveals the effect of QD window layer and core/multiple shell absorbing layer HJ-QD thin film cell structure. Increasing the volumetric concentration of selected metal cores improved the module's performance, leading to greater output power and increased efficiency through increased performance. Solar power substations can be designed more cheaply and efficiently with the increase in volumetric concentration of metal cores. The innovative HJ-QD solar cells (with a QD window and a core/multiple shell absorb layer) can be used to build solar power substations, allowing them to produce more electricity. In comparison with the solar power substations using traditional crystalline monocrystalline and multicrystalline NP solar cells, it succeeds in reducing the total number of strings and achieving higher efficiency at a much lower cost.

10.6.3 Economic Trends in Wind Energy

A number of industries have adopted AI to produce power and energy. As part of this, analytics are used to optimize operational performance, wind farms are optimized by forecasting wind speed, distributed generation is flexible, microgeneration is integrated, equipment inspections are conducted with drones, generation output is networked, demand management is proactive, generation optimization is autonomous, and renewable generation is optimized [141, 142].

It is imperative that a more efficient grid be implemented along with an increase in the use of renewable energy sources, such as solar and wind power, coupled with intermittent electricity loads, such as electric vehicles or busses, batteries, and energy storage, together with the use of decentralized renewable energy sources. It is believed that with the integration of AI into the smart grid, the smart grid will be able to learn and communicate between the nature of intermittent loads. AI is undoubtedly going to be beneficial to a variety of solar and wind stakeholder groups and stakeholders, allowing them to perform remote inspections, troubleshoot problems, perform maintenance on solar panels, optimize onshore wind farms, optimize wind farms, ensure efficient and effective solar panel inspections, and utilize AI in real-time to expedite due diligence tasks. AI is expected to help emerging markets reduce delivery failures and resolve maintenance and reliability problems; AI will also allow distributed electricity systems/networks and microgrids to operate autonomously and incorporate renewable sources into the power grid. The potential for AI to help grid operators to integrate renewable energy sources into the grid is likely to be a critical area of focus in the coming years [143].

Recently, power machine efficiency is being optimized by adjusting generation in real time with AI. Machine learning can be used to improve/increase wind turbine production by taking into account historical performance, real-time communication with other wind farms, grid networks, and wind direction and wind transitions [144].

The use of big data and AI in digital forecasting permits even more solar energy to be incorporated into the power grid; for instance, wind power and speed data are collected for forecasting purposes. A new generation of wind turbines based on AI allows more renewable energy to be deployed at lower cost than utilities ever thought possible owing to the AI-based software. Scientists are working around the world to maximize the surplus of solar power although solar energy is lagging wind power production. Through AI, there will still be the investigation on how to balance fluctuating wind energy with demand and monitor charging costs and timing effectively [145].

REFERENCES

1. US Department of Energy. *A National Vision of America's Transition to a Hydrogen Economy—to 2030 and Beyond*, 2002.
2. US Department of Energy. *Fuel Cell Report to the Congress*, February 2003.
3. Guo KW. Green nanotechnology of trends in future energy. *Recent Patents on Nanotechnology*, 5, 76–88, 2011.
4. Sanders B. Cheap, plastic solar cells may be on the horizon. *UC Berkeley Campus News*, 28, 2002.
5. Jadhav MV, Todkar AS, Gambhire VR, Sawant SY. Nanotechnology for powerful solar energy. *International Journal of Advanced Biotechnology and Research*, 2(1), 208–212, 2011. ISSN 0976-2612.
6. Rastogi S, Sharma G, Kandasubramanian B. Nanomaterials and the environment. In Hussain CM, editor. *The ELSI Handbook of Nanotechnology: Risk, Safety, ELSI and Commercialization*. Scrivener Publishing LLC, 2020, 1–24.
7. Falinski MM, Plata DL, Chopra SS, Theis TL, Gilbertson LM, Zimmerman J. A framework for sustainable nanomaterial selection and design based on performance, hazard, and economic considerations. *Nature Nanotechnology*, 13, 708–714, 2018.
8. Lan K, et al. A modeling framework to identify environmentally greener and lower-cost pathways of nanomaterials. *Green Chemistry*, 26, 3466–3478, 2024.
9. Züttel A, Materials for hydrogen storage. *Materials Today*, 6, 24–33, 2003.
10. Abe JO, Popoola API, Ajenifuja E, Popoola OM. Hydrogen energy, economy and storage: Review and recommendation. *International Journal of Hydrogen Energy*, 44, 15072–15086, 2019.
11. Aithal PS, Aithal S. Nanotechnology innovations and commercialization-opportunities, challenges & reasons for delay. *International Journal of Engineering and Manufacturing (IJEM)*, 6, 1–12, 2016.
12. Aithal PS, Aithal S. A new model for commercialization of nanotechnology production and services. *International Journal of Computational Research and Development (IJCRD)*, 1, 84–93, 2016.
13. He Y, Pang Y, Li X, Zhang M. Dynamic subsidy model of photovoltaic distributed generation in China. *Renewable Energy*, 118, 555–564, 2018.
14. Ren LZ, Zhao XG, Zhang YZ, Li YB. The economic performance of concentrated solar power industry in China. *Journal of Cleaner Production*, 205, 799–813, 2018.
15. Pingkuo L, Zhongfu T. How to develop distributed generation in China: In the context of the reformation of electric power system. *Renewable and Sustainable Energy Reviews*, 66, 10–26, 2016.
16. Fu R, Chung D, Lowder T, Feldman D, Ardani K, Margolis R. *U.S. Solar Photovoltaic System Cost Benchmark: Q1 2016*. Technical Report. National Renewable Energy Laboratory (NREL), 2016.

17. Obi M, et al. Calculation of levelized costs of electricity for various electrical energy storage systems. *Renewable and Sustainable Energy Reviews*, 67, 908–920, 2017.
18. Lai CS, McCulloch MD. Levelized cost of electricity for solar photovoltaic and electrical energy storage. *Applied Energy*, 190, 191–203, 2017.
19. Zhao X, Zeng Y, Zhao D. Distributed solar photovoltaics in China: Policies and economic performance. *Energy,* 88, 572–583, 2015.
20. Xin-gang Z, Yi-min X. The economic performance of industrial and commercial rooftop photovoltaic in China. *Energy*, 187, 115961, 2019. Elsevier.
21. Massi Pavan A, Mellit A, De Pieri D, Kalogirou SA. A comparison between BNN and regression polynomial methods for the evaluation of the effect of soiling in large scale photovoltaic plants. *Applied Energy*, 108, 392–401, 2013.
22. Moharram KA, Abd-Elhady MS, Kandil HA, El-Sherif H. Influence of cleaning using water and surfactants on the performance of photovoltaic panels. *Energy Conversion and Management*, 68, 266–272, 2013.
23. Catelani M, Ciani L, Cristaldi L, Faifer M, Lazzaroni M, Rossi M. Characterization of photovoltaic panels: The effect of dust. In *Proceedings of the 2nd IEEE International Energy Conference and Exhibition, Advances in Energy Conversion Symposium*, Florence, Italy, pp. 45–50, 2012.
24. Bianchini A, Gambuti M, Pellegrini M, Saccani C. Performance analysis and economic assessment of different photovoltaic technologies based on experimental measurements. *Renewable Energy*, 85, 1–11, 2016.
25. Olatomiwa L, Mekhilef S, Huda ASN, Ohunakin OS. Economic evaluation of hybrid energy systems for rural electrification in six geo-political zones of Nigeria. *Renewable Energy,* 83, 435–446, 2015.
26. Rezaei M, Dampage U, Das BK, Nasif O, Borowski PF, Mohamed MA. Investigating the impact of economic uncertainty on optimal sizing of grid-independent hybrid renewable energy systems. *Processes*, 9, 1468, 2021.
27. Sinha S, Chandel SS. Review of software tools for hybrid renewable energy systems. *Renewable and Sustainable Energy Reviews*, 32, 192–205, 2014.
28. Vaziri Rad MA, Ghasempour R, Rahdan P, Mousavi S, Arastounia M. Techno-economic analysis of a hybrid power system based on the cost-effective hydrogen production method for rural electrification, a case study in Iran. *Energy*, 190, 116421, 2020.
29. Singh S, Singh M, Kaushik SC. Feasibility study of an islanded microgrid in rural area consisting of PV, wind, biomass and battery energy storage system. *Energy Conversion Management,* 128, 178–190, 2016.
30. Dehghani Sanij A, et al. The influence of research and development activities and nanofab centers on product development in nanotechnology: Focusing on solar thermal energy and photovoltaic technology. *JREE* Summer, 9, 87–100, 2022.
31. Mirjalili S, Mirjalili SM, Lewis A. Grey wolf optimizer. *Advance in Engineering Software*, 69, 46–61, 2014
32. Saeed MA, Ahmed Z, Zhang W. Wind energy potential and economic analysis with a comparison of different methods for determining the optimal distribution parameters. *Renewable Energy*, 61, 1092–1109, 2020.
33. Van Cappellen LK, Croezen H, Rooijers F, Kirkels AF, *Feasibility Study into Blue Hydrogen: Technical, Economic & Sustainability Analysis*. EIRES Research, Technology, Innovation & Society, 2018.
34. Bertuccioli L, Chan A, Hart D, Lehner F, Madden B, Standen E. *Development of Water Electrolysis in the European Union*. Institution of Gas Engineers and Managers (IGEM), FCH JU 2014.
35. Parks G, Boyd R, Cornish J, Remick R. *Hydrogen Station Compression, Storage, and Dispensing Technical Status and Costs*. OSTI.GOV, Technical Report, AC36-08GO28308, 1130621, 2014.

36. Dash SK, Chakraborty S, Roccotelli M, Sahu UK, Hydrogen fuel for future mobility: Challenges and future aspects, MDPI. *Sustainability*, 14, 8285, 2022.
37. Parissis O-S, Zoulias E, Stamatakis E, Sioulas K, Alves L, Martins R, Tsikalakis A, Hatziargyriou N, Caralis G, Zervos A. Integration of wind and hydrogen technologies in the power system of Corvo island, Azores: A cost-benefit analysis. *International Journal of Hydrogen Energy*, 36, 8143–8151, 2011.
38. Yue M, Lambert H, Pahon E, Roche R, Jemei S, Hissel D. Hydrogen energy systems: A critical review of technologies, applications, trends and challenges. *Renewable and Sustainable Energy Reviews*, 146, 111180, 2021.
39. Yadav A, Kumar N. Solar resource estimation based on correlation matrix response for Indian geographical cities. *International Journal of Renewable Energy Research*, 6, 695–701, 2016.
40. Jurasz J, Canales FA, Kies A, Guezgouz M, Beluco A. A review on the complementarity of renewable energy sources: Concept, metrics, application and future research directions. *Solar Energy,* 195, 703–724, 2020.
41. Liczmańska-Kopcewicz K, Pypłacz P, Wiśniewska A. Resonance of investments in renewable energy sources in industrial enterprises in the food industry. *Energies*, 13, 4285, 2020.
42. Mehedintu A, Sterpu M, Soava G. Estimation and forecasts for the share of renewable energy consumption in final energy consumption by 2020 in the European Union. *Sustainability*, 10, 1515, 2018.
43. Cader J, Koneczna R, Olczak P. The impact of economic, energy, and environmental factors on the development of the hydrogen economy. *Energies*, 14, 4811, 2021.
44. Prince-Richard S, Whale M, Djilali N. A techno-economic analysis of decentralized electrolytic hydrogen production for fuel cell vehicles. *International Journal of Hydrogen Energy*, 30, 1159–1179, 2005.
45. Nistor S, Dave S, Fan Z, Sooriyabandara M. Technical and economic analysis of hydrogen refuelling. *Applied Energy*, 167, 211–220, 2016.
46. Hosoya Y, Fujii Y. Analysis of energy strategies to halve CO2 emissions by the year 2050 with a regionally disaggregated world energy model. *Energy Pfrocedia*, 43, 5853–5860, 2011.
47. Fujii Y, Komiyama R. Long-term energy and environmental strategies. In Ahn J, Carson C, Jensen M, Juraku K, Nagasaki S, Tanaka S, editors. *Reflections on the Fukushima Daiichi Nuclear Accident*. Cham: Springer, ISBN 978-3-319-12089-8, 2015.
48. Komiyama R, Fujii Y. Analysis of energy saving and environmental characteristics of electric vehicles in regionally disaggregated world energy model. *Electrical Engineering in Japan*, 186, 20–36, 2013.
49. Tamaki T, Nozawa W, Managi S. Evaluation of the ocean ecosystem: Climate change modelling with backstop technologies. *Applied Energy*, 205, 428–439, 2017.
50. Chapman A, Itaoka K. A review of four case studies assessing the potential for hydrogen penetration of the future energy system. *International Journal of Hydrogen Energy*, 44, 6371–6382, 2019.
51. Stanton EA, Ackerman F, Kartha S. Inside the integrated assessment models: Four issues in climate economics. *Climate and Devolopment*, 1(2), 166–184, 2009.
52. Tamaki T, Nozawa W, Managi S. Evaluation of the ocean ecosystem: Climate change modelling with backstop technologies. *Applied Energy,* 205, 428–439, 2017.
53. Webber ME. The water intensity of the transitional hydrogen economy. *Environmental Research Letters*, 2(3), 034007, 2007.
54. BEIS. *Energy Consumption in the UK*. UK Government: Department for Business, Energy & Industrial Strategy, 2016.
55. CCC. *Next Steps for UK Heat Policy*. London: Committee on Climate Change, 2016.
56. Speirs J, et al. *Agreener Gas Grid: What Are the Options?* London: Imperial College London, Sustainable Gas Institute, 2017.

57. Projected Costs of Generating Electricity. Technical Report. International Energy Agency, 2005.
58. Sheu EJ, Mitsos A, Eter AA, Mokheimer EMA, Habib MA, Al-Qutub A. A review of hybrid solar-fossil fuel power generation systems and performance metrics. *Journal of Solar Energy Engineering*, 134(4), 041006, 2012.
59. Dowling AW, Zheng T, Zavala VM. Economic assessment of concentrated solar power technologies: A review. *Renewable and Sustainable Energy Reviews*, 72, 1019–1032, 2017.
60. Kassem Y, Çamur H, Alhuoti SMA. Solar energy technology for Northern Cyprus: Assessment, statistical analysis, and feasibility study. *Energies*, 13, 940, 2020.
61. Ilkan M, Erdil E, Egelioglu F. Renewable energy resources as an alternative to modify the load curve in Northern Cyprus. *Energy* 30, 555–572, 2005.
62. Kassem Y, Gökçekuş H, Çamur H. Economic assessment of renewable power generation based on wind speed and solar radiation in urban regions. *Global Journal of Environmental Science and Management*, 4, 465–482, 2018.
63. Balghouthi M, et al. Potential of concentrating solar power (CSP) technology in Tunisia and the possibility of interconnection with Europe. *Renewable and Sustainable Energy Reviews,* 56, 1227–1248, 2016.
64. Zhao L, Wang W, Zhu L, Liu Y, Dubios A. Economic analysis of solar energy development in North Africa. *Global Energy Interconnection*, 1, 53–62, 2018.
65. Rehman S, Alam MM, Alhems LM, Rafique MM. Horizontal axis wind turbine blade design methodologies for efficiency enhancement—A review. *Energies*, 11, 506, 2018.
66. American Wind Energy Association. *Size Specifications of Common Industrial Wind Turbines. Technical Specifications of Common Large Wind Turbines,* 2018. http://aweo.org/windmodels.html (accessed on 26 June).
67. Adeyeye KA. *Design of a Wind Turbine for Prevalent Conditions in Africa for an Effective and Economic Power Generation*. Ph.D. Thesis. Kigali, Rwanda: University of Rwanda, 2021.
68. Akyuz HE, Gamgam H. Statistical analysis of wind speed data with Weibull, lognormal and Gamma distributions. *Cumhuriyet Science Journal*, 38, 68–76, 2017.
69. Ali S, Lee S, Jang C. Techno-economic assessment of wind energy potential at three locations in South Korea using long-term measured wind data. *Energies*, 10, 1442, 2017.
70. Ayodele TR, Jimoh AA, Munda JL, Agee JT. Viability and economic analysis of wind energy resource for power generation in Johannesburg, South Africa. *Internatinal Journal of Sustainable Energy*, 33, 284–303, 2014.
71. Ruslau MFV, Silubun HCA. Analysis of wind speed distribution at the Mopah airport in Merauke. *Atlantis Highlights in Engineering*, 1, 958–963, 2018.
72. Adeyeye KA, Ijumba N, Colton JS. A techno-economic model for wind energy costs analysis for low wind speed areas. *Processes*, 9, 1463, 2021.
73. Ioannou A, Wang L, Brennan F. Design implications towards inspection reduction of large scale structures. *Procedia CIRP*, 60, 434–439, 2017.
74. Ioannou A, Angus A, Brennan F. A lifecycle techno-economic model of offshore wind energy for different entry and exit instances. *Applied Energy*, 221, 406–424, 2018.
75. Scottish Government. Roadmap for renewable energy in Scotland, the 21st century technology, which is a crucial aspect of the reindustrialization of Scotland according to the plan, 2018.
76. Scottish Government. Roadmap for renewable energy in Scotland, the highest offshore wind resources in the world, making up 25% of all European offshore wind resources, 2018.
77. Connolly K. The regional economic impacts of offshore wind energy developments in Scotland. *Renewable Energy*, 160, 148–159, 2020.
78. Simic Z, Havelka JG, Vrhovcak MB. Small wind turbine—a unique segment of the wind power market. *Renewable Energy,* 50, 027–1036, 2013.

79. Gokcek M, Genc MS. Evaluation of electricity generation and energy cost of wind energy conversion systems (WECSS) in Central Turkey. *Applied Energy*, 86, 2731–2739, 2009.
80. De Oliveira WS, Fernandes AJ, Gouveia JJB. Economic metrics for wind energy projects. *International Journal of Energy and Environment,* 2(6), 1013–1038, 2011.
81. Islam N, Miyazaki K. Nanotechnology innovation system: Understanding hidden dynamics of nanoscience fusion trajectories. *Technological Forecasting and Social Change,* 76, 128–140, 2009.
82. Choi H. Ready-steady-go for emerging technologies in post catch-up countries: A longitudinal network analysis of nanotech in Korea. *Technology Analysis & Strategic Management*, 29, 946–959, 2017.
83. Choi H, Shin J, Hwang W-S. Two faces of scientific knowledge in the external technology search process. *Technological Forecasting & Social Change*, 133, 41–50, 2018.
84. Kumar G, Mathimani T, Rene ER, Pugazhendhi A. Application of nanotechnology in dark fermentation for enhanced biohydrogen production using inorganic nanoparticles. *International Journal of Hydrogen Energy*, 44, 13106–13113, 2019.
85. Zeng M, Li Y. Recent advances in heterogeneous electrocatalysts for the hydrogen evolution reaction. *Journal of Materials Chemistry A*, 3, 14942–14962, 2015.
86. Yan X, Deng D, Wu S, Li H, Xu L. Development of transition metal nitrides as oxygen and hydrogen electrocatalysts. *Chinese Journal of Structural Chemistry,* 41, 2207004–2207015, 2022.
87. Danish MSS. Exploring metal oxides for the hydrogen evolution reaction (HER) in the field of nanotechnology. *RSC Sustainability*, 1, 2180–2196, 2023.
88. Demirbas A. *Biorenewable Gaseous Fuels*, 1st ed. London, UK: Springer, ISBN978-1-84882-010-4, 231–260, 2009.
89. Gaurav N, Sivasankari S, Kiran GS, Ninawea A, Selvin J, Utilization of bioresources for sustainable biofuels: A review. *Renewable and Sustainable Energy Reviews,* 73, 205–214, 2017.
90. Olson DG, McBride JE, Shaw AJ, Lynd LR. Recent progress in consolidated bioprocessing, *Current Opinion in Biotechnology,* 23, 396–405, 2012.
91. Pandey G. Nanotechnology for achieving green-economy through sustainable energy, *Rasayan Journal of Chemistry,* 11(3), 942–950, 2018.
92. Ansari AA, Nazeeruddin MK, Tavakoli MM. Organic-inorganic upconversion nanoparticles hybrid in dye-sensitized solar cells. *Coordination Chemistry Reviews*, 436, 213805, 2021.
93. Bandaru SH, Becerra V, Khanna S, Radulovic J, Hutchinson D, Khusainov R. A review of photovoltaic thermal (Pvt) technology for residential applications: Performance indicators, progress, and opportunities. *Energies*, 14, 3853, 2021.
94. Alva G, Liu L, Huang X, Fang G. Thermal energy storage materials and systems for solar energy applications. *Renewable and Sustainable Energy Reviews,* 68, 693–706, 2017.
95. AL-Musawi AIA, Taheri A, Farzanehnia A, Sardarabadi M, Passandideh-Fard M. Numerical study of the effects of nanofluids and phase-change materials in photovoltaic thermal (PVT) systems. *Journal of Thermal Analysis and Calorimetry*, 137, 623–636, 2019.
96. Hassan A, et al. Thermal management and uniform temperature regulation of photovoltaic modules using hybrid phase change materials-nanofluids system. *Renewable Energy,* 145, 282–293, 2020.
97. Yazdanifard F, Ameri M, Taylor RA. Numerical modeling of a concentrated photovoltaic/thermal system which utilizes a PCM and nanofluid spectral splitting. *Energy Conversion and Management,* 215, 112927, 2020.
98. Hussein AK. Applications of nanotechnology in renewable energies—a comprehensive overview and understanding. *Renewable and Sustainable Energy Reviews,* 42, 460–476, 2015.

99. Hussain S, Hussein AK. Natural convection heat transfer enhancement in a differentially heated parallelogrammical enclosure filled with copper-water nanofluid. *Journal of Heat Transfer—Transactions of the American Society of Mechanical Engineering (ASME)*, 136, 082502-1–082502-8, 2014.
100. Hussein AK, Hussain S. Heatline visualization of natural convection heat transfer in an inclined wavy cavity filled with nanofluids and subjected to a discrete isoflux heating from its left sidewall. *Alexandria Engineering Journal*, 55, 169–186, 2016.
101. Hussein AK, Mustafa A. Natural convection in fully open parallel ogrammic cavity filled with Cu-water nanofluid and heated locally from its bottom wall. *Thermal Science and Engineering Progress*, 1, 66–77, 2017.
102. Hussein AK, Kolsi L, Younis O, Li D, Ali HM, Afrand M. Using of nanotechnology concept to enhance the performance of solar stills-recent advances and overview. *Journal of Engineering Science and Technology*, 15(6), 3991–4031, 2020.
103. Al-Karaghouli A, Renne D, Kazmerski LL. Solar and wind opportunities for water desalination in the Arab regions, *Renewable and Sustainable Energy Reviews*, 13, 2397–2407, 2009.
104. Singha T, Atieh MA, Al-Ansari T, Mohammad AW, McKay G. The role of nanofluids and renewable energy in the development of sustainable desalination systems: A review, MDPI. *Water*, 12, 2002, 2020.
105. García CE, Prett DM, Morari M. Model predictive control: Theory and practice—a survey, *Automatica*, 25(3), 335–348, 1989.
106. Qin SJ, Badgwell TA. A survey of industrial model predictive control technology, *Control Engineering Practice*, 11(7), 733–764, 2003.
107. Rawlings JB. Tutorial overview of model predictive control. *IEEE Control Systems Magazine*, 20(3), 38–52, 2000.
108. Yang Z, Li Y, Seem JE. Model predictive control for wind turbine load reduction under wake meandering of upstream wind turbine. In *Proceedings of the ACC*, Montréal, Canada, 3008–3013, 2012.
109. Khalid M, Savkin AV. A model predictive control approach to the problem of wind power smoothing with controlled battery storage. *Renewable Energy*, 35(7), 1520–1526, 2010.
110. Grunnet JD, Soltani M, Knudsen T, Kragelund MN, Bak T. Aeolus toolbox for dynamics wind farm model, simulation and control. In *Proceedings of the European Wind Energy Conference & Exhibition (EWEC)*, Warsaw, Poland, 20–23 April 2010.
111. Shaltout ML, Ma Z, Chen D. An adaptive economic model predictive control approach for wind turbines. *Journal of Dynamic Systems, Measurement and Control*, 140(5), 051007, 2018.
112. Filippov SP, Yaroslavtsev AB. Hydrogen energy: Development prospects and materials. *Russian Chemical Reviews*, 90(6), 627–643, 2021.
113. Pavlov VI, Gerasimova EV, Zolotukhina EV, Don GM, Dobrovolsky YA, Yaroslavtsev AB. Degradation of Pt/C electrocatalysts having different morphology in low-temperature PEM fuel cells. *Nanotechnology in Russia*, 11, 743–750, 2016.
114. Perez J, Paganin VA, Antolini E. Particle size effect for ethanol electro-oxidation on Pt/C catalysts in half-cell and in a single direct ethanol fuel cell. *Journal of Electroanalytical Chemistry*, 654, 108–115, 2011.
115. Peres-Alonso FJ, McCarthy DN, Nierhoff A, Hernandez-Fernandez P, Strebel C, Stephens IEL, Nielsen JH, Chorkendorff I. The effect of size on the oxygen electroreduction activity of mass-selected platinum nanoparticles. *Angewandte Chemie*, 124, 4719, 2012.
116. Meier JC, Galeano C, Katsounaros I, Witte J, Bongard HJ, Topalov AA, Baldizzone C, Mezzavilla S, Schüth F, Mayrhofer KJJ. Design criteria for stable Pt/C fuel cell catalysts. *Beilstein Journal of Nanotechnology*, 5, 44–67, 2014.

117. Morse J. Assessing the economic impact of nanotechnology: The role of nanomanufacturing. *NNN Newsletter,* 5(3), 2012.
118. Serrano E, Rus G, Garcia-Martinez J. Nanotechnology for sustainable energy. *Renewable and Sustainable Energy Reviews,* 13(9), 2373–2384, 2009.
119. Agrawal RC, Pandey GP. Solid polymer electrolytes: Materials designing and all-solid-state battery applications: An overview. *Journal of Physics D: Applied Physics,* 41(22), 223001, 2008.
120. Nam KT, et al. Virus-enabled synthesis and assembly of nanowires for lithium-ion battery electrodes. *Science,* 312(5775), 885–888, 2006.
121. Wang J, et al. Morphological effects on the electrical and electrochemical properties of carbon aerogels. *Journal of the Electrochemical Society,* 148(6), D75–D77, 2001.
122. Ciemleja G, Natalja L. Opportunities for sustainable development and challenges in nanotech industry in Latvia. *Journal of Security and Sustainability Issues,* 5(3), 2016.
123. Wiek A, Guston D, Leeuw SVD, Selin C, Shapira P. Nanotechnology in the city: Sustainability challenges and anticipatory governance. *Journal of Urban Technology,* 20(2), 45–62, 2013.
124. Babatunde DE, Denwigwe IH, Babatunde OM, Gbadamosi SL, Babalola IP. Agboola O. *Environmental and Societal Impact of Nanotechnology,* 8, 4640–4667, 2019.
125. Shockley W, Queisser HJ. Detailed balance limit of efficiency of P-N junction solar cells. *Journal of Applied Physics,* 32, 510–519, 1961.
126. Green MA. Prospects for photovoltaic efficiency enhancement using low-dimensional structures, *Nanotechnology,* 11, 401, 2000.
127. Frankl P, Nowak S. *Technology Roadmap: Solar Photovoltaic Energy.* Paris, France: OECD/IEA, 2010.
128. Asadpour F, Jamshidi HA. The use of solar energy using nanotechnology to reduce energy consumption in biological buildings of Parand. *Journal of Applied Environmental and Biological Sciences,* 5(12S), 430–436, 2010.
129. Pathakoti K, Manubolu M, Hwang HM. Nanotechnology applications for environmental industry. In *Handbook of Nanomaterials for Industrial Applications.* Elsevier, Chapter 48, 894–907, 2018.
130. Nejad JM, Asadpour F. A strategy for sustainable development: Using nanotechnology for solar energy in buildings (case study Parand town). *Journal of Engineering and Technological Science,* 51(1), 103–120, 2019.
131. Fumo N, Biswas MR. Regression analysis for prediction of residential energy consumption. *Renewable and Sustainable Energy Reviews,* 47, 332–343, 2015.
132. Zamo M, Mestre O, Arbogast P, Pannekoucke O. A benchmark of statistical regression methods for short-term forecasting of photovoltaic electricity production, part I: Deterministic forecast of hourly production. *Solar Energy,* 105, 792–803, 2014.
133. Gürtürk M. Economic feasibility of solar power plants based on PV module with levelised cost analysis. *Energy,* 171, 866–878, 2019.
134. Stevovi I, Mirjani D, Stevovi S. Possibilities for wider investment in solar energy implementation. *Energy,* 180, 495–510, 2019.
135. Energy efficiency mechanisms: Energy industry innovations. *NanoNewsNet,* 12 January 2011.
136. Building Connections. *Fostering Solutions: 2012–2013ANSI.* Annual Report. New York, NY: ANSI, 2013, 40.
137. Manukhina M, Riazanova N. Theoretical and methodological principles of renewable energy development based on the technologies' transfer. *SHS Web of Conferences,* 67, 06035, 2019.
138. Nagarajan PK, Subramani J, Suyambazhahan S, Sathyamurthy R. Nanofluids for solar collector applications: A review. *Energy Procedia,* 61, 2416–2434, 2014.

139. De Risi A, Milanese M, Colangelo G, Laforgia D. High efficiency nanofluid cooling system for wind turbines. *Thermal Science,* 18, 543–554, 2014.
140. Mahian O, et al. Recent advances in using nanofluids in renewable energy systems and the environmental implications of their uptake. *Nano Energy*, 86, 106069, 2021.
141. Entchev E, Yang L. Application of adaptive neuro-fuzzy inference system techniques and artificial neural networks to predict solid oxide fuel cell performance in residential microgeneration installation. *Journal of Power Sources*, 170, 122–129, 2007.
142. Antonopoulos I, et al. Artificial intelligence and machine learning approaches to energy demand-side response: A systematic review. *Renewable and Sustainable Energy Reviews,* 130, 109899, 2020.
143. Ahmad T, et al. Energetics systems and artificial intelligence: Applications of industry 4.0. *Energy Reports*, 8, 334–361, 2022.
144. Yousri D, Babu TS, Mirjalili S, Rajasekar N, Elaziz MA. A novel objective function with artificial ecosystem-based optimization for relieving the mismatching power loss of large-scale photovoltaic array. *Energy Conversion and Management,* 225, 113385, 2020.
145. Zhao X, Wang C, Su J, Wang J. Research and application based on the swarm intelligence algorithm and artificial intelligence for wind farm decision system. *Renewable Energy,* 134, 681–697, 2019.

Index

Note: Page numbers in *italics* indicate figures, and page numbers in **bold** indicate tables on the corresponding page.

For Product Safety Concerns and Information please contact our EU representative GPSR@taylorandfrancis.com Taylor & Francis Verlag GmbH, Kaufingerstraße 24, 80331 München, Germany

Batch number: 10397790

Printed by Printforce, the Netherlands